Calvin H. Bartholomew and Morris D. Argyle (Eds.)

Advances in Catalyst Deactivation

MDPI

This book is a reprint of the Special Issue that appeared in the online, open access journal, *Catalysts* (ISSN 2073-4344) from 2013–2015 (available at: http://www.mdpi.com/journal/catalysts/special_issues/catalyst-deactivation).

Guest Editors
Calvin H. Bartholomew
Brigham Young University
USA

Morris D. Argyle
Brigham Young University
USA

Editorial Office
MDPI AG
Klybeckstrasse 64
Basel, Switzerland

Publisher
Shu-Kun Lin

Managing Editor
Mary Fan

1. Edition 2016

MDPI • Basel • Beijing • Wuhan • Barcelona

ISSN 978-3-03842-187-0 (Hbk)
ISSN 978-3-03842-188-7 (PDF)

Table of Contents

List of Contributors

Ana Rita Almeida: Faculty of Science & Technology, University of Twente, PO Box 217, Meander 225, 7500 AE, Enschede, The Netherlands.

Mina Alyani: Department of Chemical & Biological Engineering, University of British Columbia, 2360 East Mall, Vancouver, BC V6T 1Z3, Canada.

Francesco Arena: Dipartimento di Ingegneria Elettronica, Chimica e Ingegneria Industriale, Università degli Studi di Messina, Viale F. Stagno D'Alcontres 31, I-98166 Messina, Italy; Istituto CNR-ITAE "Nicola Giordano", Salita S. Lucia 5, I-98126 Messina, Italy.

Morris D. Argyle: Chemical Engineering Department, Brigham Young University, Provo, UT 84602, USA.

Calvin H. Bartholomew: Chemical Engineering Department, Brigham Young University, Provo, UT 84602, USA.

Rob J. Berger: Anaproc c/o Delft University of Technology, Julianalaan 136, 2628 BL Delft, The Netherlands.

Jose Antonio Calles: Department of Chemical and Energy Technology, Rey Juan Carlos Universtity, c/Tulipán, s/n, Móstoles 28933, Spain.

Alicia Carrero: Department of Chemical and Energy Technology, Rey Juan Carlos Universtity, c/Tulipán, s/n, Móstoles 28933, Spain.

Paolo Ciambelli: Department of Industrial Engineering, University of Salerno, Via Giovanni Paolo II, 132; 84084, Fisciano, SA, Italy.

Burtron H. Davis: Center for Applied Energy Research, 2540 Research Park Drive, Lexington, KY 40511, USA.

Jorge Gascon: Catalysis Engineering, Chemical Engineering Department, Delft University of Technology, Julianalaan 136, 2628 BL Delft, The Netherlands.

Rahman Gholami: Department of Chemical & Biological Engineering, University of British Columbia, 2360 East Mall, Vancouver, BC V6T 1Z3, Canada.

Anders Holmen: Department of Chemical Engineering, Norwegian University of Science and Technology (NTNU), N-7491 Trondheim, Norway.

Gary Jacobs: Center for Applied Energy Research, 2540 Research Park Drive, Lexington, KY 40511, USA.

Freek Kapteijn: Catalysis Engineering, Chemical Engineering Department, Delft University of Technology, Julianalaan 136, 2628 BL Delft, The Netherlands.

Patricia J. Kooyman: Catalysis Engineering, Chemical Engineering Department, Delft University of Technology, Julianalaan 136, 2628 BL Delft, The Netherlands.

Michiel T. Kreutzer: Product & Process Engineering, Chemical Engineering Department, Delft University of Technology, Julianalaan 136, 2628 BL Delft, The Netherlands.

Montaña Lindo: Department of Chemical and Energy Technology, Rey Juan Carlos Universtity, c/Tulipán, s/n, Móstoles 28933, Spain.

Wenping Ma: Center for Applied Energy Research, 2540 Research Park Drive, Lexington, KY 40511, USA.

Michiel Makkee: Catalysis Engineering, Chemical Engineering Department, Delft University of Technology, Julianalaan 136, 2628 BL Delft, The Netherlands.

Jacob A. Moulijn: Catalysis Engineering, Chemical Engineering Department, Delft University of Technology, Julianalaan 136, 2628 BL Delft, The Netherlands.

Guido Mul: Faculty of Science & Technology, University of Twente, PO Box 217, Meander 225, 7500 AE, Enschede, The Netherlands.

Erling Rytter: Department of Chemical Engineering, Norwegian University of Science and Technology (NTNU), N-7491 Trondheim, Norway; SINTEF Materials and Chemistry, N-7465 Trondheim, Norway.

Diana Sannino: Department of Industrial Engineering, University of Salerno, Via Giovanni Paolo II, 132; 84084, Fisciano, SA, Italy.

Vera P. Santos: Catalysis Engineering, Chemical Engineering Department, Delft University of Technology, Julianalaan 136, 2628 BL Delft, The Netherlands.

Emmanuel Skupien: Catalysis Engineering, Chemical Engineering Department, Delft University of Technology, Julianalaan 136, 2628 BL Delft, The Netherlands.

Kevin J. Smith: Department of Chemical & Biological Engineering, University of British Columbia, 2360 East Mall, Vancouver, BC V6T 1Z3, Canada.

Giuseppe Trunfio: Dipartimento di Ingegneria Elettronica, Chimica e Ingegneria Industriale, Università degli Studi di Messina, Viale F. Stagno D'Alcontres 31, I-98166 Messina, Italy

Vincenzo Vaiano: Department of Industrial Engineering, University of Salerno, Via Giovanni Paolo II, 132; 84084, Fisciano, SA, Italy.

Arturo Javier Vizcaíno: Department of Chemical and Energy Technology, Rey Juan Carlos Universtity, c/Tulipán, s/n, Móstoles 28933, Spain.

About the Guest Editors

Calvin H. Bartholomew, Professor Emeritus of Chemical Engineering at Brigham Young University (BYU), has taught and mentored students at BYU in catalysis, materials, and catalyst deactivation for 42 years. He is an active researcher in heterogeneous catalysis and a recognized authority on Fischer-Tropsch synthesis and catalyst deactivation; he has co-authored over 140 journal articles, 20 chapters, four books, and three patents. He is co-author with Dr. Robert Farrauto of Fundamentals of Industrial Catalytic Processes, a leading handbook and textbook. Together with Professors Bill Hecker and Morris Argyle of BYU, he has taught short courses on "Heterogeneous Catalysis," "Fischer Tropsch Synthesis," and "Catalyst Deactivation" to more than 700 professionals from industry and academe. He has worked at four companies and consulted with more than 70 company clients on catalyst and support design, Fischer-Tropsch synthesis, selective catalytic reduction of nitrogen oxides, FT reactor design, BTL/GTL process design, and litigation relating to catalyst failure.

Morris D. Argyle is an Associate Professor of Chemical Engineering at Brigham Young University (BYU). After earning his bachelor's degree, he became interested in catalysis while working as the process engineer for one of the fluid catalytic cracking units at the Exxon Baytown Texas Refinery. After completing graduate school at the University of California, Berkeley, he joined the Department of Chemical and Petroleum Engineering at the University of Wyoming, where he became an Associate Professor and served as Department Head before joining the faculty at BYU in 2009. His research interests include metal oxide catalysts for oxidative dehydrogenation of light alkanes, high temperature water gas shift catalysts for hydrogen production, Fischer-Tropsch catalysis, plasma reactions, and carbon capture techniques. He also shares Professor Calvin Bartholomew's interest in catalyst deactivation. He has co-authored 38 journal articles, one book chapter, and five patents.

Preface

Catalyst deactivation, the loss over time of catalytic activity and/or selectivity, is a problem of immense and ongoing concern in the practice of industrial catalytic processes. Costs to industry for catalyst replacement and process shutdown total tens of billions of dollars per year. While catalyst deactivation is inevitable for most processes, some of its immediate, drastic consequences may be avoided, postponed, or even reversed. Accordingly, there is considerable motivation to better understand catalyst decay and regeneration. Indeed, the science of catalyst deactivation and regeneration and its practice have been expanding rapidly, as evidenced by the extensive growth of literature addressing these topics. This developing science provides the foundation for continuing, substantial improvements in the efficiency and economics of catalytic processes through development of catalyst deactivation models, more stable catalysts, and regeneration processes.

This special issue focuses on recent advances in catalyst deactivation and regeneration, including advances in: (1) scientific understanding of mechanisms; (2) development of improved methods and tools for investigation; and (3) more robust models of deactivation and regeneration. It consists mainly of topical reviews.

The editors thank Keith Hohn, Editor-in-Chief, for the opportunity to organize this special issue and Mary Fan, Senior Assistant Editor, and the staff of the *Catalysts* Editorial Office for their significant support, encouragement, and patience. We would also like to thank the reviewers of the submitted manuscripts for their invaluable recommendations, and the contributing authors for their hard work in revising their manuscripts several times in order to meet the high standards of this special issue. The quality of the published work appears to have rewarded these efforts.

Calvin H. Bartholomew and Morris D. Argyle
Guest Editors

Heterogeneous Catalyst Deactivation and Regeneration: A Review

Morris D. Argyle and Calvin H. Bartholomew

Abstract: Deactivation of heterogeneous catalysts is a ubiquitous problem that causes loss of catalytic rate with time. This review on deactivation and regeneration of heterogeneous catalysts classifies deactivation by type (chemical, thermal, and mechanical) and by mechanism (poisoning, fouling, thermal degradation, vapor formation, vapor-solid and solid-solid reactions, and attrition/crushing). The key features and considerations for each of these deactivation types is reviewed in detail with reference to the latest literature reports in these areas. Two case studies on the deactivation mechanisms of catalysts used for cobalt Fischer-Tropsch and selective catalytic reduction are considered to provide additional depth in the topics of sintering, coking, poisoning, and fouling. Regeneration considerations and options are also briefly discussed for each deactivation mechanism.

Reprinted from *Catalysts*. Cite as: Argyle, M.D.; Bartholomew, C.H. Heterogeneous Catalyst Deactivation and Regeneration: A Review. *Catalysts* **2015**, *5*, 145-269.

1. Introduction

Catalyst deactivation, the loss over time of catalytic activity and/or selectivity, is a problem of great and continuing concern in the practice of industrial catalytic processes. Costs to industry for catalyst replacement and process shutdown total billions of dollars per year. Time scales for catalyst deactivation vary considerably; for example, in the case of cracking catalysts, catalyst mortality may be on the order of seconds, while in ammonia synthesis the iron catalyst may last for 5–10 years. However, it is inevitable that all catalysts will decay.

Typically, the loss of activity in a well-controlled process occurs slowly. However, process upsets or poorly designed hardware can bring about catastrophic failure. For example, in steam reforming of methane or naphtha, great care must be taken to avoid reactor operation at excessively high temperatures or at steam-to-hydrocarbon ratios below a critical value. Indeed, these conditions can cause formation of large quantities of carbon filaments that plug catalyst pores and voids, pulverize catalyst pellets, and bring about process shutdown, all within a few hours.

While catalyst deactivation is inevitable for most processes, some of its immediate, drastic consequences may be avoided, postponed, or even reversed. Thus, deactivation issues (*i.e.*, extent, rate, and reactivation) greatly impact research, development, design, and operation of commercial processes. Accordingly, there is considerable motivation to understand and treat catalyst decay. Over the past three decades, the science of catalyst deactivation has been steadily developing, while literature addressing this topic has expanded considerably to include books [1–4], comprehensive reviews [5–8], proceedings of international symposia [9–14], topical journal issues (e.g., [15]), and more than 20,000 U.S. patents for the period of 1976–2013. (In a U.S. patent search conducted in November 2013 for the keywords catalyst and deactivation, catalyst and life, and catalyst and regeneration, 14,712, 62,945, and 22,520 patents were found respectively.) This area of research provides a critical understanding that is the foundation for modeling deactivation processes, designing stable catalysts, and optimizing processes to prevent or slow catalyst deactivation.

The purpose of this article is to provide the reader with a comprehensive overview of the scientific and practical aspects of catalyst deactivation with a focus on mechanisms of catalyst decay, prevention of deactivation, and regeneration of catalysts. Case studies of deactivation and regeneration of Co Fischer-Tropsch catalysts and of commercial catalysts for selective catalytic reduction of nitrogen oxides in stationary sources have been included.

2. Mechanisms of Deactivation

There are many paths for heterogeneous catalyst decay. For example, a catalyst solid may be poisoned by any one of a dozen contaminants present in the feed. Its surface, pores, and voids may be fouled by carbon or coke produced by cracking/condensation reactions of hydrocarbon reactants, intermediates, and/or products. In the treatment of a power plant flue gas, the catalyst can be dusted or eroded by and/or plugged with fly ash. Catalytic converters used to reduce emissions from gasoline or diesel engines may be poisoned or fouled by fuel or lubricant additives and/or engine corrosion products. If the catalytic reaction is conducted at high temperatures, thermal degradation may occur in the form of active phase crystallite growth, collapse of the carrier (support) pore structure, and/or solid-state reactions of the active phase with the carrier or promoters. In addition, the presence of oxygen or chlorine in the feed gas can lead to formation of volatile oxides or chlorides of the active phase, followed by gas-phase transport from the reactor. Similarly, changes in the oxidation state of the active catalytic phase can be induced by the presence of reactive gases in the feed.

Thus, the mechanisms of solid catalyst deactivation are many; nevertheless, they can be grouped into six intrinsic mechanisms of catalyst decay: (1) poisoning, (2) fouling, (3) thermal degradation, (4) vapor compound formation and/or leaching accompanied by transport from the catalyst surface or particle, (5) vapor–solid and/or solid–solid reactions, and (6) attrition/crushing. As mechanisms 1, 4, and 5 are chemical in nature while 2 and 6 are mechanical, the causes of deactivation are basically threefold: chemical, mechanical, and thermal. Each of the six basic mechanisms is defined briefly in Table 1 and treated in some detail in the subsections that follow, with an emphasis on the first three. Mechanisms 4 and 5 are treated together, since 4 is a subset of 5.

2.1. Poisoning

Poisoning [3,16–22] is the strong chemisorption of reactants, products, or impurities on sites otherwise available for catalysis. Thus, poisoning has operational meaning; that is, whether a species acts as a poison depends upon its adsorption strength relative to the other species competing for catalytic sites. For example, oxygen can be a reactant in partial oxidation of ethylene to ethylene oxide on a silver catalyst and a poison in hydrogenation of ethylene on nickel. In addition to physically blocking of adsorption sites, adsorbed poisons may induce changes in the electronic or geometric structure of the surface [17,21]. Finally, poisoning may be reversible or irreversible. An example of reversible poisoning is the deactivation of acid sites in fluid catalytic cracking catalysts by nitrogen compounds in the feed. Although the effects can be severe, they are temporary and are generally eliminated within a few hours to days after the nitrogen source is removed from the feed. Similar effects have been observed for nitrogen compound (e.g., ammonia and cyanide) addition to the syngas of cobalt Fischer-Tropsch catalysts, although these surface species require weeks to months before the lost activity is regained [23]. However, most poisons are irreversibly chemisorbed to the catalytic surface sites, as is the case for sulfur on most metals, as discussed in detail below. Regardless of whether the poisoning is reversible or irreversible, the deactivation effects while the poison is adsorbed on the surface are the same.

Table 1. Mechanisms of catalyst deactivation.

Mechanism	Type	Brief definition/description
Poisoning	Chemical	Strong chemisorption of species on catalytic sites which block sites for catalytic reaction
Fouling	Mechanical	Physical deposition of species from fluid phase onto the catalytic surface and in catalyst pores
Thermal degradation and sintering	Thermal Thermal/chemical	Thermally induced loss of catalytic surface area, support area, and active phase-support reactions
Vapor formation	Chemical	Reaction of gas with catalyst phase to produce volatile compound
Vapor–solid and solid–solid reactions	Chemical	Reaction of vapor, support, or promoter with catalytic phase to produce inactive phase
Attrition/crushing	Mechanical	Loss of catalytic material due to abrasion; loss of internal surface area due to mechanical-induced crushing of the catalyst particle

Many poisons occur naturally in feed streams that are treated in catalytic processes. For example, crude oil contains sulfur and metals, such as vanadium and nickel, that act as catalyst poisons for many petroleum refinery processes, especially those that use precious metal catalysts, like catalytic reforming, and those that treat heavier hydrocarbon fractions in which the sulfur concentrates and metals are almost exclusively found, such as fluid catalytic cracking and residuum hydroprocessing. Coal contains numerous potential poisons, again including sulfur and others like arsenic, phosphorous, and selenium, often concentrated in the ash, that can poison selective catalytic reduction catalysts as discussed later in Section 4.3.3.1. As a final example, some poisons may be added purposefully, either to moderate the activity and/or to alter the selectivity of fresh catalysts, as discussed as the end of this section, or to improve the performance of a product that is later reprocessed catalytically. An example of this latter case is lubricating oils that contain additives like zinc and phosphorous to improve their lubricating properties and stability, which become poisons when the lubricants are reprocessed in a hydrotreater or a fluid catalytic cracking unit.

Mechanisms by which a poison may affect catalytic activity are multifold, as illustrated by a conceptual two-dimensional model of sulfur poisoning of ethylene hydrogenation on a metal surface shown in Figure 1. To begin with, a strongly adsorbed atom of sulfur physically blocks at least one three- or fourfold adsorption/reaction site (projecting into three dimensions) and three or four topside sites on the metal surface. Second, by virtue of its strong chemical bond, it electronically modifies its nearest neighbor metal atoms and possibly its next-nearest neighbor atoms, thereby modifying their abilities to adsorb and/or dissociate reactant molecules (in this case H_2 and ethylene molecules), although these effects do not extend beyond about 5 atomic units [21]. A third effect may be the restructuring of the surface by the strongly adsorbed poison, possibly causing dramatic changes in catalytic properties, especially for reactions sensitive to surface structure. In addition, the adsorbed poison blocks access of adsorbed reactants to each other (a fourth effect) and finally prevents or slows the surface diffusion of adsorbed reactants (effect number five).

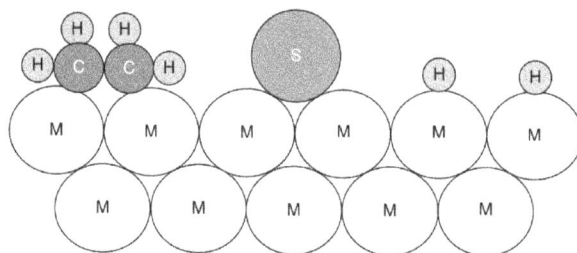

Figure 1. Conceptual model of poisoning by sulfur atoms of a metal surface during ethylene hydrogenation. Reproduced from [8]. Copyright 2006, Wiley-Interscience.

Catalyst poisons can be classified according to their chemical makeup, selectivity for active sites, and the types of reactions poisoned. Table 2 lists four groups of catalyst poisons classified according to chemical origin and their type of interaction with metals. It should be emphasized that interactions of Group VA–VIIA elements with catalytic metal phases depend on the oxidation state of the former, e.g., how many electron pairs are available for bonding and the degree of shielding of the sulfur ion by ligands [16]. Thus, the order of decreasing toxicity for poisoning of a given metal by different sulfur species is H_2S, SO_2, SO_4^{2-}, *i.e.*, in the order of increased shielding by oxygen. Toxicity also increases with increasing atomic or molecular size and electronegativity, but decreases if the poison can be gasified by O_2, H_2O, or H_2 present in the reactant stream [21]; for example, adsorbed carbon can be gasified by O_2 to CO or CO_2 or by H_2 to CH_4.

Table 2. Common poisons classified according to chemical structure.

Chemical type	Examples	Type of interaction with metals
Groups VA and VIA	N, P, As, Sb, O, S, Se, Te	Through s and p orbitals; shielded structures are less toxic
Group VIIA	F, Cl, Br, I	Through s and p orbitals; formation of volatile halides
Toxic heavy metals and ions	As, Pb, Hg, Bi, Sn, Cd, Cu, Fe	Occupy d orbitals; may form alloys
Molecules that adsorb with multiple bonds	CO, NO, HCN, benzene, acetylene, other unsaturated hydrocarbons	Chemisorption through multiple bonds and back bonding

Table 3 lists a number of common poisons for selected catalysts in important representative reactions. It is apparent that organic bases (e.g., amines) and ammonia are common poisons for acidic solids, such as silica–aluminas and zeolites in cracking and hydrocracking reactions, while sulfur- and arsenic-containing compounds are typical poisons for metals in hydrogenation, dehydrogenation, and steam reforming reactions. Metal compounds (e.g., of Ni, Pb, V, and Zn) are poisons in automotive emissions control, catalytic cracking, and hydrotreating. Acetylene is a poison for ethylene oxidation, while asphaltenes are poisons in hydrotreating of petroleum residuum.

Table 3. Poisons for selected catalysts in important representative reactions.

Catalyst	Reaction	Poisons
Silica–alumina, zeolites	Cracking	Organic bases, hydrocarbons, heavy metals
Nickel, platinum, palladium	Hydrogenation/dehydrogenation	Compounds of S, P, As, Zn, Hg, halides, Pb, NH_3, C_2H_2
Nickel	Steam reforming of methane, naphtha	H_2S, As
Iron, ruthenium	Ammonia synthesis	O_2, H_2O, CO, S, C_2H_2, H_2O
Cobalt, iron	Fischer–Tropsch synthesis	H_2S, COS, As, NH_3, metal carbonyls
Noble metals on zeolites	Hydrocracking	NH_3, S, Se, Te, P
Silver	Ethylene oxidation to ethylene oxide	C_2H_2
Vanadium oxide	Oxidation/selective catalytic reduction	As/Fe, K, Na from fly ash
Platinum, palladium	Oxidation of CO and hydrocarbons	Pb, P, Zn, SO_2, Fe
Cobalt and molybdenum sulfides	Hydrotreating of residuum	Asphaltenes, N compounds, Ni, V

Poisoning selectivity is illustrated in Figure 2, a plot of activity (the reaction rate normalized to initial rate) *versus* normalized poison concentration. "Selective" poisoning involves preferential adsorption of the poison on the most active sites at low concentrations. If sites of lesser activity are blocked initially, the poisoning is "antiselective". If the activity loss is proportional to the concentration of adsorbed poison, the poisoning is "nonselective." An example of selective poisoning is the deactivation of platinum by CO for the para-H_2 conversion (Figure 3a) [24] while Pb poisoning of CO oxidation on platinum is apparently antiselective (Figure 3b) [25], and arsenic poisoning of cyclopropane hydrogenation on Pt is nonselective (Figure 3c) [26]. For nonselective poisoning, the linear decrease in activity with poison concentration or susceptibility (σ) is defined by the slope of the activity *versus* poison concentration curve. Several other important terms associated with poisoning are defined in Table 4. Poison tolerance, the activity at saturation coverage of the poison, and resistance (the inverse of deactivation rate) are important concepts that are often encountered in discussions of poisoning including those below.

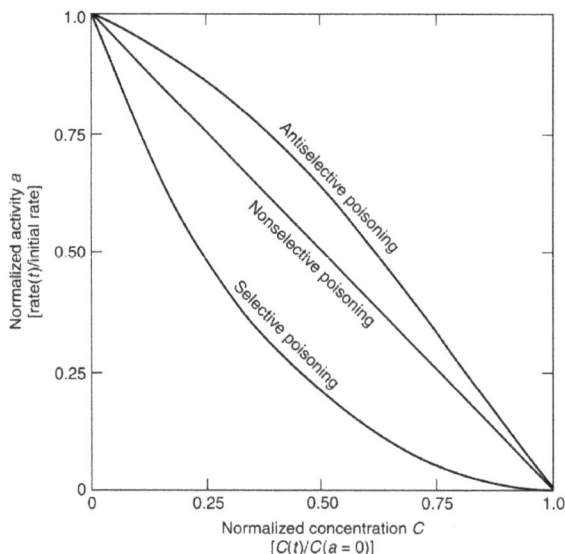

Figure 2. Three kinds of poisoning behavior in terms of normalized activity *versus* normalized poison concentration. Reproduced from [8]. Copyright 2006, Wiley-Interscience.

Table 4. Important Poisoning Parameters.

Parameter	Definition
Activity (a)	Reaction rate at time t relative to that at $t = 0$
Susceptibility (σ)	Negative slope of the activity *versus* poison concentration curve [$\sigma = (a - 1)/C(t)$]. Measure of a catalyst's sensitivity to a given poison
Toxicity	Susceptibility of a given catalyst for a poison relative to that for another poison
Resistance	Inverse of the deactivation rate. Property that determines how rapidly a catalyst deactivates
Tolerance ($a(C_{sat})$)	Activity of the catalyst at saturation coverage (some catalysts may have negligible activity at saturation coverage)

The activity *versus* poison concentration patterns illustrated in Figure 2 are based on the assumption of uniform poisoning of the catalyst surface and surface reaction rate controlling, *i.e.*, negligible pore-diffusional resistance. These assumptions, however, are rarely met in typical industrial processes because the severe reaction conditions of high temperature and high pressure bring about a high pore-diffusional resistance for either the main or poisoning reaction or both. In physical terms, this means that the reaction may occur preferentially in the outer shell of the catalysts particle, or that poison is preferentially adsorbed in the outer shell of the catalyst particle, or both. The nonuniformly distributed reaction and/or poison leads to nonlinear activity *versus* poison concentration curves that mimic the patterns in Figure 2 but do not represent truly selective or antiselective poisoning. For example, if the main reaction is limited to an outer shell in a pellet where poison is concentrated, the drop in activity with concentration will be precipitous. Pore diffusional effects in poisoning (nonuniform poison) are treated later in this review.

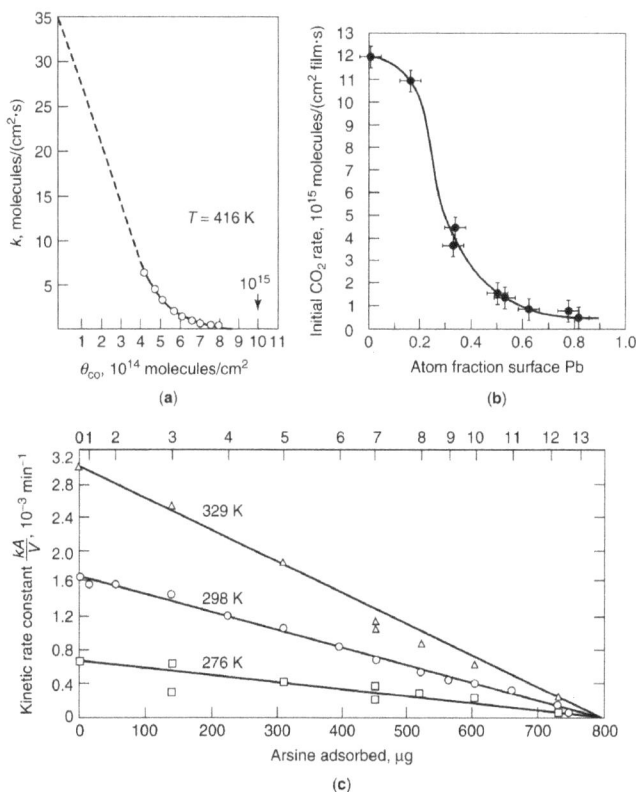

Figure 3. (a) CO poisoning of para-H_2 conversion over a Pt foil, reproduced from [24], copyright 1974, Wiley-VHC; (b) effect of lead coverage on the rate of CO oxidation of Pt film, reproduced from [25], copyright 1978, Elsevier; (c) rate constants of cyclopropane hydrogenolysis over a Pt film as a function of the amount of AsH_3 adsorbed, reproduced from [26], copyright 1970, Elsevier.

As sulfur poisoning is a difficult problem in many important catalytic processes (e.g., hydrogenation, methanation, Fischer–Tropsch synthesis, steam reforming, and fuel cell power production), it merits separate discussion as an example of catalyst poisoning phenomena. Studies of sulfur poisoning in hydrogenation and CO hydrogenation reactions have been thoroughly reviewed [8,21,27–31]. Much of the previous work focused on poisoning of nickel metal catalysts by H_2S, the primary sulfur poison in many important catalytic processes, and thus provides some useful case studies of poisoning.

Previous adsorption studies [28–30] indicate that H_2S adsorbs strongly and dissociatively on nickel metal surfaces. The high stability and low reversibility of adsorbed sulfur is illustrated by the data in Figure 4 [28], in which most of the previous equilibrium data for nickel are represented on a single plot of log (P_{H_2S}/P_{H_2}) *versus* reciprocal temperature. The solid line corresponds to the equilibrium data for formation of bulk Ni_3S_2. Based on the equation $\Delta G = RT \ln(P_{H_2S}/P_{H_2}) = \Delta H - T\Delta S$, the slope of this line is $\Delta H/R$, where $\Delta H = -75$ kJ/mol and the intercept is $-\Delta S/R$. Most of the adsorption data lie between the dashed lines corresponding to $\Delta H = -125$ and -165 kJ/mol for coverages ranging from 0.5 to 0.9, indicating that adsorbed sulfur is more stable than the bulk sulfide. Indeed, extrapolation of high temperature data to zero coverage using a Tempkin isotherm [29] yields an enthalpy of adsorption of -250 kJ/mol; in other words, at low sulfur coverages, surface nickel–sulfur bonds are a factor of 3 more stable than bulk nickel–sulfur bonds. It is apparent from Figure 4 that the absolute heat of adsorption increases with decreasing coverage and that the equilibrium partial pressure of H_2S increases with increasing temperature and increasing coverage. For instance, at 725 K (450 °C) and $\theta = 0.5$, the values of P_{H_2S}/P_{H_2} range from about 10^{-8} to 10^{-9}. In other words, half coverage occurs at 1–10 ppb H_2S, a concentration range at the lower limit of our present analytical capability. At the same temperature (450 °C), almost complete coverage ($\theta > 0.9$) occurs at values of P_{H_2S}/P_{H_2} of 10^{-7}–10^{-6} (0.1–1 ppm) or at H_2S concentrations encountered in many catalytic processes after the gas has been processed to remove sulfur compounds. These data are typical of sulfur adsorption on most catalytic metals. Thus, we can expect that H_2S (and other sulfur impurities) will adsorb essentially irreversibly to high coverage in most catalytic processes involving metal catalysts.

Two important keys to reaching a deeper understanding of poisoning phenomena include (1) determining surface structures of poisons adsorbed on metal surfaces and (2) understanding how surface structure and hence adsorption stoichiometry change with increasing coverage of the poison. Studies of structures of adsorbed sulfur on single crystal metals (especially Ni) [3,28,32–38] provide such information. They reveal, for example, that sulfur adsorbs on Ni(100) in an ordered p(2 × 2) overlayer, bonded to four Ni atoms at $S/Ni_s < 0.25$ and in a c(2 × 2) overlayer to two Ni atoms for $S/Ni_s = 0.25$–0.50 (see Figure 5; Ni_s denotes a surface atom of Ni); saturation coverage of sulfur on Ni(100) occurs at $S/Ni_s = 0.5$. Adsorption of sulfur on Ni(110), Ni(111), and higher index planes of Ni is more complicated; while the same p(2 × 2) structure is observed at low coverage, complex overlayers appear at higher coverages—for example, at $S/Ni_s > 0.3$ on Ni(111) a $(5\sqrt{3} \times 2)S$ overlayer is formed [32–34]. In more open surface structures, such as Ni(110) and Ni(210), saturation coverage occurs at $S/Ni_s = 0.74$ and 1.09 respectively; indeed, there is a trend of increasing S/Ni_s with decreasing planar density and increasing surface roughness for Ni, while the saturation sulfur concentration remains constant at 44 ng/cm^2 Ni (see Table 5).

Reported saturation stoichiometries for sulfur adsorption on polycrystalline and supported Ni catalysts (S/Ni_s) vary from 0.25 to 1.3 [28]. The values of saturation coverage greater than $S/Ni_s = 0.5$ may be explained by (1) a higher fractional coverage of sites of lower coordination number, *i.e.*, atoms located on edges or corners of rough, high-index planes (Table 5); (2) enhanced adsorption capacity at higher gas phase concentrations of H_2S in line with the observed trend of increasing saturation coverage with increasing H_2S concentration in Figure 4; and/or (3) reconstruction of planar surfaces to rougher planes by adsorbed sulfur at moderately high coverages and adsorption temperatures.

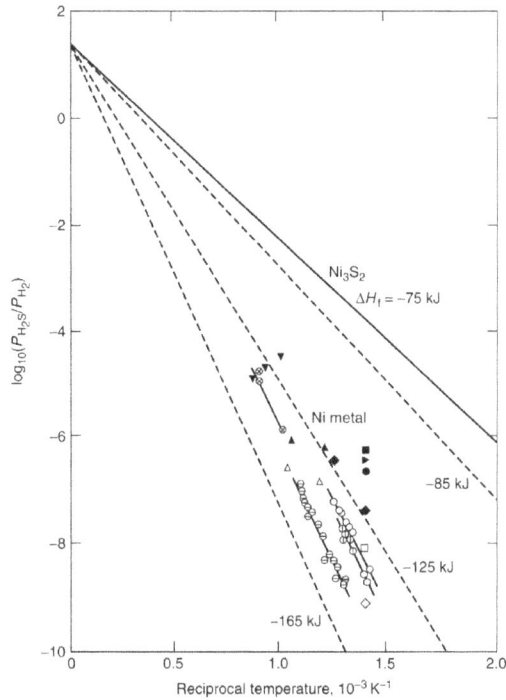

Figure 4. Equilibrium partial pressure of H₂S *versus* reciprocal temperature (values of ΔH_f based on 1 mole of H₂S); open symbols: θ = 0.5–0.6; closed symbols: θ = 0.8–0.9. Reproduced from [28]. Copyright 1982, Academic Press.

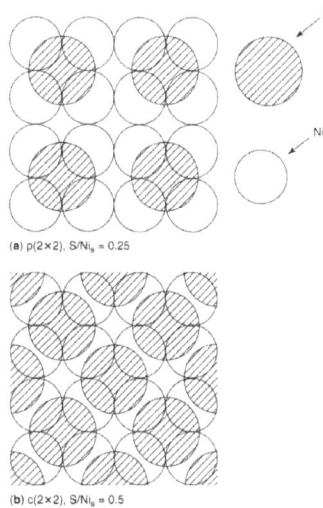

Figure 5. Schematic view of sulfur adsorbed on a Ni(100) surface at a (**a**) S/Ni_s = 0.25 in a p(2 × 2) structure and (**b**) S/Ni_s = 0.50 in a c(2 × 2) structure. Reproduced from [39]. Copyright 2001, Elsevier.

Table 5. Sulfur Adsorption Densities on Various Crystal Faces of Nickel [a].

Crystal face	Sulfur conc. at saturation, ng·S/cm^2	Number of S atoms/cm^2 ($\times 10^{15}$)	Number of Ni atoms/cm^2 ($\times 10^{15}$)	S atoms per surface Ni atom
(111)	47 ± 1	0.86	1.8	0.48
(100)	43 ± 1	0.80	1.6	0.50
(110)	44.5 ± 1	0.82	1.1	0.74
(210)	42 ± 1	0.78	0.72	1.09
Polycrystalline	44.5 ± 1	0.82	—	—

[a] Data from [35].

The first effect would be favored, and in fact is observed, for supported metals of higher dispersion [28]. The second effect may explain the typically lower observed values of S/Ni$_s$ for single crystal Ni, which are measured at extremely low pressures (high vacuum) relative to the higher values of S/Ni$_s$ for polycrystalline and supported Ni, typically measured at orders of magnitude higher pressure; thus, in the case of the single crystal studies, the surface is not in equilibrium with gas phase H$_2$S/H$_2$.

The third effect, reconstruction of nickel surfaces by adsorbed sulfur, has been reported by a number of workers [28,32,33,36–38]; for example, McCarroll and co-workers [37,38] found that sulfur adsorbed at near saturation coverage on a Ni(111) face was initially in a hexagonal pattern, but upon heating above 700 K reoriented to a distorted c(2 × 2) (100) overlayer. Oudar [36] reported that sulfur adsorbed on a Ni(810) surface caused decomposition to (100) and (410) facets. During adsorption of H$_2$S at RT, Ruan et al. [33] observed surface restructuring of Ni(111) from a p(2 × 2) at low coverage to a missing-row $(5\sqrt{3} \times 2)$S terrace structure (0.4 monolayer) sparsely covered with small, irregular islands composed of sulfur adsorbed on disordered nickel; upon annealing to 460 K for 5 min, the islands ordered to the $(5\sqrt{3} \times 2)$S phase and their size increased, suggesting further diffusion of Ni atoms from the terraces. The reconstruction of Ni (111) involving ejection and migration of Ni atoms was attributed to compressive surface stresses induced by sulfur adsorption; the role of compressive surface stress due to sulfur coverages exceeding 0.3 was confirmed by Grossmann et al. [32]. From these and similar studies, it is concluded that at moderately high temperatures (300–600 K) and coverages greater than 0.3, restructuring by sulfur of different facets of Ni to rougher, more open, stable structures is probably a general phenomenon. Thus, reconstruction probably accounts at least in part for observed increases in saturation S coverage with decreasing Ni site density.

The nature of reconstruction of a surface by a poison may depend on its pretreatment. For example, in a scanning tunneling microscopy (STM) study of room temperature H$_2$S adsorption on Ni(110), Ruan and co-workers [40] found that the S/Ni structure at saturation varied with the initial state of the surface, i.e., whether clean or oxygen covered. Beginning with a clean Ni(110) surface, oxygen adsorbs dissociatively to form a (2 × 1)O overlayer at 1/2 monolayer coverage (Figure 6a); this is accompanied by a homogeneous nucleation of low-coordinated -Ni-O- rows along the [001] direction. As the oxygen-covered surface is exposed stepwise to 3 and then 8 Langmuirs (L) of H$_2$S, oxygen atoms are removed by reaction with hydrogen to water; the surface is first roughened, after which white islands and black troughs having a c(2 × 2) structure are formed as sulfur atoms replace oxygen atoms (Figure 6b). Upon exposure to 25 L of H$_2$S, the c(2 × 2) islands dissolve, while low-coordinated rows (periodicity of 1) form in the [001] direction, developing into ordered regions with a periodicity of 4 in the [1 $\bar{1}$ 0] direction (Figure 6c). After exposure to 50 L of H$_2$S (Figure 6d), a stable, well-ordered (4 × 1)S structure appears, a surface clearly reconstructed relative to the original Ni(110). Moreover, the reconstructed surface in Figure 6d is very different from that observed upon direct exposure of the Ni(110) to H$_2$S at room temperature, *i.e.*, a c(2 × 2)S overlying the original Ni(110) (similar to Figure 5b); in other words, it appears that no reconstruction occurs by direct exposure to H$_2$S at room temperature, rather only in the presence of O$_2$ (or air). This emphasizes the complexities inherent in predicting the structure and stability of a given poison adsorbed on a given catalyst during a specified reaction as a function of different pretreatments or process disruptions, e.g., exposure to air.

In the previous discussion of Figure 4, $-\Delta H_{ads}$ was observed to decrease with increasing sulfur coverage; data in Figure 7 [41] show that $-\Delta H_{ads}$ decreases with increasing gas-phase H$_2$S concentration and coverage. However, in contrast to the data in Figure 4, those in Figure 7 [41] show that at very high H$_2$S concentrations and high adsorption temperatures, $-\Delta H_{ads}$ falls well below the $-\Delta H_{formation}$ of bulk Ni$_3$S$_2$; at the same time, the S/Ni$_s$ ratio approaches that of Ni$_2$S$_3$. This is a unique result, since all of the data obtained at lower temperatures and H$_2$S concentrations [28] show $-\Delta H_{ads}$ to be greater than $-\Delta H_{formation}$ of Ni$_3$S$_2$.

From the above discussion, the structure and stoichiometry of sulfur adsorbed on nickel evidently are complex functions of temperature, H$_2$S concentration, sulfur coverage, and pretreatment, phenomena that account at least in part for the complex nature of nickel poisoning by sulfur. Could one expect similar complexities in the poisoning of other metals? Probably, since poisoning of nickel is prototypical, *i.e.*, similar principles operate and similar poisoning behaviors are observed in other poison/metal systems, although none have been studied to the same depth as sulfur/nickel.

Since one of the necessary steps in a catalytic reaction is the adsorption of one or more reactants, investigation of the effects of adsorbed sulfur on the adsorption of other molecules, can provide useful insights into the poisoning process [21,28]. Previous investigations [28,42–48] indicate that both H$_2$ and CO adsorptions on nickel are poisoned by adsorbed sulfur. For example, thermal desorption studies of CO from presulfided Ni(100) [44] reveal a weakening of the CO adsorption bond and a rapid, nonlinear decline in the most strongly bound β_2 state (bridged CO) with increasing sulfur coverage, corresponding to a poisoning of about 8–10 Ni atoms for bridged CO adsorption per adsorbed sulfur atom at low sulfur coverage (see Figure 8); moreover, the β_2 CO species is completely poisoned at about 0.2–0.4 mL of sulfur relative to a saturation coverage of 0.5 mL. Hydrogen adsorption is poisoned in a similar nonlinear fashion. On the other hand, the coverage of the β_1 state (linear CO) is constant with increasing sulfur coverage. The sharp nonlinear drop in CO and hydrogen adsorptions at low sulfur coverages has been interpreted in terms of a combination of short-range electronic and steric effects operating over a range of less than 5 atomic units [13]. The different effects of sulfur on β_1 and β_2 states of CO have important implications for sulfur poisoning in reactions involving CO; that is, sulfur poisoning can affect reaction selectivity as well as activity [28].

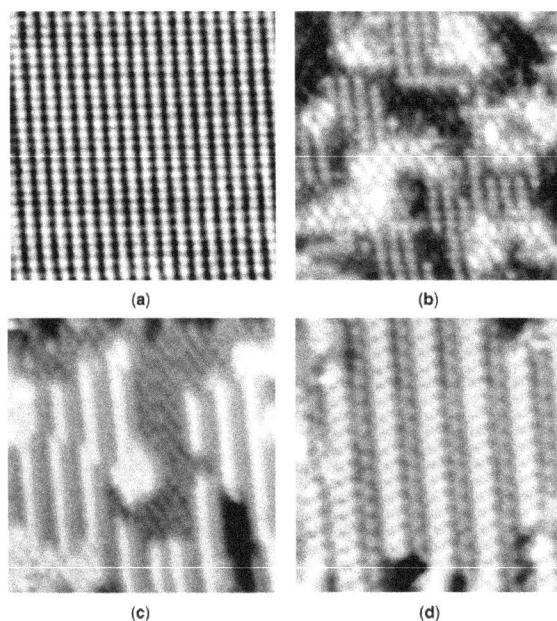

(a)

(b)

(c)

(d)

Figure 6. A series of *in situ* scanning tunneling microscope (STM) images recorded after exposure of Ni(110) to oxygen and then progressively higher exposures of H$_2$S: (**a**) (2 × 1)O overlayer; (**b**) white islands and black troughs with a c(2 × 2)S structure after exposure to 3 and 8 L of H$_2$S; (**c**) 25 L, islands transform to low-coordinated rows in the [001] direction; and (**d**) 50 L, stable, well-ordered (4 × 1)S. Reproduced from [40]. Copyright 1992, American Physical Society.

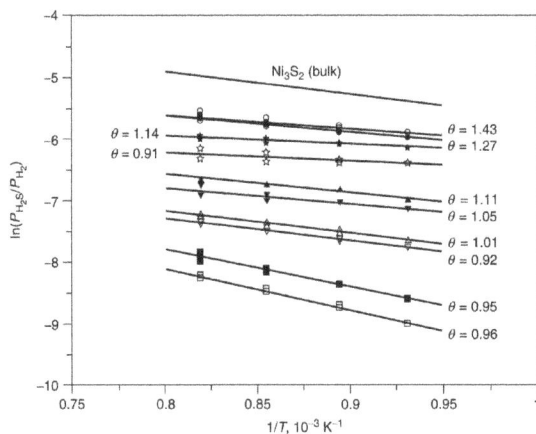

Figure 7. Sulfur chemisorption isosteres on a Ni/α-Al$_2$O$_3$ catalyst at high temperatures and high H$_2$S concentrations. Reproduced from [41]. Copyright 1999, Elsevier.

Because sulfur adsorbs so strongly on metals and prevents or modifies the further adsorption of reactant molecules, its presence on a catalyst surface usually effects substantial or complete loss of activity in many important reactions. This is illustrated by the data in Figure 9 showing the steady-state methanation activities of Ni, Co, Fe, and Ru relative to the fresh, unpoisoned surface activity as a function of gas phase H_2S concentration. These data indicate that Ni, Co, Fe, and Ru all suffer 3–4 orders of magnitude loss in activity at 15–100 ppb of H_2S, $i.e.$, their sulfur tolerances are extremely low. Moreover, the sharp drop in activity with increasing H_2S concentration suggests highly selective poisoning. Nevertheless, the rate of sulfur poisoning and hence sulfur resistance varies from catalyst to catalyst and is apparently a function of catalyst composition [28] and reaction conditions [49]. Indeed, it is possible to significantly improve sulfur resistance of Ni, Co, and Fe with catalyst additives such as Mo and B that selectively adsorb sulfur. Because the adsorption of sulfur compounds is generally rapid and irreversible, surface sulfur concentrations in catalyst particles and beds are nonuniform, e.g., H_2S adsorbs selectively at the entrance to a packed bed and on the outer surface of catalyst particles, making the experimental study and modeling of sulfur poisoning extremely difficult.

There are other complications in the study of sulfur poisoning. For example, the adsorption stoichiometry of sulfur in CO hydrogenation on Ni is apparently a function of the temperature, H_2/CO ratio, and water partial pressure [49]. Moreover, at high CO partial pressures sulfur may be removed from the surface as COS, which is not as strongly adsorbed as H_2S. At low temperature conditions, e.g., those representative of Fischer–Tropsch synthesis or liquid phase hydrogenations, the gas phase concentration of H_2S in poisoning studies must be kept very low, $i.e.$, below 0.1–5 ppm, to avoid formation of bulk metal sulfides—a phenomenon that seriously compromises the validity of the results. Thus, the importance of studying poisoning phenomena in $situ$ under realistic reaction conditions, at low process-relevant poison concentrations, and over a process-representative range of temperature and concentration conditions is emphasized.

As mentioned earlier, there are a number of industrial processes in which one intentionally poisons the catalyst in order to improve its selectivity. For example, Pt-containing naphtha reforming catalysts are often pre-sulfided to minimize unwanted cracking reactions. On basic Pt/KL zeolite catalysts, these short term, low concentration exposures are beneficial to produce Pt ensemble sizes that promote aromatization, while longer term or higher concentration exposures poison the catalyst both by forming Pt-S bonds and producing large crystallites that block pores, as shown by transmission electron microscopy (TEM) and X-ray absorption fine structure spectroscopy (EXAFS), and favor only dehydrogenation [50–53]. Other examples are sulfur added to Fischer-Tropsch catalysts that have been reported to have either beneficial or negligibly harmful effects, which are important considerations in setting the minimum gas clean-up requirements [27,30,54–56]. S and P are added to Ni catalysts to improve isomerization selectivity in the fats and oils hydrogenation industry, while S and Cu are added to Ni catalysts in steam reforming to minimize coking. In catalytic reforming, sulfided Re or Sn is added to Pt to enhance the dehydrogenation of paraffins to olefins while poisoning hydrogenolysis/coking reactions. V_2O_5 is added to Pt to suppress SO_2 oxidation to SO_3 in diesel emissions control catalysts.

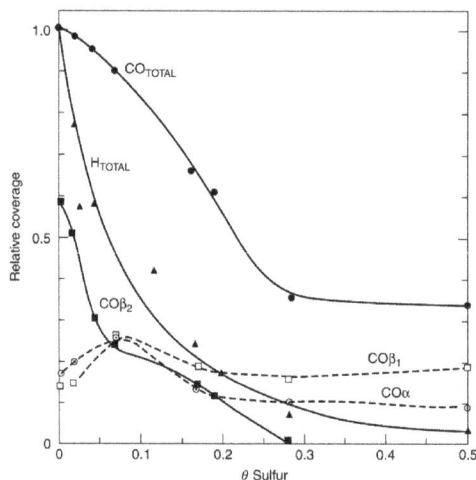

Figure 8. Area under thermal programmed desorption spectra for H_2 and the α, β_1, β_2, and total CO adsorption curves, as a function of sulfur precoverage. Reproduced from [44]. Copyright 1981, Elsevier.

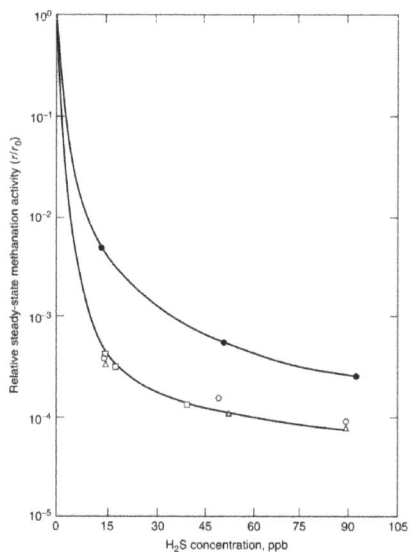

Figure 9. Relative steady-state methanation activity profiles for Ni ($\bullet$), Co (Δ), Fe ($\circ$), and Ru ($\square$) as a function of gas-phase H_2S concentration. Reaction conditions: 100 kPa, 400 °C, 1% CO/99% H_2 for Co, Fe, and Ru, 4% CO/96% H_2 for Ni. Reproduced from [28]. Copyright 1982, Academic Press.

2.2. Fouling, Coking, and Carbon Deposition

2.2.1. Fouling

Fouling is the physical (mechanical) deposition of species from the fluid phase onto the catalyst surface, which results in activity loss due to blockage of sites and/or pores. In its advanced stages, it may result in disintegration of catalyst particles and plugging of the reactor voids. Important examples include mechanical deposits of carbon and coke in porous catalysts, although carbon- and coke-forming processes also involve chemisorption of different kinds of carbons or condensed hydrocarbons that may act as catalyst poisons. The definitions of carbon and coke are somewhat arbitrary and by convention related to their origin. Carbon is typically a product of CO disproportionation while coke is produced by decomposition or condensation of hydrocarbons on catalyst surfaces and typically consists of polymerized heavy hydrocarbons. Nevertheless, coke forms may vary from high molecular weight hydrocarbons to primarily carbons such as graphite, depending upon the conditions under which the coke was formed and aged. A number of books and reviews treat the formation of carbons and coke on catalysts and the attendant deactivation of the catalysts [1,4,57–62].

The chemical structures of cokes or carbons formed in catalytic processes vary with reaction type, catalyst type, and reaction conditions. Menon [62] suggested that catalytic reactions accompanied by carbon or coke formation can be broadly classified as either coke-sensitive or coke-insensitive, analogous to Boudart's more general classification of structure-sensitive and structure-insensitive catalytic reactions. In coke-sensitive reactions, unreactive coke is deposited on active sites, leading to activity decline, while in coke-insensitive reactions, relatively reactive coke precursors formed on active sites are readily removed by hydrogen (or other gasifying agents). Examples of coke-sensitive reactions include catalytic cracking and hydrogenolysis; on the other hand, Fischer–Tropsch synthesis, catalytic reforming, and methanol synthesis are examples of coke-insensitive reactions. On the basis of this classification, Menon [62] reasoned that the structure and location of a coke are more important than its quantity in affecting catalytic activity.

Consistent with Menon's classification, it is also generally observed that not only structure and location of coke vary but also its mechanism of formation varies with catalyst type, e.g., whether it is a metal or metal oxide (or sulfide, sulfides being similar to oxides). Because of these significant differences in mechanism, formation of carbon and coke is discussed below separately for supported metals and for metal oxides and sulfides.

2.2.2. Carbon and Coke Formation on Supported Metal Catalysts

Possible effects of fouling by carbon (or coke) on the functioning of a supported metal catalyst are illustrated in Figure 10. Carbon may (1) chemisorb strongly as a monolayer or physically adsorb in multilayers and in either case block access of reactants to metal surface sites, (2) totally encapsulate a metal particle and thereby completely deactivate that particle, and (3) plug micro- and mesopores such that access of reactants is denied to many crystallites inside these pores. Finally, in extreme cases, strong carbon filaments may build up in pores to the extent that they stress and fracture the support material, ultimately causing the disintegration of catalyst pellets and plugging of reactor voids. For example, in steam methane reforming (SMR) catalysts, which are typically nickel supported on alumina with alkaline earth oxides, the carbon can diffuse through and begin to grow filaments from the back side of the nickel particles (structural type 3 in Table 6) especially at high reaction temperatures and low steam to methane ratios, which push the nickel particles off the support surface. Thermal or mechanical shock can then cause the carbon filaments to fall off the support, thus permanently deactivating the catalyst [8,60]. However, the behavior is complex because for other reaction conditions and other metals, the filaments may grow from the top surface of the metal particles or the carbon may diffuse into the metal and form bulk carbides [8].

An example of recent interest for biomass reactions that points to the complex interaction between the active metal and the support during carbon deposition is the steam reforming of light alcohols and other oxygenates, in which deactivation occurs primarily through coking. For traditional SMR catalysts (e.g., $Ni/MgAl_2O_4$) the coke is believed to originate primarily from alkene formation [63,64]. However, for the case of Ni/La_2O_3 catalysts, carbon appears to form at the interface between the active metal and the support to block the active phase [65].

Mechanisms of carbon deposition and coke formation on metal catalysts from carbon monoxide and hydrocarbons, including methane during SMR for hydrogen production [4,57–61], are illustrated in Figures 11 and 12. Different kinds of carbon and coke that vary in morphology and reactivity are formed in these reactions (see Tables 6 and 7). For example, CO dissociates on metals to form C_α, an adsorbed atomic carbon; C_α can react to C_β, a polymeric carbon film. The more reactive, amorphous forms of carbon formed at low temperatures (e.g., C_α and C_β) are converted at high temperatures over a period of time to less reactive, graphitic forms [60]

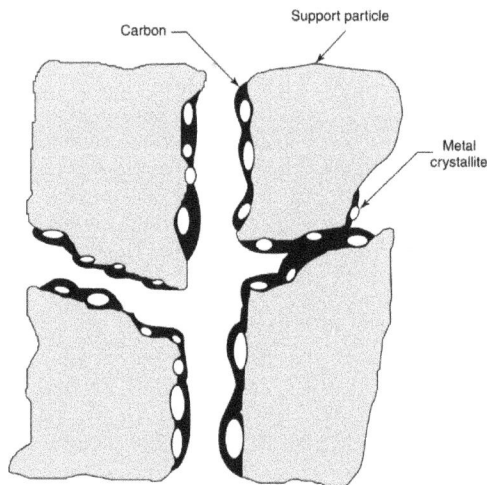

Figure 10. Conceptual model of fouling, crystallite encapsulation, and pore plugging of a supported metal catalyst owing to carbon deposition. Reproduced from [8]. Copyright 2006, Wiley-Interscience.

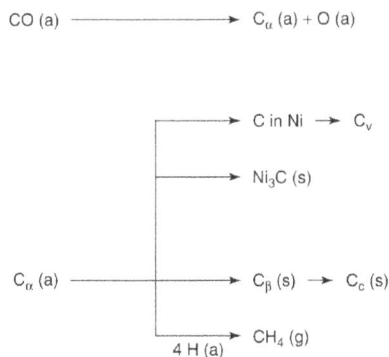

CO (a) ──────────────→ C_α (a) + O (a)

C in Ni → C_v

Ni_3C (s)

C_α (a) ──────────────→ C_β (s) → C_c (s)

CH₄ (g)

4 H (a)

Figure 11. Formation, transformation, and gasification of carbon on nickel (a, g, s refer to adsorbed, gaseous, and solid states respectively). Reproduced from [60]. Copyright 1983, Elsevier.

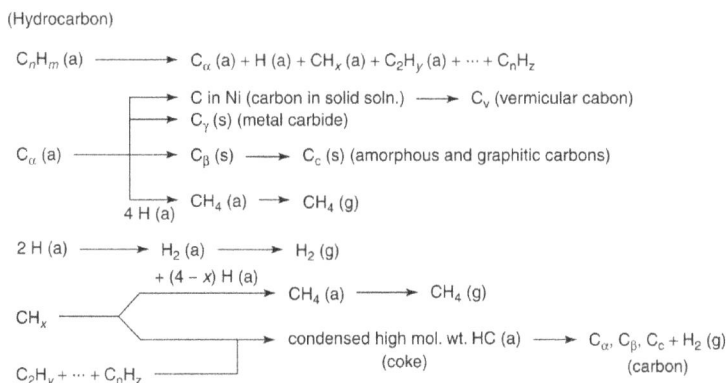

(Hydrocarbon)

C_nH_m (a) ──────→ C_α (a) + H (a) + CH_x (a) + C_2H_y (a) + ⋯ + C_nH_z

C in Ni (carbon in solid soln.) ──────→ C_v (vermicular cabon)
C_γ (s) (metal carbide)

C_α (a) ──────→ C_β (s) ──────→ C_c (s) (amorphous and graphitic carbons)

4 H (a)   CH₄ (a) → CH₄ (g)

2 H (a) ──────→ H_2 (a) ──────→ H_2 (g)

+ (4 − x) H (a)

CH_x ──────→ CH₄ (a) ──────→ CH₄ (g)

──────→ condensed high mol. wt. HC (a) ──────→ C_α, C_β, C_c + H_2 (g)
(coke)   (carbon)

C_2H_y + ⋯ + C_nH_z

Figure 12. Formation and transformation of coke on metal surfaces (a, g, s refer to adsorbed, gaseous, and solid states respectively); gas phase reactions are not considered. Reproduced from [60]. Copyright 1983, Elsevier.

Table 6. Forms and Reactivities of Carbon Species Formed by Decomposition of CO on Nickel [a].

Structural type	Designation	Temp. formed, °C	Peak temp. for reaction with H_2, °C
1. Adsorbed, atomic (surface carbide)	C_α	200–400	200
2. Polymeric, amorphous films or filaments	C_β	250–500	400
3. Vermicular filaments, fibers, and/or whiskers	C_v	300–1000	400–600
4. Nickel carbide (bulk)	$C\gamma$	150–250	275
5. Graphitic (crystalline) platelets or films	C_c	500–550	550–850

[a] Ref. [60].

Table 7. Carbon Species Formed in Steam Reforming of Hydrocarbons on Nickel Catalysts [a].

Attribute	Encapsulating film	Whisker-like	Pyrolytic carbon
Formation	Slow polymerization of C_nH_m radicals on Ni surface, into encapsulating film	Diffusion of C through Ni crystal, nucleation and whisker growth with Ni crystal at top	Thermal cracking of hydrocarbon; deposition of C precursors on catalyst
Effects	Progressive deactivation	No deactivation of Ni surface. Breakdown of catalyst and increasing ΔP	Encapsulation of catalyst particle; deactivation and increasing ΔP
Temp. range, °C	<500	>450	>600
Critical parameters	Low temperature, low H_2O/C_nH_m, low H_2/C_nH_m, aromatic feed	High temperature, low H_2O/C_nH_m, no enhanced H_2O adsorption, low activity, aromatic feed	High temperature, high void fraction, low H_2O/C_nH_m, high pressure, acidic catalyst

[a] Ref. [60].

It should also be emphasized that some forms of carbon result in loss of catalytic activity and some do not. For example, at low reaction temperatures (<300–375 °C) condensed polymer or β-carbon films and at high temperatures (>650 °C) graphitic carbon films encapsulate the metal surfaces of methanation and steam reforming catalysts [60]. Deactivation of steam reforming catalysts at high reaction temperatures (500–900 °C) may be caused by precipitation of atomic (carbidic) carbon dissolved in the Ni surface layers to a depth of more than 50–70 nm [62,66]. If it accumulates on the metal surface (at high or low temperatures), adsorbed atomic carbon can deactivate metal sites for adsorption and/or reaction. For example, Durer and co-workers [67] demonstrated that carbon atoms residing in the fourfold hollow sites of Rh(100) block the adsorption of hydrogen (and hence could block sites for hydrogenation). In the intermediate temperature range of 375–650 °C, carbon filaments (Figure 13) are formed by precipitation of dissolved carbon at the rear side of metal crystallites, causing the metal particles to grow away from the support [57]. Filament growth ceases when sufficient carbon accumulates on the free surface to cause encapsulation by a carbon layer; however, encapsulation of the metal particles does not occur if H_2/CO or H_2O/hydrocarbon ratios are sufficiently high. Thus, carbon filaments sometimes formed in CO hydrogenation or steam reforming of hydrocarbons would not necessarily cause a loss of intrinsic catalyst activity unless they are formed in sufficient quantities to cause plugging of the pores [60] or loss of metal occurs as the carbon fibers are removed during regeneration [68,69]. However, in practice, regions of carbon forming potential in steam reforming must be carefully avoided, since once initiated, the rates of filamentous carbon formation are sufficiently high to cause catastrophic pore plugging and catalyst failure within a few hours to days.

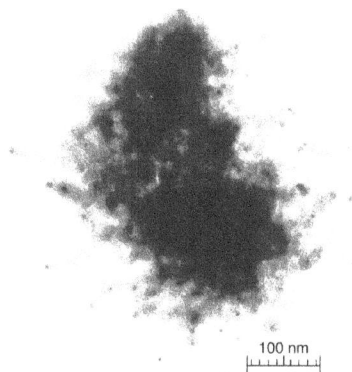

Figure 13. Electron micrograph of 14% Ni/Al_2O_3 having undergone extensive carbon deposition during CO disproportionation at 673 K, P_{CO} = 4.55 kPa (magnification of 200,000). Courtesy of the BYU Catalysis Laboratory.

The rate at which deactivation occurs for a given catalyst and reaction depends greatly on reaction conditions—especially temperature and reactant composition. A fundamental principle for coke-insensitive reactions on metals (e.g., methanation, Fischer–Tropsch synthesis, steam reforming, catalytic reforming, and methanol synthesis) is that deactivation rate depends greatly on the difference in rates of formation and gasification of carbon/coke precursors, *i.e.*, $r_d = r_f - r_g$. If the rate of gasification, r_g, is equal to or greater than that of formation, r_f, carbon/coke is not deposited. Rates of carbon/coke precursor formation and gasification both increase exponentially with temperature, although the difference between them varies a great deal with temperature because of differences in preexponential factors and activation energies. Thus, carbon/coke formation is avoided in regions of temperature in which precursor gasification rate exceeds deposition rate. This is illustrated in Figure 14, an Arrhenius plot for rates of formation and hydrogenation of alpha and beta carbons on nickel during CO methanation. Since at temperatures below 600 K ($1/T > 1.66 \times 10^{-3}$ K^{-1}) the rate of C_α gasification exceeds that of C_α formation, no carbon is deposited. However above 600 K, C_α accumulates on the surface since the rate of C_α formation exceeds that of C_α gasification. As C_α accumulates (at 600–700 K), it is converted to a C_β polymeric chain or film that deactivates the nickel catalyst; however, above 700 K ($1/T < 1.43 \times 10^{-3}$ K^{-1}), the rate of C_β hydrogenation exceeds that of formation and no deactivation occurs. Thus, the "safe" regions of methanation for avoiding deactivation by carbon are below 600 K and above 700 K; of course, these regions will vary somewhat with reactant concentrations and catalyst activity. A similar principle operates in steam reforming, *i.e.*, at a sufficiently low reaction temperature, the rate of hydrocarbon adsorption exceeds the rate of hydrocracking and a deactivating polymer film is formed [70]; accordingly, it is necessary to operate above this temperature to avoid deactivation.

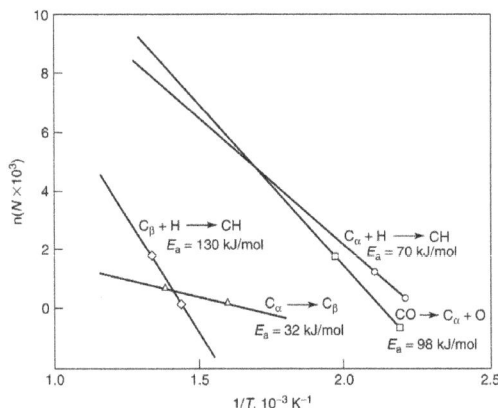

Figure 14. Rates of formation (log scale) and hydrogenation of C_α and C_β *versus* reciprocal temperature. Reproduced from [60]. Copyright 1983, Elsevier.

In steam reforming, filamentous carbon formation rate is a strong function of reactant hydrocarbon structure; for example, it decreases in the order acetylenes, olefins, paraffins, *i.e.*, in order of decreasing reactivity, although activation energies for nickel are in the same range (125–139 kJ) independent of hydrocarbon structure and about the same as those observed for formation of filamentous carbon from decomposition of CO [60]. This latter observation suggests that the reactions of CO and different hydrocarbons to filamentous carbon proceed by a common mechanism and rate-determining step—probably the diffusion of carbon through the metal crystallites [60].

The rate at which a carbon or coke is accumulated in a given reaction under given conditions can vary significantly with catalyst structure, including metal type, metal crystallite size, promoter, and catalyst support. For example, supported Co, Fe, and Ni are active above 350–400 °C for filamentous carbon formation from CO and hydrocarbons; the order of decreasing activity is reportedly Fe > Co > Ni [60]. Pt, Ru, and Rh catalysts, on the other hand, while equally or more active than Ni, Co, or Fe in steam reforming, produce little or no coke or carbon. This is attributed to reduced mobility and/or solubility of carbon in the noble metals, thus retarding the nucleation process. Thus, it is not surprising that addition of noble metals to base metals retards carbon formation; for example, addition of Pt in Ni lowers carbon deposition rate during methanation, while addition of Cu or Au to Ni substantially lowers carbon formation in steam reforming [60,71]. In contrast to the moderating effects of noble metal additives, addition of 0.5% Sn to cobalt substantially increases the rate of carbon filament formation from ethylene [72], an effect desirable in the commercial production of carbon filament fibers.

Since carbon formation and gasification rates are influenced differently by modifications in metal crystallite surface chemistry, which are in turn a function of catalyst structure, oxide additives or oxide supports may be used to moderate the rate of undesirable carbon or coke accumulation. For example, Bartholomew and Strasburg [73] found the specific rate (turnover frequency) of filamentous carbon deposition on nickel during methanation at 350 °C to decrease in the order Ni/TiO_2 > $NiAl_2O_3$ > Ni/SiO_2, while Vance and Bartholomew [74] observed C_α

hydrogenation rates at 170 °C to decrease in this same order (the same as for methanation at 225 °C). This behavior was explained in terms of promotional or inhibiting effects due to decoration of metal crystallites by the support, for example silica, inhibiting both CO dissociation and carbon hydrogenation. This hypothesis is consistent with observations [75,76] that silica evaporated on metal surfaces and supported metals inhibits formation of filamentous carbon. Similarly Bitter and co-workers [77] observed rates of carbon formation in CO_2/CH_4 reforming to decrease in the order $Pt/\gamma\text{-}Al_2O_3 \rightarrow Pt/TiO_2 > Pt/ZrO_2$; while 90% of the carbon deposited on the support, the authors linked deactivation to carbon accumulated on the metal owing to an imbalance between carbon formed by methane dissociation and oxidation by chemisorbed CO_2. The rate of formation of coke in steam reforming is delayed and occurs at lower rates in nickel catalysts promoted with alkali or supported on basic MgO [78].

Since formation of coke, graphite, or filamentous carbon involves the formation of C-C bonds on multiple atoms sites, one might expect that coke or carbon formation on metals is structure-sensitive, *i.e.*, sensitive to surface structure and metal crystallite size. Indeed, Bitter and co-workers [77] found that catalysts containing larger Pt crystallites deactivate more rapidly than those containing small crystallites. Moreover, a crystallite size effect, observed in steam reforming of methane on nickel [60,78], appears to operate in the same direction, *i.e.*, formation of filamentous carbon occurs at lower rates in catalysts containing smaller metal crystallites.

In summary, deactivation of supported metals by carbon or coke may occur chemically, owing to chemisorption or carbide formation, or physically and mechanically, owing to blocking of surface sites, metal crystallite encapsulation, plugging of pores, and destruction of catalyst pellets by carbon filaments. Blocking of catalytic sites by chemisorbed hydrocarbons, surface carbides, or relatively reactive films is generally reversible in hydrogen, steam, CO_2, or oxygen. Further details of the thermodynamics, kinetics, and mechanisms of carbon and coke formation in methanation and steam reforming reactions are available in reviews by Bartholomew [60] and Rostrup-Nielsen [70,78]. In recent reviews addressing deactivation of Co catalysts by carbon during Fischer-Tropsch synthesis [79,80], the same or similar carbon species, e.g., α, β, polymeric, and graphitic carbons, are observed on Co surfaces as on Ni; moreover, poisoning or fouling of the Co surfaces with β, polymeric, and graphitic carbon layers are found to be major causes of deactivation.

2.2.3. Coke Formation on Metal Oxide and Sulfide Catalysts

In reactions involving hydrocarbons, coke may be formed in the gas phase and on both noncatalytic and catalytic surfaces. Nevertheless, formation of coke on oxides and sulfides is principally a result of cracking reactions involving coke precursors (typically olefins or aromatics) catalyzed by acid sites [81,82]. Dehydrogenation and cyclization reactions of carbocation intermediates formed on acid sites lead to aromatics, which react further to higher molecular weight polynuclear aromatics that condense as coke (see Figure 15). Reactions 1–3 in Figure 15 illustrate the polymerization of olefins, reactions 4–8 illustrate cyclization from olefins, and reactions 9–14 illustrate chain reaction formation of polynuclear aromatics that condense as coke on the catalyst surface. Because of the high stability of the polynuclear carbocations (formed in reactions 10–13), they can continue to grow on the surface for a relatively long time before a termination reaction occurs through the back donation of a proton.

From this mechanistic scheme (Figure 15), it is clear that olefins, benzene and benzene derivatives, and polynuclear aromatics are precursors to coke formation. However, the order of reactivity for coke formation is clearly structure dependent, *i.e.*, decreases in the order polynuclear aromatics > aromatics > olefins > branched alkanes > normal alkanes. For example, the weight percent coke formed on silica–alumina at 500 °C is 0.06, 3.8, 12.5, and 23% for benzene, naphthalene, fluoranthene, and anthracene respectively [83].

Coking reactions in processes involving heavy hydrocarbons are very complex; different kinds of coke may be formed and they may range in composition from CH to C and have a wide range of reactivities with oxygen and hydrogen, depending upon the time on stream and temperature to which they are exposed. For example, coke deposits occurring in hydrodesulfurization of residuum have been classified into three types [84]:

(1) Type I deposits are reversibly adsorbed normal aromatics deposited during the first part of the cycle at low temperature.

(2) Type II deposits are reversibly adsorbed asphaltenes deposited early in the coking process.

(3) Type III deposits result from condensation of aromatic concentrates into clusters and then crystals that constitute a "mesophase." This crystalline phase is formed after long reaction times at high temperature. This hardened coke causes severe deactivation of the catalyst [84].

(a) Polymerization of Olefins

Step 1: Reaction of olefin with Brønsted acid form secondary carbenium ion:

$$H_2C=CHCH_3 + HX \rightleftharpoons CH_3-\overset{+}{C}HCH_3 + X^- \tag{1}$$

Step 2: Condensation reaction of a C_3 cabocation with a C_3 olefin to form a condensed, branched C_6 product with a carbenium ion:

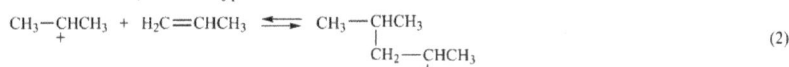

$$CH_3-\overset{+}{C}HCH_3 + H_2C=CHCH_3 \rightleftharpoons \begin{array}{l} CH_3-CHCH_3 \\ | \\ CH_2-\overset{+}{C}HCH_3 \end{array} \tag{2}$$

Step 3: Reaction of carbenium ion with Brønsted base to form olefin:

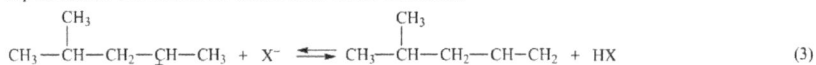

$$\begin{array}{cc} CH_3 & CH_3 \\ | & | \\ CH_3-CH-CH_2-\overset{+}{C}H-CH_3 + X^- \rightleftharpoons CH_3-CH-CH_2-CH=CH_2 + HX \end{array} \tag{3}$$

(b) Cyclization from Olefins

Step 1: Formation of an allylic carbocation by reaction of a diene with a primary carbocation:

$$R_1^+ + R_2-CH=CH-CH=CH-CH-CH_2CH_3 \rightleftharpoons R_1H + (R_2-CH\cdots CH\cdots CH\cdots CH\cdots CH-CH_2CH_3)^+ \tag{4}$$

Step 2: Reaction of an allylic carbocation with a Brønsted base to form a triene:

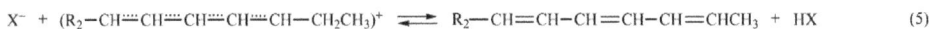

$$X^- + (R_2-CH\cdots CH\cdots CH\cdots CH\cdots CH-CH_2CH_3)^+ \rightleftharpoons R_2-CH=CH-CH=CH-CH=CHCH_3 + HX \tag{5}$$

Step 3: Cyclization of a triene to form a substituted cyclohexadiene:

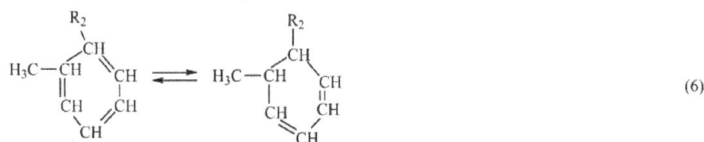

$$\tag{6}$$

Figure 15. *Cont.*

23

Step 4: Formation of a tertiary carbocation:

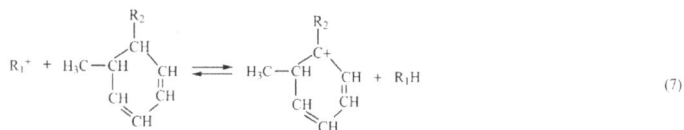

(7)

Step 5: Reaction of a tertiary carbocation with Brønsted base to form substituted benzene:

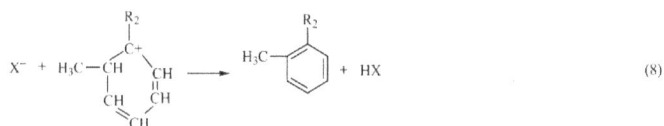

(8)

(c) Formation of Polynuclear Aromatics from Benzene

Step 1: Initiation (protonation of benzene):

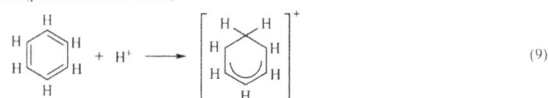

(9)

Step 2: Propagation (condensation reaction of carbocation with benzene, followed by H abstraction):

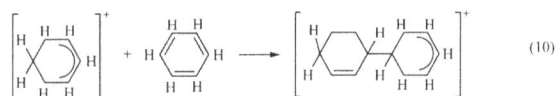

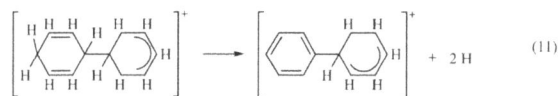

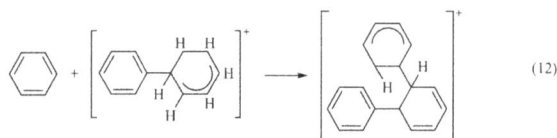

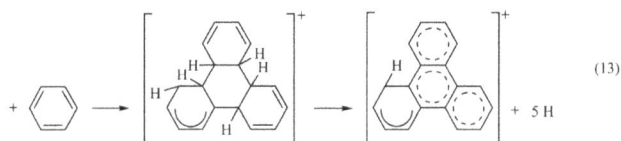

(10)

(11)

(12)

(13)

and so forth.

Step 3: Termination (reaction of carbocation with Brønsted base):

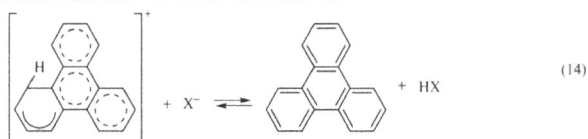

(14)

Figure 15. Coke-forming reactions of alkenes and aromatics on oxide and sulfide catalysts: (**a**) polymerization of alkenes, (**b**) cyclization from alkenes, and (**c**) formation of polynuclear aromatics from benzene. Reproduced from [8], Copyright 2006, Wiley-Interscience.

In addition to hydrocarbon structure and reaction conditions, extent and rate of coke formation are also a function of the acidity and pore structure of the catalyst. Generally, the rate and extent of coke formation increase with increasing acid strength and concentration. Coke yield decreases with decreasing pore size (for a fixed acid strength and concentration); this is especially true in zeolites where shape selectivity plays an important role in coke formation. For example, coke yield in fluid catalytic cracking is only 0.4% for ZSM-5 (pore diameters of 0.54×0.56 nm) compared to 2.2% for Y-faujasite (aperture diameter of 0.72 nm) [82]. However, in pores of molecular diameter, a relatively small quantity of coke can cause substantial loss of activity. It should be emphasized that coke yield can vary considerably into the interior pores of a catalyst particle or along a catalyst bed, depending upon the extent to which the main and deactivation reactions are affected by film mass transport and pore diffusional resistance.

The mechanisms by which coke deactivates oxide and sulfide catalysts are, as in the case of supported metals, both chemical and physical. However, some aspects of the chemistry are quite different. The principal chemical loss of activity in oxides and sulfides is due to the strong adsorption of coke molecules on acidic sites. However, as discussed earlier, strong acid sites also play an important role in the formation of coke precursors, which subsequently undergo condensation reactions to produce large polynuclear aromatic molecules that physically coat catalytic surfaces. Physical loss of activity also occurs as coke accumulates, ultimately partially or completely blocking catalyst pores as in supported metal catalysts. For example, in isomerization of *cis*-butene on SiO_2/Al_2O_3 [85] catalyst deactivation occurs by rapid, selective poisoning of strong acid sites; coke evolved early in the reaction is soluble in dichloromethane and pyridine and is slightly aromatic. Apparently, the blocking of active sites does not significantly affect porosity or catalyst surface area, as SiO_2/Al_2O_3 contains relatively large mesopores.

In the case of supported bifunctional metal/metal oxide catalysts, different kinds of coke are formed on the metal and the acidic oxide support, e.g., soft coke (high H/C ratio) on Pt or Pt–Re metals and hard coke (low H/C ratio) on the alumina support in catalytic reforming [86]. In this case, coke precursors may be formed on the metal via hydrogenolysis, following which they migrate to the support and undergo polymerization and cyclization reactions, after which the larger molecules are dehydrogenated on the metal and finally accumulate on the support, causing loss of isomerization activity. Mild sulfiding of these catalysts (especially Pt–Re/alumina) substantially reduces the rate of hydrogenolysis and the overall formation of coke on both metal and support; it especially reduces the hard coke, which is mainly responsible for deactivation.

Several recent studies [82,87–97] have focused on coke formation during hydrocarbon reactions in zeolites including (1) the detailed chemistry of coke precursors and coke molecules formed in zeolite pores and pore intersections (or supercages) and (2) the relative importance of adsorption on acid sites *versus* pore blockage. The principal conclusions from these studies can be summarized as follows: (1) the formation of coke and the manner in which it deactivates a zeolite catalyst are shape-selective processes, (2) deactivation is mainly due to the formation and retention of heavy aromatic clusters in pores and pore intersections, and (3) while both acid-site poisoning and pore blockage participate in the deactivation, the former dominates at low coking rates, low coke coverages (e.g., in Y-zeolite below 2 wt%), and high temperatures, while the latter process dominates at high reaction rates, high coke coverages, and low temperatures. Thus, pore size and pore structure are probably more important than acid strength and density under typical commercial process conditions. Indeed, deactivation is typically more rapid in zeolites having small pores or apertures and/or a monodimensional structure [95]. Figure 16 illustrates four possible modes of deactivation of HZSM-5 by carbonaceous deposits with increasing severity of coking [95].

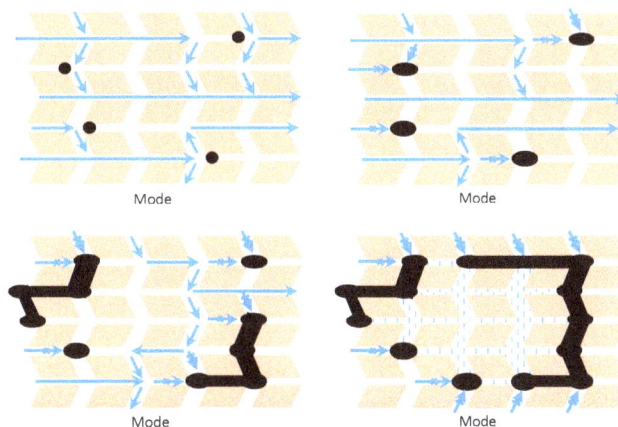

Figure 16. Schematic of the four possible modes of deactivation by carbonaceous deposits in HZSM-5: (1) reversible adsorption on acid sites, (2) irreversible adsorption on sites with partial blocking of pore intersections, (3) partial steric blocking of pores, and (4) extensive steric blocking of pores by exterior deposits. Adapted from [95].

These conclusions (in the previous paragraph) are borne out, for example, in the study by Cerqueira and co-workers [97] of USHY zeolite deactivation during methylcyclohexane transformation at 450 °C, showing the following:

(1) Coke is probably mainly formed by rapid transformation of toluenic C_7 carbenium ions with lesser contributions from reactions of cyclopentadiene, C_3–C_6 olefins, and aromatics.

(2) Soluble coke consists of polynuclear aromatic clusters containing three to seven five- and six-membered rings having a typical compositions of $C_{30}H_{40}$ to $C_{40}H_{44}$ and having dimensions of 0.9×1.1 nm to 1.1×1.5 nm, *i.e.*, sizes that would cause them to be trapped in the supercages of Y-zeolite.

(3) At short contact times, coking is relatively slow and deactivation is mainly due to acid-site poisoning, while at long contact times, coking is much faster because of the high concentrations of coke precursors; under these latter conditions coke is preferentially deposited at the outer pore openings of zeolite crystallites and deactivation is dominated by pore-mouth blockage.

That coke formed at large contact times not only blocks pores and/or pore intersections inside the zeolite, but also migrates to the outside of zeolite crystallites, where it blocks pore entrances, has been observed in several studies [91,93,94,97]. However, the amount, structure, and location of coke in ZSM-5 depends strongly on the coke precursor, e.g., coke formed from mesitylene is deposited on the external zeolite surface, whereas coking with isobutene leads to largely paraffinic deposits inside pores; coke from toluene, on the other hand, is polyaromatic and is deposited both on external and internal zeolite surfaces [91].

2.3. Thermal Degradation and Sintering

2.3.1. Background

Thermally induced deactivation of catalysts results from (1) loss of catalytic surface area due to crystallite growth of the catalytic phase, (2) loss of support area due to support collapse and of catalytic surface area due to pore collapse on crystallites of the active phase, and/or (3) chemical transformations of catalytic phases to noncatalytic phases. The first two processes are typically referred to as "sintering". The third is discussed in the next section under solid–solid reactions. Sintering processes generally take place at high reaction temperatures (e.g., > 500 °C) and are generally accelerated by the presence of water vapor.

Most of the previous sintering and redispersion work has focused on supported metals. Experimental and theoretical studies of sintering and redispersion of supported metals published before 1997 have been reviewed fairly extensively [8,98–107]. Three principal mechanisms of metal crystallite growth have been advanced: (1) crystallite migration, (2) atomic migration, and (3) (at very high temperatures) vapor transport. The processes of crystallite and atomic migration are illustrated in Figure 17. Crystallite migration involves the migration of entire crystallites over the support surface, followed by collision and coalescence. Atomic migration involves detachment of metal atoms or molecular metal clusters from crystallites, migration of these atoms over the support surface, and ultimately, capture by larger crystallites. Redispersion, the reverse of crystallite growth in the presence of O_2 and/or Cl_2, may involve (1) formation of volatile metal oxide or metal chloride complexes that attach to the support and are subsequently decomposed to small crystallites upon reduction and/or (2) formation of oxide particles or films that break into small crystallites during subsequent reduction.

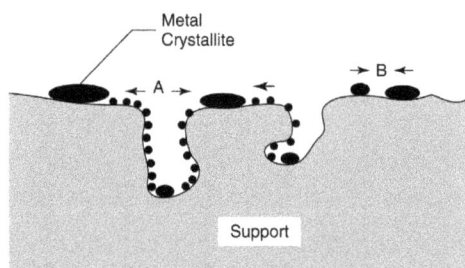

Figure 17. Two conceptual models for crystallite growth due to sintering by (**A**) atomic migration or (**B**) crystallite migration. Reproduced from [8], Copyright 2006, Wiley-Interscience.

There is controversy in the literature regarding which mechanism of sintering (or redispersion) operates at a given set of conditions. Logically, atomic migration would be favored at lower temperatures than crystallite migration, since the higher diffusivities of atoms or small cluster would facilitate their migration, whereas the thermal energy necessary to induce motion of larger crystallites would only be available at higher temperatures. Moreover, migration of small crystallites might be favorable early in the sintering process but unfavorable as crystallites become larger. However, focusing on only one of the three sintering mechanisms (and two dispersion mechanisms) is a simplification that ignores the possibility that all mechanisms may occur simultaneously and may be coupled with each other through complex physicochemical processes, including the following: (1) dissociation and emission of metal atoms or metal-containing molecules from metal crystallites, (2) adsorption and trapping of metal atoms or metal-containing molecules on the support surface, (3) diffusion of metal atoms, metal-containing molecules and/or metal crystallites across support surfaces, (4) metal or metal oxide particle spreading, (5) support surface wetting by metal particles, (6) metal particle nucleation, (7) coalescence of, or bridging between, two metal particles, (8) capture of atoms or molecules by metal particles, (9) liquid formation, (10) metal volatilization through volatile compound formation, (11) splitting of crystallites in O_2 atmosphere owing to formation of oxides of a different specific volume, and (12) metal atom vaporization. Depending upon reaction or redispersion conditions, a few or all of these processes may be important; thus, the complexity of sintering/redispersion processes is emphasized.

In general, thermal sintering processes are kinetically slow (at moderate reaction temperatures) and irreversible or difficult to reverse. Thus, sintering is more easily prevented than cured.

2.3.2. Factors Affecting Metal Particle Growth and Redispersion in Supported Metals

Temperature, atmosphere, metal type, metal dispersion, promoters/impurities, and support surface area, texture, and porosity are the principal parameters affecting rates of sintering and redispersion (see Table 8) [8,103–107]. Sintering rates increase exponentially with temperature. Metals sinter relatively rapidly in oxygen and relatively slowly in hydrogen, although depending upon the support, metal redispersion can be facilitated by exposure at high temperature (e.g., 500–550 °C for Pt/Al_2O_3) to oxygen and chlorine, followed by reduction. Water vapor also increases the sintering rate of supported metals, likely through chemical-assisted sintering effects similar to those described in Section 2.4.3.

Normalized dispersion (percentage of metal exposed at any time divided by the initial percentage exposed) *versus* time data in Figure 18 show that at temperatures of 650 °C or higher, rates of metal surface area loss (measured by hydrogen chemisorption) due to sintering of Ni/silica in hydrogen atmosphere are significant, causing 70% loss of the original metal surface area within 50 h at 750 °C. In reducing atmosphere, metal crystallite stability generally decreases with decreasing metal melting temperature, *i.e.*, in the order Ru > Ir > Rh > Pt > Pd > Ni > Cu > Ag, although this order may be affected by relatively stronger metal–support interactions, e.g., the observed order of decreasing stability of supported platinum in vacuum is Pt/Al_2O_3 > Pt/SiO_2 > Pt/C. In oxidizing atmospheres, metal crystallite stability depends on the volatility of metal oxides and the strength of the metal–oxide–support interaction. For noble metals, metal stability in air decreases in the order Rh > Pt > Ir > Ru; formation of volatile RuO_4 accounts for the relative instability of ruthenium.

Table 8. Effects of Important Reaction and Catalyst Variables on Sintering Rates of Supported Metals Based on General Power-Law Expression (GPLE) Data [a].

Variable	Effect
Temperature	Sintering rates are exponentially dependent on T; E_{act} varies from 30 to 150 kJ/mol. E_{act} decreases with increasing metal loading; it increases in the following order with atmosphere: $NO < O_2 < H_2 < N_2$
Atmosphere	Sintering rates are much higher for noble metals in O_2 than in H_2 and higher for noble and base metals in H_2 relative to N_2. Sintering rate decreases for supported Pt in atmospheres in the following order: $NO > O_2 > H_2 > N_2$
Metal	Observed order of decreasing thermal stability in H_2 is Ru > Ir $\cong$ Rh > Pt; thermal stability in O_2 is a function of (1) volatility of metal oxide and (2) strength of metal oxide–support interaction
Support	Metal–support interactions are weak (bond strengths of 5–15 kJ/mol); with a few exceptions, thermal stability for a given metal decreases with support in the following order: $Al_2O_3 > SiO_2$ > carbon
Promoters	Some additives decrease atom mobility, e.g., C, O, CaO, BaO, CeO_2, GeO_2; others increase atom mobility, e.g., Pb, Bi, Cl, F, or S. Oxides of Ba, Ca, or Sr are "trapping agents" that decrease sintering rate
Pore size	Sintering rates are lower for porous *versus* nonporous supports; they decrease as crystallite diameters approach those of the pores

[a] Refs. [8,103–107]. For the definition of a GPLE, see Equation 2 later in this section.

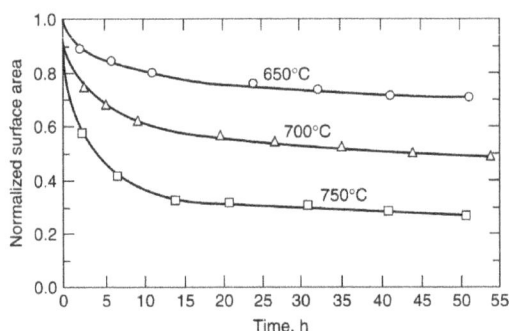

Figure 18. Normalized nickel surface area (based on H_2 adsorption) *versus* time data during sintering of 13.5% Ni/SiO_2 in H_2 at 650, 700, and 750 °C. Reproduced from [108]. Copyright 1983, Elsevier.

The effect of temperature on sintering of metals and oxides can be understood physically in terms of the driving forces for dissociation and diffusion of surface atoms, which are both proportional to the fractional approach to the absolute melting point temperature (T_{mp}). Thus, as temperature increases, the mean lattice vibration of surface atoms increases; when the Hüttig temperature ($0.3T_{mp}$) is reached, less strongly bound surface atoms at defect sites (e.g., edges and corner sites) dissociate and diffuse readily over the surface, while at the Tamman temperature ($0.5T_{mp}$), atoms in the bulk become mobile. Accordingly, sintering rates of a metal or metal oxide are significant above the Hüttig temperature and very high near the Tamman temperature; thus, the relative thermal stability of metals or metal oxides can be correlated in terms of the Hüttig or Tamman temperatures [109]. This can be illustrated from values of the melting and Tamman temperatures for noble and base metals and their compounds listed in Table 9. For example, sintering of copper catalysts for methanol synthesis is promoted by traces of chlorine in the feed, which react at about 225 °C (500 K) with the active metal/metal oxide surface to produce a highly mobile copper chloride phase having a Tamman temperature of only 79–174 °C (352–447 K) relative to 405–527 °C (678–800 K) for copper metal or metal oxides [110].

Table 9. Values of Melting and Tamman Temperatures (°C) for Common Catalytic Metals and Their Compounds [a].

Compound	T_{mp}, K	T_{Tamman}, K	$T_{Hüttig}$, K
Ag	1233	617	370
Au	1336	668	401
Co	1753	877	526
Cu	1356	678	407
CuO	1599	800	480
Cu_2O	1508	754	452
$CuCl_2$	893	447	268
Cu_2Cl_2	703	352	211
Fe	1808	904	542
Mo	2883	1442	865
MoO_3	1068	534	320
MoS_2	1458	729	437
Ni	1725	863	518
NiO	2228	1114	668
$NiCl_2$	1281	641	384
$Ni(CO)_4$	254	127	76
Rh	2258	1129	677
Rh_2O_3	1373	687	412
Ru	2723	1362	817
Pd	1828	914	548
PdO	1023	512	307
Pt	2028	1014	608
PtO	823	412	247
PtO_2	723	362	217
$PtCl_2$	854	427	256
$PtCl_4$	643	322	193
Zn	693	347	208
ZnO	2248	1124	674

[a] Adapted from Ref. [109].

Promoters or impurities affect sintering and redispersion by either increasing (e.g., chlorine and sulfur) or decreasing (e.g., oxygen, calcium, cesium) metal atom mobility on the support; in the latter case, this is due to their high resistance to dissociation and migration due to high melting points, as well as their hindering dissociation and surface diffusion of other atoms. Similarly, support surface defects or pores impede surface migration of metal particles—especially micropores and mesopores with pore diameters about the same size as the metal crystallites.

Historically, sintering rate data were fitted to a simple power-law expression (SPLE) of the form

$$-\frac{d(\frac{D}{D_0})}{dt} = k_s \left(\frac{D}{D_0}\right)^n \tag{1}$$

where k_s is the sintering rate constant, D_0 the initial dispersion, and n is the sintering order, which for typical catalyst systems may vary from 3 to 15; unfortunately, the SPLE is, in general, not valid for sintering processes because it assumes that surface area or dispersion ultimately reaches zero, given sufficient time, when in fact, for a given temperature and atmosphere, a nonzero or limiting dispersion is observed after long sintering times. Moreover, the use of the SPLE is further questionable because variations in sintering order are observed as a function of time and temperature for a given catalyst in a fixed atmosphere [105–107]; thus, data obtained for different samples and different reaction conditions cannot be quantitatively compared. Nevertheless, it has been shown by Fuentes [111,112] and Bartholomew [104–106] that the effects of temperature, atmosphere, metal, promoter, and support can be quantitatively determined by fitting sintering kinetic data to the general power-law expression (GPLE)

$$-\frac{d\left(\frac{D}{D_0}\right)}{dt} = k_s \left(\frac{D}{D_0} - \frac{D_{eq}}{D_0}\right)^m \tag{2}$$

which adds a term $-D_{eq}/D_0$ to account for the observed asymptotic approach of the typical dispersion *versus* time curve to a limiting dispersion D_{eq} at infinite time; m, the order of sintering, is found to be either 1 or 2. A recently compiled, comprehensive quantitative treatment of previous sintering rate data based on the GPLE with an order m of 2 [104–106] quantitatively addresses the effects of catalyst properties and reaction conditions on sintering rate. Some of these data are summarized in Table 10 [108,113–115]. These data show, for example, that the rate constant, and hence the rate of sintering, is less for Ni/Al_2O_3 than for Pt/Al_2O_3, an unexpected result in view of the lower heat of vaporization for Ni. This result is possibly explained by a greater metal support interaction for Ni with alumina.

Table 10. Comparison of Second-Order Sintering Rate Constants and Activation Energies for Pt, Ni, and Ag Catalysts [a].

Catalyst	Atm.	D_0 [b]	k_s [c] (400 °C)	k_s (650 °C)	k_s (700 °C)	k_s (750 °C)	E_{act} [d] kJ/mol	Ref.
0.6% Pt/γ-Al$_2$O$_3$	H$_2$	~0.85	0.007	0.310	0.530	1.32	79	[113]
5% Pt/γ-Al$_2$O$_3$	H$_2$	0.10	0.420	0.76	0.84	0.97	13	[114]
15% Ni/γ-Al$_2$O$_3$	H$_2$	0.16	0.004	0.083	0.13	0.27	66	[108]
0.6% Pt/γ-Al$_2$O$_3$	Air	~0.85	0.024	0.29	0.41	0.75	52	[113]
5% Pt/γ-Al$_2$O$_3$	Air	0.10	0.014	1.46	2.79	8.51	97	[114]
1.8% Ag/η-Al$_2$O$_3$	Air	0.36	0.69	-	-	-	-	[115]

[a] Refs. [105,106]; [b] Initial metal dispersion or percentage exposed; [c] Second-order sintering rate constant from general power-law expression (GPLE) with units of h^{-1}; [d] Sintering activation energy for GPLE, $-d(D/D_0)/dt = k_s[D/D_0 - D_{eq}/D_0]^m$, where $m = 2$.

Sintering studies of supported metals are generally of two types: (1) studies of commercially relevant supported metal catalysts and (2) studies of model metal–support systems. The former type provides useful rate data that can be used to predict sintering rates, while the latter type provides insights into the mechanisms of metal particle migration and sintering, although the results cannot be quantitatively extrapolated to predict behavior of commercial catalysts. There is direct evidence from the previous studies of model-supported catalysts [104,107] for the occurrence of crystallite migration (mainly in well-dispersed systems early in the sintering process), atomic migration (mainly at longer sintering times), and spreading of metal crystallites (mainly in oxygen atmosphere). There is also evidence that under reaction conditions, the surface is dynamic, *i.e.*, adsorbates and other adatoms rapidly restructure the surface and slowly bring about faceting; moreover, thermal treatments cause gradual changes in the distribution of coordination sites to minimize surface energy. There is a trend in increasing sophistication of spectroscopic tools used to study sintering and redispersion. In the next decade, we might expect additional insights into atomic and molecular processes during reaction at the atomic scale using STM, analytical high resolution transmission electron microscopy (HRTEM), and other such powerful surface science tools.

2.3.3. Sintering of Catalyst Carriers

Sintering of carriers (supports) has been reviewed by Baker and co-workers [103] and Trimm [116]. Single-phase oxide carriers sinter by one or more of the following processes: (1) surface diffusion, (2) solid-state diffusion, (3) evaporation/condensation of volatile atoms or molecules, (4) grain boundary diffusion, and (5) phase transformations. In oxidizing atmospheres, γ-alumina and silica are the most thermally stable carriers; in reducing atmospheres, carbons are the most thermally stable carriers. Additives and impurities affect the thermal properties of carriers by occupying defect sites or forming new phases. Alkali metals, for example, accelerate sintering; while calcium, barium, nickel, and lanthanum oxides form thermally stable spinel phases with alumina. Steam accelerates support sintering by forming mobile surface hydroxyl groups that are subsequently volatilized at higher temperatures. Chlorine also promotes sintering and grain growth in magnesia and titania during high temperature calcination. This is illustrated in Figure 19 [117]. By contrast, sulfuric acid treatment of hydrated alumina (gibbsite) followed by two-step calcination, results in a very stable transitional alumina with needle-like particle morphology [116]. Dispersed metals in supported metal catalysts can also accelerate support sintering; for example, dispersed nickel accelerates the loss of Al_2O_3 surface area in Ni/Al_2O_3 catalysts.

As an important example of support sintering through phase transformations, Al_2O_3 has a rich phase behavior as a function of temperature and preparation. A few among the many important phases that are stable or metastable, include boehmite, γ-alumina, and α-alumina [8,118,119]. Other phases are possible and the temperatures at which the phase transitions occur depend on crystal size and moisture content of the starting material, but as an example, as temperature is raised, boehmite, which is a hydrated or hydroxyl form of alumina, transforms to γ-alumina between 300 and 450 °C, then to δ-alumina at ~850°C, θ-alumina at ~1000°C, and finally α-alumina at ~1125 °C. The corresponding crystal structures for these five phases are orthorhombic, cubic defective spinel, orthorhombic, deformed monoclinic spinel, and hexagonal close pack (hcp with ABAB stacking) [8,118,119]. The approximate surface areas of these respective phases, as measured by nitrogen physisorption using Brunauer, Emmett, Teller (BET) analysis, are approximately 400, 200, 120, 50, and 1 m^2/g [8]. The dramatic drop in surface area during the transition from θ to α is associated with collapse of the microporous structure and formation of the dense hcp phase.

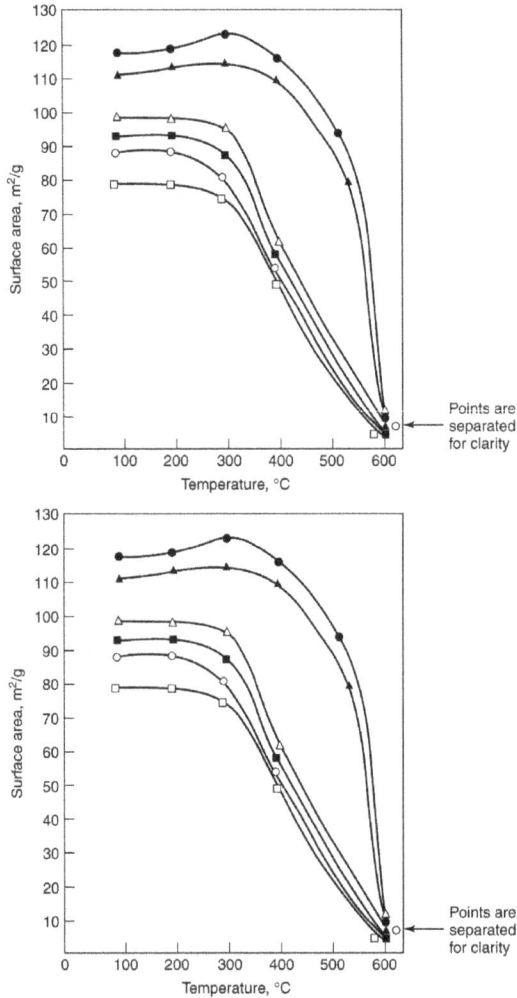

Figure 19. BET surface area of titania as a function of thermal treatment and chlorine content of fresh samples (before pretreatment). Samples were treated at the temperature indicated for 2 h. Reproduced from [117]. Copyright 1985, Elsevier. • = Blank TiO_2; ▲ = TiO_2 soaked in H_2O; △ = TiO_2 soaked in HCl/H_2O (2.06 wt% Cl); ■ = TiO_2 soaked in HCl/H_2O (2.40 wt% Cl); ○ = TiO_2 soaked in HCl/H_2O (2.55 wt% Cl); □ = TiO_2 soaked in HCl/H_2O (2.30 wt% Cl).

2.3.4. Effects of Sintering on Catalyst Activity

Baker and co-workers [103] have reviewed the effects of sintering on catalytic activity. Specific activity (based on catalytic surface area) can either increase or decrease with increasing metal crystallite size during sintering if the reaction is structure-sensitive, or it can be independent of changes in metal crystallite size if the reaction is structure-insensitive. Thus, for a structure-sensitive reaction, the impact of sintering may be either magnified or moderated; while for a structure insensitive-reaction, sintering has in principle no effect on specific activity (per unit surface area). In the latter case, the decrease in mass-based activity is proportional to the decrease in metal surface area. Ethane hydrogenolysis and ethane steam reforming are examples of structure-sensitive reactions, while CO hydrogenation on supported cobalt, nickel, iron, and ruthenium is largely structure-insensitive in catalysts of moderate loading and dispersion.

2.4. Gas/Vapor–Solid and Solid-State Reactions

In addition to poisoning, there are a number of chemical routes leading to catalyst deactivation: (1) reactions of the vapor phase with the catalyst surface to produce (a) inactive bulk and surface phases (rather than strongly adsorbed species), (b) volatile compounds that exit the catalyst and reactor in the vapor phase, or (c) sintering due to adsorbate interactions, that we call chemical-assisted sintering to distinguish it from thermal sintering previously discussed; (2) catalytic solid-support or catalytic solid-promoter reactions, and (3) solid-state transformations of the catalytic phases during reaction. Each of these routes is discussed in some detail below.

2.4.1. Gas/Vapor–Solid Reactions

2.4.1.1. Reactions of Gas/Vapor with Solid to Produce Inactive Phases

Dispersed metals, metal oxides, metal sulfides, and metal carbides are typical catalytic phases, the surfaces of which are similar in composition to the bulk phases. For a given reaction, one of these catalyst types is generally substantially more active than the others, e.g., only Fe and Ru metals are active for ammonia synthesis, while the oxides, sulfides, and carbides are inactive. If, therefore, one of these metal catalysts is oxidized, sulfided, or carbided, it will lose essentially all of its activity. While these chemical modifications are closely related to poisoning, the distinction here is that rather than losing activity owing to the presence of an adsorbed species, the loss of activity is due to the formation of a new phase altogether.

Examples of vapor-induced chemical transformations of catalysts to inactive phases are listed in Table 11 [8,120–127]. These include the formation of $RhAl_2O_4$ in the three-way Pt–Rh/Al_2O_3 catalyst during high temperature operation in an auto exhaust; oxidation of Fe by low levels of O_2 during ammonia synthesis or by H_2O during regeneration; dealumination (migration of Al from the zeolite framework) of Y-zeolite during high temperature catalytic cracking and regeneration in steam; reaction of SO_3 with the alumina support to form aluminum sulfate leading to support breakdown and catalyst pore plugging in several processes, including CO oxidation in a gas turbine exhaust, conversion of CO and hydrocarbons in a diesel exhaust converter, and selective catalytic reduction (SCR) of NO_x in utility boiler flue gases [8,122–124,127]; oxidation of Fe_5C_2 to Fe_3O_4 and of Co metal supported on alumina or silica to Co surface aluminates or silicates during Fischer–Tropsch synthesis at high conversions and hence high P_{H_2O}; and formation of $NiAl_2O_4$ during reaction and steam regeneration of Ni/Al_2O_3 in a slightly oxidizing atmosphere above about 500 °C, especially if more reactive aluminas, e.g., γ, δ, or θ forms, are used as supports. Because reaction of SO_3 with γ-Al_2O_3 to produce $Al_2(SO_4)_3$ is a serious cause of deactivation of alumina-supported catalysts in several catalytic processes (e.g., diesel exhaust abatement and SCR), TiO_2 or SiO_2 carriers are used rather than Al_2O_3 or in the diesel or automotive exhaust the alumina catalyst is stabilized by addition of BaO, SrO, or ZrO_2 [8,122–127].

Table 11. Examples of Reactions of Gases/Vapors with Catalytic Solids to Produce Inactive Phases.

Catalytic process	Gas/vapor composition	Catalytic solid	Deactivating chemical reaction	Ref.
Auto emissions control	N_2, O_2, HCs, CO, NO, H_2O, SO_2	Pt–Rh/Al_2O_3	2 Rh_2O_3 + γ-$Al_2O_3 \rightarrow RhAl_2O_4$ + 0.5 O_2	[120,121]
Ammonia synthesis and regeneration	H_2, N_2	Fe/K/Al_2O_3	Fe→FeO at >50 ppm O_2	[8]
	Traces O_2, H_2O		Fe→FeO at >0.16 ppm H_2O/H_2	
Catalytic cracking	HCs, H_2, H_2O	La-Y-zeolite	H_2O induced Al migration from zeolite framework causing zeolite destruction	[8]
CO oxidation, gas turbine exhaust	N_2, O_2, 400 ppm CO, 100–400 ppm SO_2	Pt/Al_2O_3	2 SO_3 + γ-$Al_2O_3 \rightarrow Al_2(SO_4)_3$ which blocks catalyst pores	[8]
Diesel HC/soot emissions control	N_2, O_2, HCs (gas and liquid), CO, NO, H_2O, soot, SO_2	Pt/Al_2O_3 and β-zeolite; oxides of CaCuFeVK on TiO_2	Formation of $Al_2(SO_4)_3$ or sulfates of Ca, Cu, Fe, or V, which block catalysts pores and lower activity for oxidation; Al_2O_3 stabilized by BaO	[122–124]
Fischer–Tropsch	CO, H_2, H_2O, CO_2, HCs	Fe/K/Cu/SiO_2	$Fe_5C_2 \rightarrow Fe_3O_4$ due to oxidation at high X_{CO} by product H_2O, CO_2	[125]
Fischer–Tropsch	CO, H_2, H_2O, HCs	Co/SiO_2	Co + $SiO_2 \rightarrow CoO \cdot SiO_2$ and collapse of SiO_2 by product H_2O	[126]
Selective catalytic reduction (SCR), stationary	N_2, O_2, NO, PM [a], H_2O, SO_2	$V_2O_5/WO_3/TiO_2$	Formation of $Al_2(SO_4)_3$ if Al_2O_3 is used	[127]
Steam reforming and regeneration in H_2O	CH_4, H_2O, CO, H_2, CO_2	Ni/Al_2O_3	Ni + $Al_2O_3 \rightarrow NiAl_2O_4$	[8]

[a] Particulate matter.

2.4.1.2. Reactions of Gas/Vapor with Solid to Produce Volatile Compounds

Metal loss through direct vaporization is generally an insignificant route to catalyst deactivation. By contrast, metal loss through formation of volatile compounds, e.g., metal carbonyls, oxides, sulfides, and halides in CO, O_2, H_2S, and halogen-containing environments, can be significant over a wide range of conditions, including relatively mild conditions. Classes and examples of volatile compounds are listed in Table 12. Carbonyls are formed at relatively low temperatures but high pressures of CO; halides can be formed at relatively low temperatures and low concentration of the halogens. However, the conditions under which volatile oxides are formed vary considerably with the metal; for example, RuO_3 can be formed at room temperature, while PtO_2 is formed at measurable rates only at temperatures exceeding about 500 °C.

Table 12. Types and Examples of Volatile Compounds Formed in Catalytic Reactions.

Gaseous environment	Compound type	Example of compound
CO, NO	Carbonyls and nitrosyl carbonyls	$Ni(CO)_4$, $Fe(CO)_5$ (0–300 °C) [a]
O_2	Oxides	RuO_3 (25 °C), PbO (>850 °C), PtO_2 (>700 °C)
H_2S	Sulfides	MoS_2 (>550 °C)
Halogens	Halides	$PdBr_2$, $PtCl_4$, PtF_6, $CuCl_2$, Cu_2Cl_2

[a] Temperatures of vapor formation are listed in parentheses.

While the chemical properties of volatile metal carbonyls, oxides, and halides are well known, there is surprisingly little information available on their rates of formation during catalytic reactions. There have been no reviews on this subject and relatively few reported studies to define the effects of metal loss on catalytic activity [28,128–141]. Most of the previous work has focused on volatilization of Ru in automotive converters [128–131]; nickel carbonyl formation in nickel catalysts during methanation of CO [133,139] or during CO chemisorption at 25 °C [28,135], and formation of Ru carbonyls during Fischer–Tropsch synthesis [136,137]; volatilization of Pt during ammonia oxidation on Pt–Rh gauze catalysts [140,141]; and volatilization of Cu from methanol synthesis and diesel soot oxidation catalysts, leading to sintering in the former and better catalyst–soot contact but also metal loss in the latter case [109].

Results of selected studies are summarized in Table 13. Bartholomew [131] found evidence of significant (50%) Ru loss after testing of a Pd–Ru catalyst in an actual reducing automobile exhaust for 100 h, which he attributed to formation of a volatile ruthenium oxide and which was considered responsible at least in part for a significant loss (20%) of NO reduction activity.

Table 13. Documented Examples of Reactions of Vapor with Solid to Produce Volatile Compounds.

Catalytic process	Catalytic solid	Vapor formed	Comments on deactivation process	Ref.
Automotive converter	Pd–Ru/Al_2O_3	RuO_4	50% loss of Ru during 100-h test in reducing automotive exhaust	[131]
Methanation of CO	Ni/Al_2O_3	$Ni(CO)_4$	$P_{CO} > 20$ kPa and $T < 425$ °C due to $Ni(CO)_4$ formation, diffusion and decomposition on the support as large crystallites	[133]
CO chemi-sorption	Ni catalysts	$Ni(CO)_4$	$P_{CO} > 0.4$ kPa and $T > 0$ °C due to $Ni(CO)_4$ formation; catalyzed by sulfur compounds	[134]
Fischer–Tropsch synthesis (FTS)	Ru/NaY zeolite, Ru/Al_2O_3, Ru/TiO_2	$Ru(CO)_5$, $Ru_3(CO)_{12}$	Loss of Ru during FTS (H_2/CO = 1, 200–250 °C, 1 atm) on Ru/NaY zeolite and Ru/Al_2O_3; up to 40% loss while flowing CO at 175–275 °C over Ru/Al_2O_3 for 24 h. Rate of Ru loss less on titania-supported Ru and for catalysts containing large metal crystallites (3 nm) relative to small metal crystallites (1.3 nm). Surface carbon lowers loss	[136,137]
Ammonia oxidation	Pt–Rh gauze	PtO_2	Loss: 0.05–0.3 g Pt/ton HNO_3; recovered with Pd gauze; loss of Pt leads to surface enrichment with inactive Rh	[8,142]
HCN synthesis	Pt–Rh gauze	PtO_2	Extensive restructuring and loss of mechanical strength	[8,143]
Methanol synthesis	CuZnO	$CuCl_2$, Cu_2Cl_2	Mobile copper chloride phase leads to sintering at reaction temperature (225 °C)	[109]
Diesel soot oxidation	Oxides of K, Cu, Mo, and trace Cl	$CuCl_2$, Cu_2Cl_2	Mobile copper chloride improves catalyst–soot contact; catalyst evaporation observed	[109]

Shen and co-workers [133] found that Ni/Al$_2$O$_3$ methanation catalysts deactivate rapidly during methanation at high partial pressures of CO (>20 kPa) and temperatures below 425 °C because of Ni(CO)$_4$ formation, diffusion, and decomposition on the support as large crystallites; under severe conditions (very high P_{CO} and relatively low reaction temperatures) loss of nickel metal occurs. Thus, loss of nickel and crystallite growth could be serious problems at the entrance to methanation reactors where the temperature is low enough and P_{CO} high enough for metal carbonyl formation. Agnelli and co-workers [139] investigated kinetics and modeling of sintering due to formation and migration of nickel carbonyl species. They found that the initially sharp crystallite size distribution evolved during several hours of sintering under low temperature (230 °C) reaction conditions to a bimodal system consisting of small spherical crystallites and large faceted crystals favoring (111) planes. The sintering process was modeled in terms of an Ostwald-ripening mechanism coupled with mass transport of mobile subcarbonyl intermediates. Long-term simulations were found to predict reasonably well the ultimate state of the catalyst. On the basis of their work, they proposed two solutions for reducing loss of nickel: (1) increasing reaction temperature and decreasing CO partial pressure in order to lower the rate of carbonyl formation, and (2) changing catalyst composition, e.g., alloying nickel with copper or adding alkali to inhibit carbonyl species migration.

Of note, Kuo and Hwang have shown that the particle morphology itself affects the rate of Ostwald ripening due to different relative chemical potential energies of the surfaces [144]. Using silver nanoparticles, they found that atoms at sharp edges and corners were removed first, resulting in more rounded particles for all starting geometries. Thus, initial particle geometry appears to have an effect in addition to the chemical atmosphere experienced by the particles.

Loss of nickel metal during CO chemisorption on nickel catalysts at temperatures above 0 °C is also a serious problem; moreover, this loss is catalyzed by sulfur poisoning [28]. In view of the toxicity of nickel tetracarbonyl, the rapid loss of nickel metal, and the ill-defined adsorption stoichiometries, researchers are advised to avoid using CO chemisorption for measuring nickel surface areas; instead, hydrogen chemisorption, an accepted ASTM method with a well-defined adsorption stoichiometry, is recommended [145]. Figure 20 illustrates a mechanism for the formation of Ni(CO)$_4$ on a crystallite of nickel in CO atmosphere.

Goodwin and co-workers [136,137] studied the influence of reaction atmosphere, support, and metal particle size on the loss of Ru due to carbonyl formation. They found that the loss of Ru during CO hydrogenation (H$_2$/CO = 1, 200–250 °C, 1 atm) on Ru/NaY zeolite and Ru/Al$_2$O$_3$ for extended periods of time was significant (e.g., up to 40% while flowing CO at 175–275 °C over Ru/Al$_2$O$_3$ for 24 h). The loss of Ru was significantly less on titania-supported Ru; moreover, the rate of loss was lower for catalysts containing large metal crystallites (3 nm) relative to those containing small metal crystallites (1.3 nm). Metal loss was inhibited in part at higher reaction temperatures as a result of carbon deposition. Thus, while it is clear that loss of ruthenium could be a serious problem in Fischer–Tropsch synthesis, there are measures in terms of catalyst design and choice of reaction conditions that can be taken to minimize loss.

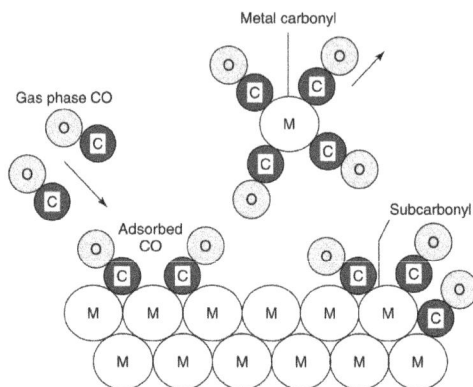

Figure 20. Formation of volatile tetra-nickel carbonyl at the surface of nickel crystallite in CO atmosphere. Reproduced from [8]. Copyright 2006, Wiley-Interscience.

One of the most dramatic examples of vapor phase loss of the catalyst occurs during NH_3 oxidation on Pt–Rh gauze, an important reaction in the manufacture of nitric acid [8,140,141]. At the high reaction temperature (~900 °C), formation of a volatile platinum oxide (PtO_2) occurs at a very significant rate; in fact, the rate of loss of 0.05–0.3 g Pt/ton of HNO_3 is high enough to provide a substantial economic incentive for Pt recovery [8]. The most effective recovery process involves placing a woven Pd-rich alloy gauze immediately below the Pt–Rh gauze to capture the Pt through formation of a Pd–Pt alloy. Pt loss is also the most significant cause of catalyst deactivation as the gauze surface becomes enriched in nonvolatile but inactive rhodium oxide [142], requiring shutdown and catalyst replacement every 3–12 months [8].

Decomposition of volatile platinum oxide species formed during high temperature reaction may (similar to the previously discussed formation of large crystallites of Ni from $Ni(CO)_4$) lead to formation of large Pt crystallites and/or substantial restructuring of the metal surface. For example, Wu and Phillips [146–148] observed surface etching, enhanced sintering, and dramatic surface restructuring of Pt thin films to faceted particles during ethylene oxidation over a relatively narrow temperature range (500–700 °C). The substantially higher rate of sintering and restructuring in O_2/C_2H_4 relative to that in nonreactive atmospheres was attributed to the interaction of free radicals such as HO_2, formed homogeneously in the gas phase, with the metal surface to form metastable mobile intermediates. Etching of Pt–Rh gauze in a H_2/O_2 mixture under the same conditions as Pt surfaces (600 °C, $N_2/O_2/H_2$ = 90/7.5/2.5) has also been reported [143]. A significant weight loss was observed in a laminar flow reactor with little change in surface roughness, while in an impinging jet reactor, there was little weight loss, but substantial restructuring of the surface to particle-like structures, 1–10 μm in diameter; these particles were found to have the same Pt–Rh composition as the original gauze. The nodular structures of about 10-μm diameter formed in these experiments are strikingly similar to those observed on Pt–Rh gauze after use in production of HCN at 1100 °C in 15% NH_3, 13% CH_4, and 72% air (see Figure 21). Moreover, because of the high space velocities during HCN production, turbulent rather than laminar flow would be expected, as in the impinging jet reactor. While little Pt is volatilized from the Pt–Rh gauze catalyst during HCN synthesis, the extensive restructuring leads to mechanical weakening of the gauze [8].

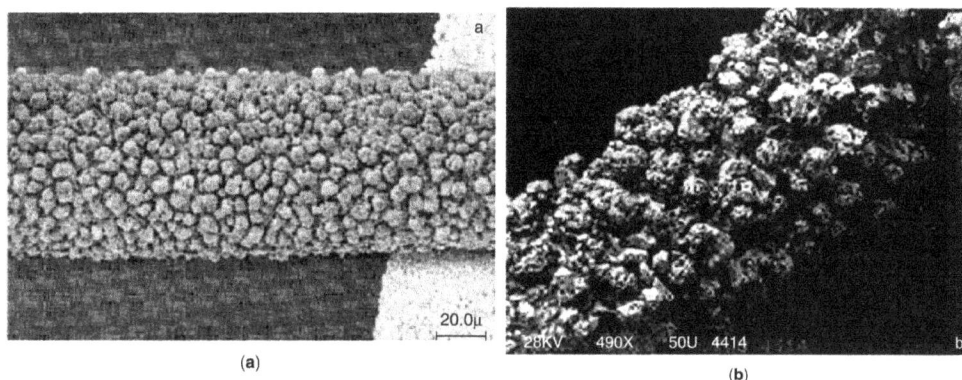

(a)   (b)

Figure 21. (a) SEM of Pt–Rh gauze after etching in $N_2/O_2/H_2 = 90/7.5/2.5$ at 875 K for 45 h. Reproduced from [143]. Copyright 1992, Elsevier. (b) SEM of Pt–Rh gauze after use in production of HCN (magnification 1000×). Photograph courtesy of Ted Koch at DuPont, personal correspondence to the author.

Other examples of catalyst deactivation due to volatile compound formation include (1) loss of the phosphorus promoter from the VPO catalyst used in the fluidized-bed production of maleic anhydride, with an attendant loss of catalyst selectivity [8], (2) vapor-phase loss of the potassium promoter from steam-reforming catalysts in the high temperature, steam-containing environment [8], and (3) loss of Mo from a 12-Mo-V-heteropolyacid due to formation of a volatile Mo species during oxydehydrogenation of isobutyric acid to methacrylic acid [138].

While relatively few definitive studies of deactivation by volatile compound formation have been reported, the previous work does provide the basis for enumerating some general principles. A generalized mechanism of deactivation by formation of volatile metal compounds can be postulated (see Figure 22). In addition, the roles of kinetics and thermodynamics can be stated in general terms:

(1) At low temperatures and partial pressures of the volatilization agent (VA), the overall rate of the process is limited by the rate of volatile compound formation.

(2) At intermediate temperatures and partial pressures of the VA, the rate of formation of the volatile compound exceeds the rate of decomposition. Thus, the rate of vaporization is high, the vapor is stable, and metal loss is high.

(3) At high temperatures and partial pressures of the VA, the rate of formation equals the rate of decomposition, *i.e.*, equilibrium is achieved. However, the volatile compound may be too unstable to form or may decompose before there is an opportunity to be transported from the system. From the previous work, it is also evident that besides temperature and gas phase composition, catalyst properties (crystallite size and support) can play an important role in determining the rate of metal loss.

Generalized Mechanism:

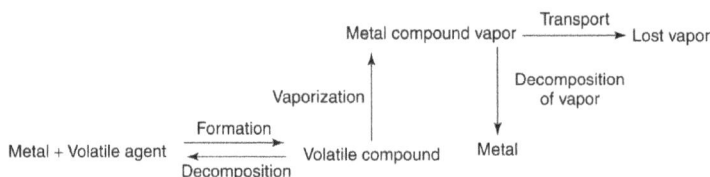

Metal compound vapor —————Transport————→ Lost vapor

Vaporization ↑ | ↓ Decomposition of vapor

Formation ————→

Metal + Volatile agent ←———————— Volatile compound   Metal
              Decomposition

Generalized Kinetics:

(a) rate of volatile compound formation = rate of formation – rate of decomposition

(b) rate of metal loss = rate of vaporization – rate of vapor decomposition

Figure 22. Generalized mechanisms and kinetics for deactivation by metal loss. Reproduced from [8]. Copyright 2006, Wiley-Interscience.

2.4.2. Solid-State Reactions

Catalyst deactivation by solid-state diffusion and reaction appears to be an important mechanism for degradation of complex multicomponent catalysts in dehydrogenation, synthesis, partial oxidation, and total oxidation reactions [8,149–160]. However, it is difficult in most of these reactions to know the extent to which the solid-state processes, such as diffusion and solid-state reaction, are affected by surface reactions. For example, the rate of diffusion of Al_2O_3 to the surface to form an aluminate may be enhanced by the presence of gas-phase oxygen or water or the nucleation of a different phase may be induced by either reducing or oxidizing conditions. Recognizing this inherent limitation, the focus here is nevertheless on processes in which formation of a new bulk phase (and presumably the attendant surface phase) leads to substantially lower activity. There is probably some overlap with some of the examples given under Gas/Vapor–Solid Reactions involving reactions of gas/vapor with solid to produce inactive phases.

Examples from the literature of solid-state transformations leading to catalyst deactivation are summarized in Table 14. They include (1) the formation of $KAlO_2$ during ammonia synthesis at the $Fe/K/Al_2O_3$ catalyst surface, (2) decomposition of the active phase PdO to inactive Pd metal during catalytic combustion on PdO/Al_2O_3 and PdO/ZrO_2 catalysts, (3) transformation of active carbides to inactive carbides in Fischer–Tropsch synthesis on Fe/K/Cu catalysts, (4) formation of inactive V(IV) compounds in SO_2 oxidation, and (5) reductive transformation of iron molybdate catalysts during partial oxidation of benzene, methanol, propene, and isobutene.

Table 14. Examples of Solid-State Transformations Leading to Catalyst Deactivation.

Catalytic process	Catalytic solid	Deactivating chemical reaction	Ref.
Ammonia synthesis	$Fe/K/Al_2O_3$	Formation of $KAlO_2$ at catalyst surface	[159]
Catalytic combustion	PdO/Al_2O_3, PdO/ZrO_2	PdO→Pd at $T > 800$ °C	[152]
Catalytic combustion	Co/K on MgO, CeO_2, or La_2O_3	Formation of CoO–MgO solid soln., $LaCoO_3$, or K_2O film on CeO_2	[160]

Table 14. *Cont.*

Catalytic process	Catalytic solid	Deactivating chemical reaction	Ref.
Dehydrogenation of ethyl benzene to styrene	$Fe_2O_3/Cr_2O_3/K_2O$	K migration to center of pellet caused by thermal gradient	[8]
Fischer–Tropsch	Fe/K, Fe/K/CuO	Transformation of active carbides to inactive carbides	[157,158]
Oxidation of SO_2 to SO_3	$V_2O_5/K_2O/Na_2O/$	Formation of inactive V(IV) compounds at $T < 420$–430 °C	[155]
Partial oxidation of benzene to maleic anhydride	V_2O_5–MoO_3	Decreased selectivity due to loss of MoO_3 and formation of inactive vanadium compounds	[149]
Partial oxidation of methanol to formaldehyde	$Fe_2(MoO_4)_3$ plus MoO_3	Structural reorganization to β-$FeMoO_4$; reduction of MoO_3	[150,156]
Partial oxidation of propene to acrolein	$Fe_2(MoO_4)_3$	Reductive transformation of $Mo_{18}O_{52}$ to Mo_4O_{11}	[153,156]
Partial oxidation of isobutene to methacrolein	$Fe_2(MoO_4)_3$	Reduction to $FeMoO_4$ and MoO_{3-x}	[151,154]

There are basic principles underlying most solid-state reactions in working catalysts that have been enumerated by Delmon [156]: (1) the active catalytic phase is generally a high-surface-area defect structure of high surface energy and as such a precursor to more stable, but less active phases and (2) the basic reaction processes may itself trigger the solid-state conversion of the active phase to an inactive phase; for example, it may involve a redox process, part of which nucleates the inactive phase.

A well-documented example of these principles occurs in the partial oxidation of propene to acrolein on a $Fe_2(MoO_4)_3$ catalyst [153,156]. This oxidation occurs by the "*Mars van Krevelen*" *mechanism*, *i.e.*, a redox mechanism in which lattice oxygen reacts with the adsorbed hydrocarbon to produce the partially oxygenated product; the reduced catalyst is restored to its oxidized state through reaction with gaseous oxygen. In propene oxidation, two atoms of oxygen from the catalyst are used, one for removing two hydrogen atoms from the olefin and the other one in forming the unsaturated aldehyde. The fresh, calcined catalyst MoO_3 consists of corner-sharing MoO_6 octahedra (with Mo at the center and six oxygen atoms at the corners); but upon reduction to MoO_2, octahedra share edges as shown in Figure 23. However, it has been reported [153,156] that only slightly reduced (relative to MoO_3), open structures such as $Mo_{18}O_{52}$ and Mo_8O_{23} are the most active, selective phases; more complete reduction of either of these structures leads to formation of Mo_4O_{11} (see Figure 24) having substantially lower selectivity. Delmon and co-workers [154,156] have shown that addition of an oxygen donor such as Sb_2O_4 facilitates spillover of oxygen and thereby prevents overreduction and deactivation of the catalyst.

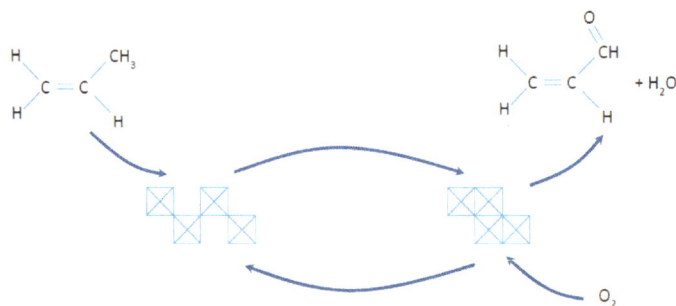

Figure 23. Schematic representation of the cyclic reduction/oxidation of twin pairs of MoO_6 octahedra between the corner and the edge-sharing arrangements (boxes represent MoO_6 octahedra with sharing of oxygen atoms at corners for MoO_3 or edges for MoO_2). The figure is not completely accurate, because it cannot take into account the fact that the arrangements are not perpendicular to the main axes of the lattice. Adapted from [156].

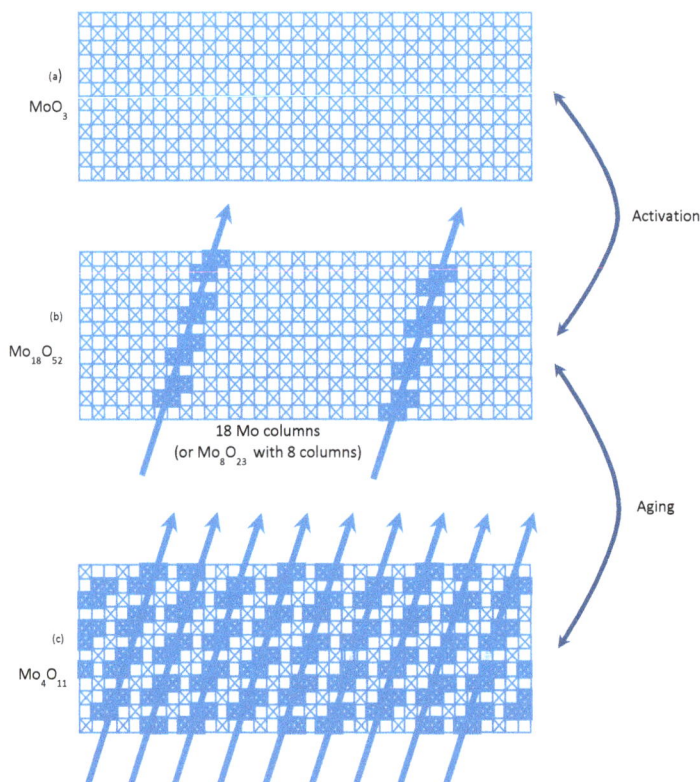

Figure 24. Schematic representation of the structure of MoO_3, $Mo_{18}O_{52}$, and Mo_4O_{11}. The shear planes in $Mo_{18}O_{52}$ and Mo_4O_{11} are represented by the oblique arrows (boxes with an "X" represent MoO_5 octahedra). Adapted from [156].

2.4.3. Reactions of Gas/Vapor with Solid to Restructure the Surface by Chemical
Assisted Sintering

The surfaces of metallic catalysts can be greatly roughened by interactions with the reactants and/or
products. However, as opposed to forming volatile species that are transported out of the reactor as
discussed in the previous section, these interactions lead to a restructuring of the surface that is similar
to that which occurs during thermal sintering, but at temperatures which are below the Tamman or
Huttig temperatures, respectively defined as 0.5 and 0.3 of the melting point (T_m) of the material, at
which thermal sintering might be expected. Therefore, this surface restructuring must be attributed to
the interaction of the gas phase with the solid. The following three examples from the literature
highlight the chemical-assisted sintering process caused by adsorbate-surface interactions on Ni, Co,
and Pd surfaces.

Chemical sintering of Ni/alumina catalysts in methanation due to formation of volatile $Ni(CO)_4$
followed by its decomposition downstream to large Ni crystallites has been well documented [8,105].
Moreover, deactivation of Ni/alumina by Ni aluminate formation is also observed at the exit of
methanators where temperature is moderately high (T = 450 °C) and steam partial pressure is
maximum [105].

Wilson and de Groot [161] reported that under high pressure (4 bar, H_2/CO = 2) and moderate
temperature (523 K) conditions, single crystal Co (0001) surfaces restructured significantly due to
interaction with the CO, which they attributed to an etch-regrowth mechanism. The left hand panel
of Figure 25a shows the scanning tunneling microscopy (STM) image of the single crystal surface,
while Figure 25b shows the same location after exposure to the H_2/CO atmosphere for 1 h The
surface restructuring and roughening is profound, with the peaks approximately four atoms high
relative to the previously smooth surface that had only well-defined steps interrupting the (0001)
planar surface.

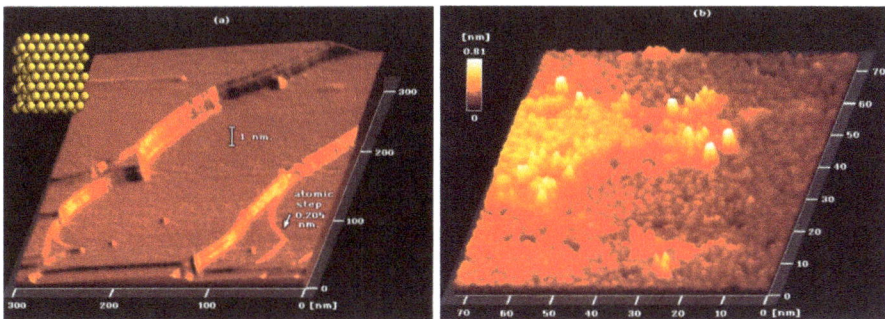

Figure 25. STM images of the Co (0001) surface (**a**) before and (**b**) after 1 h exposure to
4 MPa 2:1 H_2:CO atmosphere at 523 K. Reproduced from [161]. Copyright 1995,
American Chemical Society.

More recently, Parkinson *et al.* [162] have shown that chemical-assisted sintering occurs at
room temperature for palladium supported on magnetite under ultra high vacuum conditions with
CO partial pressures of only 5×10^{-10} mbar. Figure 26 shows four STM images from a movie that

demonstrates the surface mobility of the Pd at these low CO partial pressures. Figure 26a is the surface prior to CO exposure, while Figures 26b–d show the surface as a function of time up to about an hour of exposure. The authors note that hydroxyl-Pd groups (OH-Pd), identified by the ×'s in the images, serve as anchoring points for the coalescence of larger Pd clusters. The full movie, available with the supplementary material for this article [162], is recommended to fully appreciate the unexpectedly high atomic mobility under these conditions.

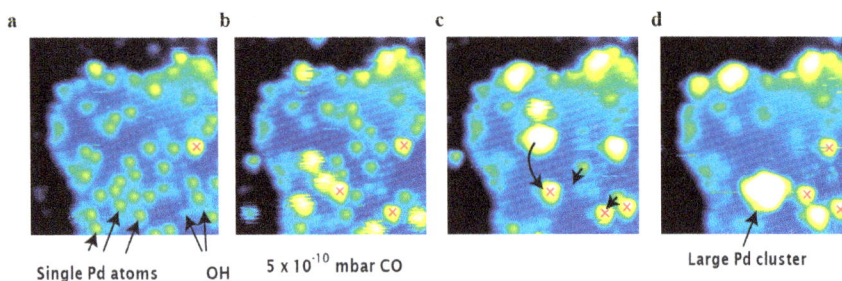

Figure 26. "The CO-induced formation of a large Pd cluster. a–d, Four STM images (14×14 nm^2, +1 V, 0.2 nA) selected from a 36-frame STM movie (duration 1 h 50 min) following the deposition of 0.2 ML Pd [on Fe$_3$O$_4$] at RT. Initially (**a**), isolated Pd atoms are present, together with hydroxyl groups and one OH–Pd (red cross). After three frames the background pressure of CO is raised to 5×10^{-10} mbar. Thirty minutes later (frame **b**), several mobile 'fuzzy' Pd carbonyl species, trapped at other Pd atoms, have formed. Shortly afterwards (**c**), three Pd carbonyls and four adatoms have formed a large cluster. Twenty-five minutes later (**d**), the cluster has captured another Pd carbonyl, and diffused to merge with an OH–Pd species.". Reproduced from [162]. Copyright 2013, MacMillan Publishers.

2.5. Mechanical Failure of Catalysts

2.5.1. Forms and Mechanisms of Failure

Mechanical failure of catalysts is observed in several different forms that depend on the type of reactor, including (1) crushing of granular, pellet, or monolithic catalyst forms due to a load in fixed beds; (2) attrition, the size reduction, and/or breakup of catalyst granules or pellets to produce fines, especially in fluid or slurry beds; and (3) erosion of catalyst particles or monolith coatings at high fluid velocities in any reactor design. Attrition is evident by a reduction in the particle size or a rounding or smoothing of the catalyst particle easily observed under an optical or electron microscope. Washcoat loss is observed by scanning a cross section of the honeycomb channel with either an optical or an electron microscope. Large increases in pressure drop in a catalytic process are often indicative of fouling, masking, or the fracturing and accumulation of attritted catalyst in the reactor bed.

Commercial catalysts are vulnerable to mechanical failure in large part because of the manner in which they are formed; that is, catalyst granules, spheres, extrudates, and pellets ranging in diameter from 50 µm to several millimeters are in general prepared by agglomeration of 0.02–2 µm aggregates of much smaller primary particles having diameters of 10–100 nm by means of precipitation or gel formation, followed by spray drying, extrusion, or compaction. These agglomerates have, in general, considerably lower strengths than the primary particles and aggregates of particles from which they are formed.

Two principal mechanisms are involved in mechanical failure of catalyst agglomerates: (1) fracture of agglomerates into smaller agglomerates of approximately $0.2d_0$–$0.8d_0$ and (2) erosion (or abrasion) from the surface of the agglomerate of aggregates of primary particles having diameters ranging from 0.1 to 10 µm [163]. While erosion is caused by mechanical stresses, fracture may be due to mechanical, thermal, and/or chemical stresses. Mechanical stresses leading to fracture or erosion in fluidized or slurry beds may result from (1) collisions of particles with each other or with reactor walls or (2) shear forces created by turbulent eddies or collapsing bubbles (cavitation) at high fluid velocities. Thermal stresses occur as catalyst particles are heated and/or cooled rapidly; they are magnified by temperature gradients across particles and by differences in thermal expansion coefficients at the interface of two different materials, e.g., catalyst coating/monolith interfaces; in the latter case the heating or cooling process can lead to fracture and separation of the catalyst coating. Chemical stresses occur as phases of different density are formed within a catalyst particle via chemical reaction; for example, carbiding of primary iron oxide particles increases their specific volume and micromorphology leading to stresses that break up these particles [164]. A further example occurs in supported metal catalysts when large quantities of filamentous carbon (according to reaction mechanisms discussed earlier) overfill catalysts pores, generating enormous stresses that can fracture primary particles and agglomerates.

2.5.2. Role of Physical and Chemical Properties of Ceramic Agglomerates in Determining Strength and Attrition Resistance

2.5.2.1. Factors Affecting the Magnitude of Stress Required for Agglomerate Breakage and the Mechanisms by Which It Occurs

The extent to which a mechanism, *i.e.*, fracture or erosion, participates in agglomerate size reduction depends upon several factors: (1) the magnitude of a stress, (2) the strength and fracture toughness of the agglomerate, (3) agglomerate size and surface area, and (4) crack size and radius. Erosion (abrasion) occurs when the stress (e.g., force per area due to collision or cavitation pressure) exceeds the agglomerate strength, *i.e.*, the strength of bonding between primary particles. Erosion rate is reportedly [163] proportional to the external surface area of the catalyst; thus, erosion rate increases with decreasing agglomerate size.

2.5.2.2. Fracture Toughness of Ceramic Agglomerates

Most heterogeneous catalysts are complex, multiphase materials that consist, in large part, of porous ceramic materials, *i.e.*, are typically oxides, sulfides, or metals on an oxide carrier or support. When a tensile stress of a magnitude close to the yield point is applied, ceramics almost always undergo brittle fracture before plastic deformation can occur. Brittle fracture occurs through formation and propagation of cracks through the cross section of a material in a direction perpendicular to the applied stress. Agglomerate fracture due to a tensile stress occurs by propagation of internal and surface flaws; these flaws, created by external stresses or inherent defects, are stress multipliers, *i.e.*, the stress is multiplied by $2(a/r)^{0.5}$, where a is the crack length and r is the radius of curvature of the crack tip; since a/r can vary from 2 to 1000, the effective stress at the tip of a crack can be 4–60 times the applied stress. Tensile stress multipliers may be microcracks, internal pores, and grain corners.

The ability of a material to resist fracture is termed *fracture toughness*. The plane strain fracture toughness, K_{Ic}, is defined as

$$K_{Ic} = Y\sigma(\pi a)^{0.5} \tag{3}$$

where Y is a dimensionless parameter (often close to 1.0–2.0), the magnitude of which depends upon both specimen and crack geometries, σ is the applied stress, and a is the length of a surface crack or half the length of an internal crack. Crack propagation and fracture are likely if the right hand side of Equation 3 exceeds the experimental value of plane strain fracture toughness (left-hand side of Equation 3). Plane strain fracture toughness values for ceramic materials are significantly smaller than for metals and typically below 10 MPa(m)$^{0.5}$; reported values for nonporous, crystalline alumina (99.9%), fused silica, and zirconia (3 mol% Y_2O_3) are 4–6, 0.8, and 7–12 MPa(m)$^{0.5}$, respectively; flexural strengths (analogous to yield strengths for metals) for the same materials are 280–550, 100, and 800–1500 MPa [165]. Thus, on the basis of both fracture toughness and flexural strength, nonporous, crystalline zirconia is much stronger toward fracture than alumina, which in turn is much stronger than fused silica.

2.5.2.3. Effects of Porosity on Ceramic Agglomerate Strength

The introduction of porosity to crystalline or polycrystalline ceramic materials will, on the basis of stress amplification, significantly decrease elastic modulus and flexural strength for materials in tension. This is illustrated by data in Figure 27, showing that elastic modulus and flexural strength of a ceramic alumina (probably alpha form) are reduced by 75 and 85% respectively as porosity is increased from 0 to 50% [166]. Thus, according to Figure 27b, the flexural strength of typical porous aluminas used as catalyst supports might lie in the range of 30–40 MPa. However, yield strengths for γ-Al_2O_3 (shown in the next section) are factors of 3–50 lower. Nevertheless, the data in Figure 27b suggest that higher strengths may be possible.

2.5.2.4. Compressive Strengths of Ceramic Materials

Thus far, the discussion has focused mainly on tensile strength, the extent of which is greatly reduced by the presence of cracks or pores. However, for ceramic materials in compression, there is no stress amplification due to flaws or pores; thus ceramic materials (including catalytic materials) in compression are much stronger (approximately a factor of 10) than in tension. In addition, the strength of ceramic materials can be dramatically enhanced by imposing a residual compressive stress at the surface through thermal or chemical tempering. Moreover, introduction of binders, such as graphite, enables agglomerates of ceramic powders to undergo significant plastic deformation before fracture.

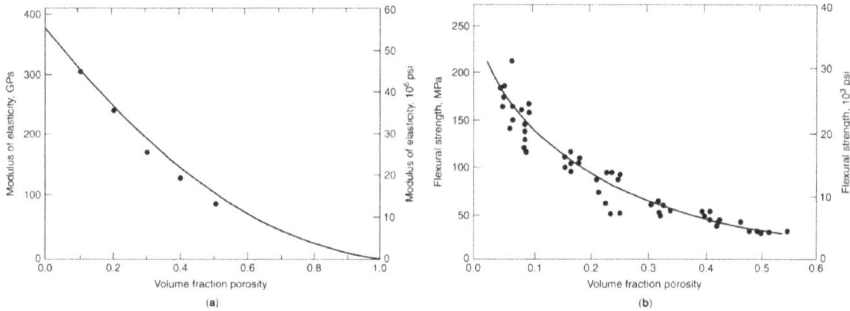

Figure 27. The influence of porosity on (**a**) the modulus of elasticity for aluminum oxide at room temperature and (**b**) the flexural strength for aluminum oxide at room temperature. Reproduced from [166]. Copyright 1956, Wiley.

2.5.3. Tensile Strengths and Attrition Resistance of Catalyst Supports and Catalysts

2.5.3.1. Tensile Strength Data for Catalyst Support Agglomerates

The strengths cited above for nonporous, annealed crystalline or polycrystalline materials do not necessarily apply to porous catalyst agglomerates, even under compression; rather, agglomerate strength is dependent upon the strengths of chemical and physical bonds, including the cohesive energy, between primary particles. Agglomerate strength would depend greatly on the preparation of the compact. Representative data for catalyst agglomerates (see Table 15) suggest they are generally substantially weaker than polycrystalline ceramic materials prepared by high temperature sintering, such as the alumina cited in Figure 27 [163,165,167–171]. For example, Pham and co-workers [163] found that the breaking strength of a VISTA B alumina agglomerate during uniaxial compaction is in the range of 5–10 MPa—substantially lower than the reported values for heat-treated polycrystalline alumina of 280–550 MPa [165]. A large part of this difference (about 85–95%) can be attributed to porosity; however, the remaining 5–15% must be due to differences in bonding between primary particles. In other words, the bonds between primary particles in catalyst agglomerates (and some ceramic agglomerates prepared by similar methods) are typically physical in nature (e.g., involve van der Waals forces) while those in sintered polycrystalline ceramic agglomerates are principally chemical because of solid bridging of primary particles. Thus, there appears to be considerable potential for strengthening catalyst agglomerates, since their strengths are typically factors of 3–50 lower than for conventional, heat-treated ceramics of similar porosity.

Table15. Mechanical Strengths and Attrition Rates of Catalyst Supports Compared to Those of Sintered Ceramic Agglomerates.

Catalyst support or ceramic	Preparation/pretreatment/properties	Strength, MPa	Attrition index, wt%/h	Ref.
High surface area catalyst supports				
γ-Al$_2$O$_3$, 1.2–4.25-mm spheres	Sol–gel granulation/dried 10 h at 40 °C, calcined 3 h at 450 °C/389 m^2/g, d_{pore} = 3.5 nm	11.6 ± 1.9	0.033	[167]
γ-Al$_2$O$_3$, 4.25-mm spheres	Alcoa LD-350	0.7	0.177	[167]
γ-Al$_2$O$_3$, 100 µm	VISTA-B-965-500C	6.2 ± 1.3	-	[163]
TiO$_2$ (anatase), 30 µm	Thermal hydrolysis/dried 110 °C, calcined 2 h at 500 °C/ 92 m^2/g, <10-nm primary crystallites	28[a]	-	[168]
TiO$_2$ (anatase), 90 µm	Basic precipitation/dried 110 °C, calcined 2 h at 500 °C/81 m^2/g, 10–14-nm primary crystallites	15[a]	-	[168]
TiO$_2$ (75% anatase, 25% rutile)	Degussa P25, fumed/4-mm extrudates/48 m^2/g, V_{pore} = 0.34 cm^3/g, d_{pore} = 21 nm	0.9	-	[169]
High surface area catalyst supports (cont.)				
TiO$_2$ (anatase)	Rhone-Poulenc DT51, ppt./4 mm extrudates/92 m^2/g, V_{pore} = 0.40 cm^3/g, d_{pore} = 8, 65 nm	0.9	-	[169]
Low surface area ceramics				
Al$_2$O$_3$	Spray dried with organic binder; plastic deformation observed	2.3	-	[170]
Al$_2$O$_3$	Heat treated (sintered), 99.9%	282–551	-	[165]
TiO$_2$ (Rutile)	Partially sintered	194	-	[170]
ZrO$_2$ (yttria additive)	Commercial samples from three companies, spray-dried	0.035–0.43	-	[171]
ZrO$_2$ (3% Y$_2$O$_3$)	Heat treated (sintered)	800–1500	-	[165]

[a]Rough estimates from break points on relative density *versus* log[applied pressure] curves; data are consistent with mass distribution *versus* pressure curves from ultrasonic tests.

2.5.3.2. Effects of Preparation and Pretreatment on Catalyst Agglomerate Strength

From the data in Table 15 it is evident that even subtle differences in preparation and pretreatment also affect agglomerate strength. For example, spheres of γ-Al$_2$O$_3$ prepared by sol–gel granulation are substantially (17 times) stronger than commercial γ-Al$_2$O$_3$ spheres [166]. Moreover, 30- and 90- µm diameter particles of TiO$_2$ prepared by thermal hydrolysis or basic precipitation are 30 and 15 times stronger than commercially available 4-mm extrudates [169].

2.5.4. Attrition of Catalyst Agglomerates: Mechanisms, Studies, and Test Methods

Catalyst attrition is a difficult problem in the operation of moving-bed, slurry-bed, or fluidized-bed reactors. Generally, stronger materials have greater attrition resistance; this conclusion is supported by representative data in Table 15 for γ-Al_2O_3, showing that the strength of the alumina prepared by sol–gel granulation is 17 times higher, while its attrition rates is 5 times lower.

The mechanism by which attrition occurs (erosion or fracture) can vary with catalyst or support preparation, crush strength, and with reactor environment; it can also vary with the mechanical test method. There is some evidence in the attrition literature, supporting the hypothesis that in the presence of a large stress, weaker oxide materials are prone to failure by fracture, while stronger materials tend to erode. For example, in the fluid catalytic cracking process, as new silica–alumina/zeolite catalyst in the form of 50–150-μm spherical agglomerates is added to replace catalyst lost by attrition, the weaker agglomerates break up fairly rapidly by fracture into smaller subagglomerates, following which the stronger agglomerates are slowly abraded to produce fine particles of 1–10 μm [172]. However, there is also contrary evidence from Thoma and co-workers [168], showing that fracture may be the preferred mechanism for strong TiO_2 agglomerates, while abrasion is favored for weaker agglomerates. That is, when subjected to ultrasonic stress, 30-μm-diameter agglomerates of amorphous anatase (TiO_2) prepared by thermal hydrolysis were observed to undergo fracture to 5–15-μm fragments, while 90-μm agglomerates of polycrystalline anatase prepared by basic precipitation were found to break down by erosion to 0.1–5-μm fragments [168]; in this case, the amorphous anatase was apparently stronger by a factor of 2 (see Table 15). Supporting a third trend, data from Pham and co-workers [163] show that attrition mechanism and rate are independent of agglomerate strength, but depend instead on the type of material. That is, 100-μm-diameter agglomerates of precipitated Fe/Cu/K Fischer–Tropsch catalyst (prepared by United Catalyst Incorporated) and having nearly the same strength shown in Table 15 for Vista-B Al_2O_3 (6.3 vs. 6.2 MPa), were found to undergo substantial fracture to 5–30-μm fragments (an increase from 45 to 85%; see Figure 28) as well as substantial erosion to 1 μm or less fragments (increase from 2 to 50%). By comparison, under the same treatment conditions, 90-μm-diameter agglomerates of Vista-B Al_2O_3 underwent much less attrition, mainly by erosion (20% increase in 0.1–5-μm fragments). The very low attrition resistance of the Fe/Cu/K Universal Catalysts, Inc. (UCI) catalyst is further emphasized by the unsatisfactory outcome of a test of this catalyst by the U.S. Department of Energy (DOE) in a pilot-scale slurry-phase bubble-column reactor in LaPorte, TX.; following one day of operation, the filter system was plugged with catalyst fines, preventing catalyst–wax separation and forcing shutdown of the plant [173].

Thus, based on these three representative examples, it follows that which of the two attrition mechanisms predominates depends much more on material composition and type than on agglomerate strength. However, irrespective of mechanism, the rate of attrition is usually greater for the weaker material.

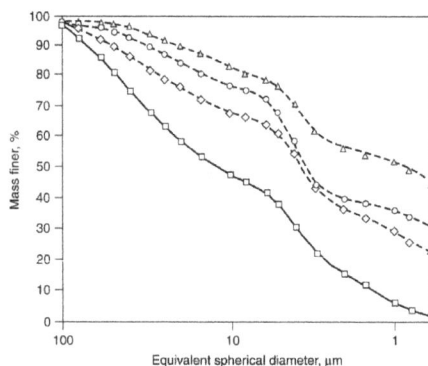

Figure 28. Sedigraph particle size distribution for a United Catalysts, Inc. (UCI) Fischer-Tropsch catalyst (designated as UCI-LAPI-COMP-DRUMC), used previously in Department of Energy (DOE) pilot-plant tests. There is considerable particle breakdown and generation of fine particles after 15 min of ultrasonic irradiation. Reproduced from [163]. Copyright 1999, Elsevier. —□— 0 min; – –◇– –5 min; - -○- - 10 min; – – Δ – –15 min.

Figure 29 illustrates the large effect that catalyst preparation method can have on the attrition resistance of an Fe/Cu Fischer–Tropsch catalyst [174]. This catalyst, prepared by precipitation, undergoes severe attrition during a 25-min treatment with ultrasonic radiation; indeed, the mass fraction finer than 0.1–5 μm increases from 0 to 65%. However, after a spray drying treatment of the same catalyst, an increase of only 0 to 10% in the same fractions is evident.

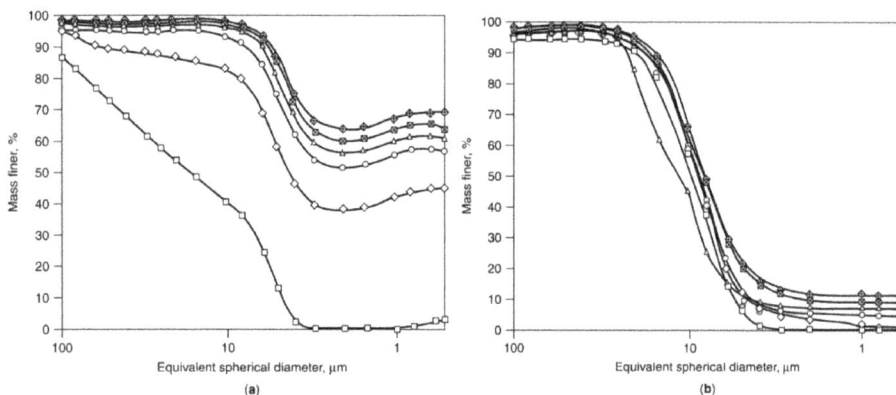

Figure 29. Sedigraph particle size distributions of a precipitated Fe–Cu catalyst, as-prepared and after spray-drying. The as-prepared catalyst (**a**) is weak and breaks down easily after 25 min of ultrasonic irradiation, while spray-drying (**b**) improves its attrition resistance. Reproduced from [174]. Copyright 2000, Elsevier. —□— 0 min; ♦ 5 min; —○— 10 min; —Δ— 15 min; —☒— 20 min; — ⊕— 25 min.

In their review of attrition and attrition test methods, Bemrose and Bridgewater [175] discuss how attrition varies with reactor type, e.g., involves mainly particle–wall impacts in moving pellet bed reactors and particle–particle impacts in fluidized-bed reactors of high fluid velocity. In fact, jet attrition of catalyst particles in a gas fluidized-bed involving principally abrasion due to collision of high-velocity particles has been modeled in some detail [172,176]. Thus, given such important differences in attrition mechanism, realistic attrition test methods should attempt to model reactor operation as closely as possible. In addition, the ideal test would require only a small catalyst sample, a simple, inexpensive apparatus, and a few minutes to complete the test. Relatively quick, inexpensive single-particle crushing tests have been devised [175]; however, properties of a single particle are rarely representative of those for the bed; moreover, it is difficult to relate the results of this crushing test to the actual abrasion process. Realistic tests have been devised for two reactor types involving a moving catalyst, i.e., an air-jet test for fluidized-bed catalysts [177,178], and a rotating drum apparatus for moving-bed catalysts [179]; however, the air-jet test requires a large quantity (e.g., 50 g) of catalyst, an expensive apparatus, and about 20 h to run. In the past decade, a new jet-cup test has been developed for testing of fluidized-bed catalysts [177,178], which requires only a 5-g sample and about 1 h to complete; comparisons of results for the jet-cup and air-jet tests indicate that the two tests give comparable results [177,178]. Nevertheless, the mechanisms for the two tests are different, i.e., the air-jet (fluid-bed) test is abrasion-(erosion-) dominant, while the jet-cup test includes both abrasion and fracture mechanisms [178]. A 30-min, 10-g ultrasonic attrition test based on cavitation has also been developed in the past decade [168,174,180]; while it likewise involves both abrasion and fracture mechanisms, the results appear to correlate with other methods. For example, particle size distributions for the same Co/silica catalyst after ultrasonic, jet-cup, and laboratory-scale, slurry-bed column reactor (SBCR) tests are very similar (see Figure 30), indicating that both fracture and abrasion mechanisms operate in the small-scale SBCR. Moreover, the good agreement among the three methods suggests that both the jet-cup and ultrasonic tests may provide data representative of the attrition process in laboratory-scale SBCR reactors. It is evident that these two small-scale methods are especially useful for screening of a series of catalysts to determine relative strength.

Nevertheless, the more realistic large-scale tests are probably needed for accurately determining design attrition rates of a commercial catalyst to be used in a full-scale process. The observation that attrition of a fluid catalytic cracking (FCC) catalyst initially involves fracture of weak agglomerates followed by abrasion of strong agglomerates emphasizes the need to collect and analyze the particle size distribution of attrited fines as a function of time in order to define which mechanism (or mechanisms) operates at startup as well as in the steady-state process. Because the mechanism may be time dependent, rapid, small-scale tests may produce misleading results.

While realistic laboratory-scale tests have been developed for simulating attrition in large moving-bed and fluidized-bed reactors, no such laboratory test has been developed and demonstrated yet for simulation of large-scale SBCR reactors, although recent research has focused on the development of such tests. For example, in laboratory-scale, SBCR tests of supported cobalt catalysts over several days [180], the attrition resistance decreases in the order $Co/Al_2O_3 > Co/SiO_2 >$

Co/TiO₂ (especially the anatase form underwent attrition at a high rate); attrition resistance was observed to increase with increasing cobalt loading from 10 to 40 wt%.

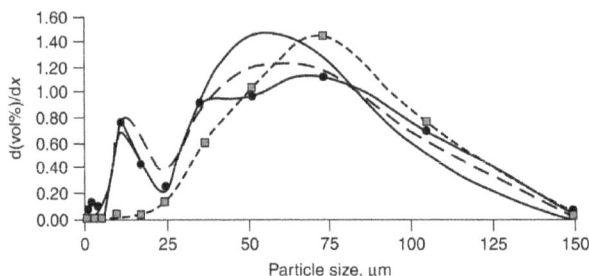

Figure 30. Particle size distributions of Co/SiO₂ catalyst. Adapted from [178]. ———— Ultrasound 250 W (>10 μm);- - - jet cup L/min (>10 μm); —●— Co/SiO₂ after SBCR; —■— Co/SiO₂ fresh.

2.5.5. Implications of Mechanistic Knowledge of Attrition for Catalyst Design

The understanding of mechanisms important in attrition of catalyst supports and catalysts, the relationship between strength and attrition rate for a given material, and test data can be used to great advantage in the design of attrition resistant catalysts. Several alternatives follow from the previous discussion for increasing attrition resistance: (1) increasing aggregate/agglomerate strength by means of advanced preparation methods, e.g., sol–gel granulation, spray drying, and carefully controlled precipitation methods (see Table 15 and Figure 29 for examples), (2) adding binders to improve strength and toughness, e.g., the addition of a polyvinylpyrrolidone binder to agglomerates of quartz sand increases agglomerate strength from 0.1 to 3 MPa [181], (3) coating aggregates with a porous but very strong material such as ZrO₂, e.g., embedding a fluidized-bed catalyst for partial oxidation of *n*-butane to maleic anhydride in a strong, amorphous matrix of zirconium hydrogen phosphate significantly improves its attrition resistance [182], and (4) chemical or thermal tempering of agglomerates to introduce compressive stresses that increase strength and attrition resistance, e.g., heating and cooling particles rapidly by passing them through a low-residence-time, high-temperature furnace to harden the agglomerate exterior, while preventing significant sintering of or phase changes in the porous interior. The subject of preventing mechanical degradation and other forms of catalyst deactivation is addressed in greater detail under Prevention of Catalyst Decay.

2.6. Summary of Deactivation Mechanisms for Solid Catalysts

Causes of solid (heterogeneous) catalyst deactivation are basically threefold: (1) chemical, (2) mechanical, and (3) thermal. Mechanisms of heterogeneous catalyst deactivation can be classified into five general areas: (1) chemical degradation including volatilization and leaching, (2) fouling, (3) mechanical degradation, (4) poisoning, and (5) thermal degradation. Poisoning and thermal degradation are generally slow processes, while fouling and some forms of chemical and mechanical degradation can lead to rapid, catastrophic catalyst failure. Some forms of poisoning and many forms of fouling are reversible; hence, reversibly poisoned or fouled catalysts are relatively easily regenerated. On the other hand, chemical, mechanical, and thermal forms of catalyst degradation are rarely reversible.

3. Prevention of Catalyst Decay

It is often easier to prevent rather than cure catalyst deactivation. Many poisons and foulants can be removed from feeds using guard beds, scrubbers, and/or filters. Fouling, thermal degradation, and chemical degradation can be minimized through careful control of process conditions, e.g., lowering temperature to lower sintering rate or adding steam, oxygen, or hydrogen to the feed to gasify carbon or coke-forming precursors. Mechanical degradation can be minimized by careful choice of carrier materials, coatings, and/or catalyst particle forming methods.

While treating or preventing catalyst deactivation is facilitated by an understanding of the mechanisms, additional perspectives are provided by examining the route by which each of the mechanisms causes loss of catalytic activity, *i.e.*, how it influences reaction rate [109]. Thus, catalytic activity can be defined in terms of the observed site-based rate constant k_{obs}, which is equal to the product of the active site density σ (number of sites per area of surface), the site-based intrinsic rate constant k_{intr}, and the effectiveness factor η, *i.e.*,

$$k_{obs} = \sigma k_{intr} \eta \qquad (4)$$

Loss of catalytic activity may be due to a decrease in any of the three factors in Equation 4, whose product leads to k_{obs}. Thus, catalyst deactivation can be caused by (1) a decrease in the site density σ, (2) a decrease in intrinsic activity (*i.e.*, decrease in k_{intr}), and/or (3) lowered access of reactants to active sites (decrease in η). Poisoning, for example, leads to a loss of active sites, *i.e.*, $\sigma = \sigma_0(1 - \alpha)$, where α is the fraction of sites poisoned; sintering causes loss of active sites through crystallite growth and reduction of active surface area. Fouling can cause both loss of active sites due to blocking of surface sites as well as plugging of pores, causing a decrease in the effectiveness η. Moreover, poisoning, as discussed earlier, can also lead to a decrease in intrinsic activity by influencing the electronic structure of neighboring atoms. Thus, each of the deactivation mechanisms affects one or more of the factors comprising observed activity (see Table 16); all of the mechanisms, however, can effect a decrease in the number of catalytic sites.

3.1. General Principles of Prevention

The age-old adage that says "an ounce of prevention is worth a pound of cure" applies well to the deactivation of catalysts in many industrial processes. The catalyst inventory for a large plant may entail a capital investment of tens of millions of dollars. In such large-scale processes, the economic return on this investment may depend on the catalyst remaining effective over a period of up to 3–5 years. This is particularly true of those processes involving irreversible or only partially reversible deactivation (e.g., sulfur poisoning or sintering). Some typical industrial catalysts, approximate catalyst lifetimes, and factors that determine their life are listed as examples in Table 17. It is evident that in many processes more than one mechanism limits catalyst life. Moreover, there is a wide variation in catalyst lifetimes among different processes, *i.e.*, from 10^{-6} to 15 years. While there is clearly greater interest in extending catalyst lifetimes in processes where life is short, it should be emphasized that great care must be exercised in protecting the catalyst in any process from process upsets (e.g., temperature runaway, short-term exposure to impure feeds, or changes in reactant composition) that might reduce typical catalyst life by orders of magnitude, e.g., from years to hours.

Table 16. How Deactivation Mechanisms Affect the Rate of a Catalyzed Reaction and the Rapidity and Reversibility of Deactivation Process.

Deactivation mechanism	Effects on reaction rate			Deactivation process	
	Decrease in number of active sites	Decrease in intrinisic activity (k_{intr})	Decrease in effectiveness factor (η)	Fast or slow [a]	Reversible
Chemical degradation	×	×	× [b,c]	Varies	No
Fouling	×	×	-	Fast	Yes
Mechanical degradation	×	-	-	Varies	No
Poisoning	×	×	-	Slow	Usually
Thermal degradation/Sintering	×	× [b,d]	× [b,e]	Slow	Sometimes
Vaporization/leaching	×	× [b,f]	-	Fast	Sometimes

[a] Generally; [b] In some cases; [c] Chemical degradation can cause breakdown of support, pore plugging, and loss of porosity; [d] If the reaction is structure-sensitive, sintering could either increase or decrease intrinsic activity; [e] Sintering of the support may cause support collapse and loss of porosity; it may also increase average pore diameter. [f] Leaching of aluminum or other cations from zeolites can cause buildup of aluminum or other oxides in zeolite pores.

Table 17. Typical Lifetimes and Factors Determining the Life of Some Important Industrial Catalysts [a].

Reaction	Operating conditions	Catalyst	Typical life (years)	Process affecting life of catalyst charge	Catalyst property affected
Ammonia synthesis $N_2 + 3\ H_2 \rightarrow 2\ NH_3$	450–470 °C 200–300 atm	Fe with promoters (K_2O) and stabilizer (Al_2O_3)	10–15	Slow sintering	Activity
Methanation (ammonia and hydrogen plants) $CO/CO_2 + H_2 \rightarrow CH_4 + H_2O$	250–350 °C 30 atm	Supported nickel	5–10	Slow poisoning by S, As, K_2CO_3 from plant upsets	Activity and pore blockage
Acetylene hydrogenation ("front end") $C_2H_2 + H_2 \rightarrow C_2H_4$	30–150 °C 20–30 atm	Supported palladium	5–10	Slow sintering	Activity/selectivity and temperature
Sulfuric acid manufacturing $2\ SO_2 + O_2 \rightarrow 2\ SO_3$	420–600 °C 1 atm	Vanadium and potassium sulfates on silica	5–10	Inactive compound formation; pellet fracture; plugging by dust	Activity, pressure drop, and mass transfer
Methanol synthesis $CO + 2\ H_2 \rightarrow CH_3OH$	200–300 °C 50–100 atm	Copper on zinc and aluminum oxides	2–5	Slow sintering; poisoning by S, Cl, and carbonyls	Activity
Low temperature water gas shift $CO + H_2O \rightarrow CO_2 + H_2$	200–250 °C 10–30 atm	Copper on zinc and aluminum oxides	2–4	Slow poisoning and accelerated sintering by poisons	Activity
Hydrocarbon hydrodesulfurization $R_2S + 2\ H_2 \rightarrow H_2S + R_2$	300–400 °C 30 atm	Cobalt and molybdenum sulfides on aluminum oxide	1–10	Slow coking, poisoning by metal deposits in residuum	Activity, mass transfer, and pressure drop
High temperature water gas shift $CO + H_2O \rightarrow H_2 + CO_2$	350–500 °C 20–30 atm	Fe_3O_4 and chromia	1–4	Slow sintering, pellet breakage due to steam	Activity and pressure drop

Table 17. *Cont.*

Reaction	Operating conditions	Catalyst	Typical life (years)	Process affecting life of catalyst charge	Catalyst property affected
Steam reforming, natural gas $CH_4 + H_2O \rightarrow CO + 3\ H_2$	500–850 °C 30 atm	Nickel on calcium aluminate or α-alumina	1–3	Sintering, sulfur-poisoning, carbon formation, and pellet breakage due to plant upsets	Activity and pressure drop
Ethylene partial oxidation $2\ C_2H_4 + O_2 \rightarrow 2\ C_2H_4O$	200–270 °C 10–20 atm	Silver on α-alumina with alkali metal promoters	1–3	Slow sintering, poisoning by Cl, S	Activity and selectivity
Butane oxidation to maleic anhydride $C_4H_{10} + 3.5\ O_2 \rightarrow C_4H_2O_3 + 4\ H_2O$	400–520 °C 1–3 atm	Vanadium phosphorus oxide with transition metal additives	1–2	Loss of P; attrition or pellet breakage; S, Cl poisoning	Activity and selectivity
Reduction of aldehydes to alcohols $RCHO + H_2 \rightarrow RCH_2OH$	220–270 °C 100–300 atm	Copper on zinc oxide	0.5–1	Slow sintering, pellet breakage (depends on feedstock)	Activity or pressure drop
Ammonia oxidation $2\ NH_3 + 5/2\ O_2 \rightarrow 2\ NO + 3\ H_2O$	800–900 °C 1–10 atm	Pt–Rh alloy gauze	0.1–0.5	Surface roughness, loss of platinum	Selectivity, fouling by Fe
Oxychlorination of ethylene to ethylene dichloride $2\ C_2H_4 + 4\ HCl + O_2 \rightarrow 2\ C_2H_4Cl_2 + 2\ H_2O$	230–270 °C 1–10 atm	Copper chlorides on alumina (fluidized bed)	0.2–0.5	Loss by attrition and other causes resulting from plant upsets	Fluidized state and activity
Catalytic hydrocarbon reforming	460–525 °C 8–50 atm	Platinum alloys on treated alumina	0.01–0.5	Coking, frequent regeneration	Activity and mass transfer
Catalytic cracking of oils	500–560 °C 2–3 atm	Synthetic zeolites (fluidized bed)	0.000002	Very rapid coking, continuous regeneration	Activity and mass transfer

Adapted from Ref. [9].

While complete elimination of catalyst deactivation is not possible, the rate of damage can be minimized in many cases through understanding of the mechanisms, thereby enabling control of the deactivation process, *i.e.*, prevention is possible through control of catalyst properties, process conditions (*i.e.*, temperatures, pressures), feedstock impurities, methods of contacting, and process design. Figure 31 illustrates general approaches to eliminating or moderating deactivation through modifications in catalyst and/or process. Examples of how deactivation can be prevented are discussed below in connection with the most important causes of deactivation: chemical degradation, fouling by coke and carbon, poisoning, sintering, and mechanical degradation. Principles for preventing deactivation by these mechanisms are summarized in Table 18, while representative results from studies focusing on prevention or minimization of catalyst deactivation are summarized in Table 19.

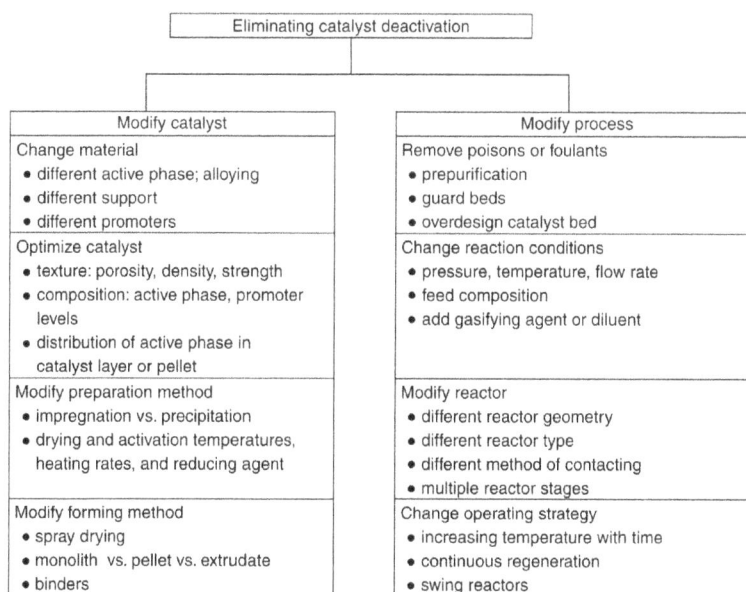

Figure 31. Approaches to eliminating catalyst deactivation.

Table 18. Methods for Preventing Catalyst Decay [8,41].

Basic mechanism	Problem	Cause	Methods of minimization
Chemical degradation	Oxidation of metal catalysts to inactive oxides	Oxidation of metal by contaminant O_2 or reactant/product water	(1) Purify feed of oxidants (2) Minimize reactant/product water by recycle/separation, staged reactors, and otherwise limiting conversion (3) Incorporate additives that facilitate resistance to oxidation
	Transformation of active phase to stable, inactive phase	Solid-state reaction of active phase with support or promoters	(1) Avoid conditions (e.g., oxidizing condition, high steam pressures, and high temperatures) that favor solid-state reactions (2) Select combinations of active phase and promoters/supports that are noninteracting
		Overreduction of active oxide phases	(1) Stabilize oxidation state using promoters that induce resistance to reduction or that serve as oxygen donors (2) Add steam to the reactants to prevent overreduction
		Free radical reactions in gas phase	(1) Avoid formation of free radicals, lower temperature (2) Minimize free space (3) Free radical traps, diluents (4) Add gasifying agents (e.g., H_2, H_2O)
		Free radical reactions at reactor walls	(1) Coat reactor with inert material
Fouling by coke or carbon	Loss of catalytic surface sites due to formation of carbon or coke films	Formation and growth on metal surfaces	(1) Avoid accumulation of coke precursors (e.g., atomic carbon, olefins) through careful choice of reactant conditions or membranes (2) Add gasifying agents (e.g., H_2, H_2O), diluents (3) Incorporate catalyst additives to increase rate of gasification or to change ensemble size (4) Passivate metal surfaces with sulfur (5) Decrease dispersion (6) Recycle inerts to flush surface of heavy oligomers and to moderate temperature

Table 18. *Cont.*

Basic mechanism	Problem	Cause	Methods of minimization
Fouling by coke or carbon (cont.)	Loss of catalytic surface sites due to formation of carbon or coke films	Formation and growth on metal oxides, sulfides	(1) Utilize measures 1, 2, 3, and 6 for metal surfaces (2) Design catalyst for optimum pore structure and acidity (3) Use shape-selective, coke-resistant molecular sieves
	Loss of catalyst effectiveness; plugging of pores; destruction of catalyst	Formation of gas phase coke, vermicular carbons, and liquid or solid cokes in massive quantities	(1) Minimize formation of free radicals or coke precursors as above (2) Use gasifying agents (3) Incorporate catalyst additives that lower solubility of carbon in metal or that change ensemble size (4) Use supports with large pores; large pellets
		Hot spots in pellet or bed	(1) Use wash coat or small pellets (2) Use slurry- or fluid-bed reactor, gas diluents
	Crushing of granules, pellets, or monoliths in a fixed bed	Brittle fracture due to a mechanical load	(1) Minimize porosity of pellets or monoliths (2) Improve bonding of primary particles in agglomerates that make up pellets or monoliths using advanced forming methods, e.g., spray drying and controlled thermal treatments (3) Add binders such as carbon to the support material, which facilitate plastic deformation and thus protect against brittle fracture (4) Chemically or thermally temper agglomerates
Mechanical failure	Attrition and/or erosion in fixed or moving beds	Abrasion of catalyst coatings or particles due to mechanical, thermal, or chemical stresses	(1) Avoid highly turbulent shear flows and/or cavitation, leading to high erosion rates (2) Avoid thermal stresses in the preparation and use of catalysts that lead to fracture or separation of coatings (3) Avoid formation of chemical phases of substantially different densities or growth of carbon filaments that cause fracture of primary particles and agglomerates. Choose supports, support additives, and coating materials, such as titanates, zirconia, and zirconates, having high fracture toughness

Table 18. *Cont.*

Basic mechanism	Problem	Cause	Methods of minimization
Poisoning	Loss of catalytic surface sites	Blockage of sites by strong adsorption of impurity	(1) Purify feed and/or use guard bed to adsorb poison (2) Employ additives that selectively adsorb poison (3) Choose reaction conditions that lower adsorption strength (4) Optimize pore structure and choose mass transfer regimes that minimize adsorption of poison on active sites (5) Apply coating that serves as diffusion barrier to poison
Thermal degradation, sintering	Loss of metal area	Metal particle or subparticle migration at high temperatures	(1) Lower or limit reaction temperature while facilitating heat transfer (2) Add thermal stabilizers to catalyst; and (3) avoid water
	Loss of support area	Crystallization and/or structural modification or collapse	Same as for avoiding loss of metal area

Table 19. Representative Results from Studies Focusing on Prevention/Minimization of Catalyst Deactivation.

Deactivation mechanism Process/Reaction Catalyst	Problem/cause	Method(s) of minimization	Ref.
		Chemical degradation	
Auto emissions control Pt– or Pd–Rh/Al₂O₃	In three-way catalyst, Rh is very active for NO reduction, but it forms a solid solution with Al₂O₃ that has no activity and alloys with Pt or Pd that reduce its activity	Place Rh in a separate catalyst layer from Pt or Pd to prevent alloying; support Rh on ZrO_2, which is a noninteracting support for Rh. In general, multilayer strategies (up to 6 layers) are used to prevent undesirable interactions between different components of the catalyst	[183–185]

Table 19. *Cont.*

Deactivation mechanism Process/Reaction Catalyst	Problem/cause	Method(s) of minimization	Ref.
Chemical degradation			
Fischer–Tropsch synthesis Co supported on Al_2O_3, SiO_2, TiO_2, and $Fe/Cu/K/SiO_2$	Oxidation of active Co metal crystallites to inactive Co oxides, aluminates, and silicates and of active iron carbides to inactive Fe_3O_4 or Fe_3C in the presence of high pressure steam at high conversion	(1) Employ two- or three-stage process that enables lower conversion and lower concentrations of steam product in the first stage. Treat gaseous stream leaving the first or second stage to remove water and liquid hydrocarbons (2) Add noble metal promoters that facilitate and maintain high reducibility of the metal or metal carbide phases (3) Stabilize silica and alumina supports with coatings of hydrothermally stable materials such as ZrO_2 and $MgAl_2O_4$	[8,126,186,187]
Partial oxidation of isobutene to methacrolein $Fe_2(MoO_4)_3$, $Mo_{12}Bi_xCe_yO_z$	Overreduction of the catalyst during reaction leads to activity decrease	(1) Stabilize reduction state of iron molybdate catalyst using an oxygen donor such as α-Sb_2O_4; the oxygen donor dissociates molecular oxygen to atomic oxygen that readily spills over to the catalyst (2) $Mo_{12}Bi_xCe_yO_z$ catalyst promoted with Co, Mg, Rb, and/or Cs oxides is highly resistant to reduction, highly selective to methacrolein, and long-lived	[154,156,188]
Steam reforming and steam-oxygen conversion of propane Pd/Al_2O_3	In the absence of steam, PdO is reduced to less active, less thermally stable Pd metal	Adding steam to the reactants inhibits oxidation of propane at lower reaction temperatures while preventing reduction of PdO at higher temperatures (up to 700–900 °C)	[189]

Table 19. *Cont.*

Deactivation mechanism Process/Reaction Catalyst	Problem/cause	Method(s) of minimization	Ref.
Fouling by coke, carbon			
Alkene oligomerization Zeolites, esp. ZSM-5, −22, −23, beta-zeolite, ferrierite	Catalyst fouling by condensation of heavy oligomers to coke	(1) Recycle of heavy paraffins flushes the surface of heavy oligomers while moderating temperature, thereby decreasing the rate of coke formation (2) Addition of steam improves conversion and catalyst life—probably by cleaning the catalyst surface of coke precursors	[190–192]
Alkylation of isoparaffins on solid catalysts Sulfated zirconia, USY [a], Nafion	Rapid catalyst deactivation due to coke formation; unacceptable product quality, and thermal degradation of catalyst during regeneration	(1) Near critical operation favors desorption and removal of coke precursors from pores while enabling lower reaction temperature (2) Remove oxygen, oxygenates, diolefins, and aromatics from feed; passivate stainless steel surfaces with silicon or bases (3) Design catalyst for optimum pore structure and acidity (4) Use stirred-slurry or fluid-bed reactor while minimizing olefin concentration	[193,194]
Catalytic reforming of naphtha Pt/Al$_2$O$_3$ promoted with Re, Sn, Ge, or Ir	Poisoning and fouling by coke produced by condensation of aromatics and olefins	(1) Use bimetallic catalyst, e.g., sulfided Pt–Re/Al$_2$O$_3$, which is substantially more resistant to coke formation and longer-lived than is Pt/Al$_2$O$_3$. Re sulfide sites break up large Pt ensembles that produce coke. Sn and Ge have a similar effect; Sn and Ir also improve selectivity (2) Optimize reaction conditions and reactor design, e.g., moving bed and low pressure; maintain optimum Cl and S contents of catalyst throughout the bed (3) Near critical reaction mixtures provide an optimum combination of solvent and transport properties for maximizing isomerization rates while minimizing coking	[8,195–198]

Table 19. *Cont.*

Deactivation mechanism Process/Reaction Catalyst	Problem/cause	Method(s) of minimization	Ref.
Fouling by coke, carbon			
Dehydrogenation of propane and butane Cr_2O_3/Al_2O_3, Cr_2O_3/ZrO_2, FeO/K/MgO, Pt/Al$_2$O$_3$, Pt–Sn/Al$_2$O$_3$, Pt–Sn/KL-zeolite	Catalyst activity is low owing to equilibrium limitations and buildup of product H$_2$; rapid loss of activity occurs owing to coke formation	(1) Add Sn and alkali metals to Pt/Al$_2$O$_3$—additives reduce coke coverage of active sites; Sn decreases Pt ensemble size and enhances reactivity of hydrogen with coke (2) Use H$_2$-selective silica membrane to remove product H$_2$, which increases propane conversion; catalyst deactivation is slowed and catalyst life increases, probably due to a lowering of surface coverage of reaction intermediates, including coke precursors, thereby reducing the rate of coke formation	[8,199–203]
Hydrocracking of heavy naphtha CoMo, NiW, MoW on Al$_2$O$_3$ or SiO$_2$–Al$_2$O$_3$; Pt or Pd on Y-zeolite, mordenite or ZSM-5	Loss of activity due to poisoning of sites and blocking of small zeolite pores by coke	(1) Optimize metals loading and porosity of catalyst; use coke-resistant zeolites; incorporate amorphous silica–alumina, which prevents build up of bulky compounds in shape-selective zeolites (2) Design process to prevent build up of polynuclear aromatics, e.g., through distillation, bleeding, flashing, precipitation, and adsorption (3) Decouple aromatics saturation and hydrocracking reactions to improve selectivity, controllability, and catalyst life, while decreasing H$_2$ consumption	[8,198,204]

Table 19. *Cont.*

Deactivation mechanism Process/Reaction Catalyst	Problem/cause	Method(s) of minimization	Ref.
Fouling by coke, carbon			
Methane reforming CO$_2$/Co/SiO$_2$, Pt/SiO$_2$, Pt/ZrO$_2$, MgO-supported noble metals, NiO·MgO solid solution	High rates of carbon formation, which rapidly deactivate catalyst	(1) Add MgO or CaO to reduce carbon deposition on Co or Ni catalysts. CO$_2$ adsorbs strongly on these basic oxides, possibly providing oxygen atoms that gasify coke precursors (2) Adding Sn to Pt catalysts increases stability; ZrO$_2$ support promotes activity and selectivity by aiding dissociation of CO$_2$ (3) Add water or H$_2$ or increase pressure to decrease carbon deposition rate	[205–208]
Methanol to olefins or gasoline Silica–alumina, Y-zeolite, ZSM-5, other zeolites, and aluminophosphate molecular sieves	Severe coking and deactivation of silica–alumina and Y-zeolite catalysts observed during high conversions of MeOH; also substantial coking of ZSM-5, other zeolites, and alumino-phosphate molecular sieves	(1) Maintain a positive methanol concentration through the reactor (e.g., CSTR) to decrease olefin concentration, favor olefin–MeOH reaction to higher olefins over olefin–olefin reactions to coke precursors, substantially decrease coking and deactivation rates, and thereby greatly improve activity and selectivity (2) Increase concentration of water, which attenuates coke formation on SAPO-34 by competing with coke precursors for active sites (3) Treat SAPO-34 above 700 °C in steam to lower acidity, increase catalyst life, and increase selectivity for C$_2$–C$_3$ olefins. Addition of diluent to feed is also beneficial (4) Silanation decreases activity but improves life of zeolites, e.g., HY, HZSM-5	[209–236]

Table 19. *Cont.*

Deactivation mechanism Process/Reaction Catalyst	Problem/cause	Method(s) of minimization	Ref.
Fouling by coke, carbon			
Steam reforming of light hydrocarbons or naphtha Ni on MgO, MgAl₂O₄ or CaAl₂O₄ promoted with S, Cu, or Au	High rates of carbon and coke formation, which rapidly deactivate catalyst	(1) Use basic supports or oxide promoters, which lower carbon deposition rate by preventing hydrocracking and by facilitating adsorption of water, which facilitates gasification of surface carbon (2) Promote with S, Cu, or Au, which lower rate of graphite formation on Ni by decreasing ensemble size (since ensemble size for C-C bond breaking is smaller than for graphite formation)	[8,60,70,71]
Poisoning			
Auto emissions control Pt–Rh/Al₂O₃ or Pd/Al₂O₃	Poisoning of noble metal catalyst by P and S compounds and large hydrocarbons from lube oil	Optimize pore structure of alumina, deposit noble metals in layers below the support surface, or provide a diffusion barrier coating of zeolite or alumina; these measures prevent access of large poison molecules to catalyst layer	[18,212]
Fischer–Tropsch synthesis Co/Al₂O₃	100 ppb of HCN and NH₃ poisons cobalt slurry catalyst within 4 days	Remove HCN and NH₃ to less than 50 ppb total by (1) catalytic hydrolysis of HCN to NH₃, followed by scrubbing with water or (2) guard bed containing acidic solid absorbent	[213]
Fluidized catalytic cracking (FCC) USY or REO-Y [b] in silica matrix	(1) Poisoning of acid sites by *N*-containing compounds. (2) Deposition of Ni and V metals that change selectivity and decrease activity	(1) FCC matrix serves as a coating to remove *N*-containing compounds before they reach zeolites (2) Add Group 13–15 compounds to passivate metals (Sb and Bi for Ni and In for V) and/or trap V with MgO or SrO	[8,198]

Table 19. *Cont.*

Deactivation mechanism Process/Reaction Catalyst	Problem/cause	Method(s) of minimization	Ref.
Poisoning			
Hydrotreating of gas oil; deep HDS Al_2O_3- supported CoMo, noble metals	Noble metal hydrogenation and high-activity HDS catalysts are poisoned by H_2S	(1) Two-stage operation with removal of H_2S between stages (2) Split feed into light and heavy streams; desulfurize light and hydrocrack heavy streams, combine, and conduct deep hydrogenation/HDS	[198]
Hydrotreating of residuum Al_2O_3- supported Mo and CoMo	Pore-mouth poisoning and blockage by Ni, V, and Fe sulfides present in feed as organometallics	(1) Use guard bed or multistage bed to remove metals with first stage containing large-pore, low-activity catalyst for removal of metals and subsequent stages containing progressively smaller-pore, higher-activity catalysts (2) Use catalysts with bimodal pore distributions	[8,214]
Thermal degradation			
Auto emissions control PdO/δ- or θ-Al_2O_3 doped with BaO, La_2O_3, Pr_2O_3, CeO_2, and ZrO_2	In close-couple, fast-warm-up converters, exhaust temperatures reach 1000–1100 °C; conventional Pt–Rh/ γ-Al_2O_3 catalysts sinter rapidly under these conditions; CeO_2 used as oxygen storage material also sinters rapidly	(1) Use δ- or θ-Al_2O_3 and Al_2O_3 spinels having a higher thermal stability than γ-alumina (2) Thermally stabilize Al_2O_3 with BaO, La_2O_3, Pr_2O_3, CeO_2, and ZrO_2; stabilize CeO_2 with ZrO_2 or Pr_2O_3 and ZrO_2 with Y (3) Employ PdO that interacts more strongly than Pt with oxide supports and is hence more stable against sintering	[8,215–222]
Catalytic combustion of methane and LNG PdO/La_2O_3, Pr_2O_3, CeO_2, and ZrO_2	Reaction temperatures ranging up to 1400 °C cause rapid sintering of most catalytic materials. Conversion above 800 °C of PdO to Pd metal is followed by rapid sintering of Pd and loss of activity	(1) Develop PdO/REO catalysts that resist sintering and decomposition of PdO to Pd up to 1300 °C (2) Maintain catalyst temperature below 1000 °C by (a) using lean mixtures, followed by post-catalyst injection of most of the fuel, or (b) employing a metal monolith with heat exchange and gradient of catalyst through bed	[8,224–227]

Table 19. *Cont.*

Deactivation mechanism Process/Reaction Catalyst	Problem/cause	Method(s) of minimization	Ref.
		Thermal degradation	
Dehydrogenation of butene to butadiene Cr_2O_3/Al_2O_3, Pt–Sn/Al_2O_3	Permanent loss of catalytic activity by sintering at high reaction temperatures (550–650 °C)	Optimize the operation of staged catalytic reactors (cycle time between regenerations, temperature, and composition of the feed as variables) while placing a limit on the upper temperature	[8,228]
Fischer–Tropsch synthesis Co/Al_2O_3, Co/SiO_2, Co/TiO_2	Sintering in hot spots and loss of hydrocarbon selectivity at higher reaction temperatures due to highly exothermic reaction	(1) Employ two-stage process that enables lower conversion, better heat removal, and thereby a smaller temperature increase in the first reactor (2) Employ slurry reactor with superior heat transfer efficiency	[186,229]
Fluid catalytic cracking (FCC) USY, REO-Y	Dealumination and destruction of zeolite crystallinity and loss of surface area/pore volume during high-temperature (650–760 °C, 3 atm) regeneration in steam/air	(1) Carry out controlled dealumination or silanization of Y-zeolite to produce USY (2) Use of REO-Y to improve thermal stability (3) Limit steam partial pressure during regeneration	[8,198]
Methane steam reforming Ni on MgAl₂O₄ or CaAl₂O₄	Sintering of Ni and support during high-temperature reaction (800–1000 °C) in high-pressure steam (20 atm)	(1) Design relatively low surface area catalyst with rugged, hydrothermally stable spinel carrier of about 5 m²/g (2) Form catalyst into rings to facilitate heat transfer and prevent overheating at the heated tube-wall	[8,70]

Table 19. *Cont.*

Deactivation mechanism Process/Reaction Catalyst	Problem/cause	Method(s) of minimization	Ref.
		Mechanical degradation	
Partial oxidation of n-butane to maleic anhydride VPO	Attrition in fluidized-bed process	Imbed catalyst particles in a strong, amorphous matrix of zirconium hydrogen phosphate	[182]
Fischer–Tropsch synthesis in a bubble-column slurry reactor Co/Al$_2$O$_3$, Co/SiO$_2$, Co/TiO$_2$	Attrition in bubble column slurry reactor	(1) Spray drying improves density and attrition resistance. (Attrition resistance improves with higher particle density; attrition resistance decreases in the order Co/Al$_2$O$_3$ > Co/SiO$_2$ > Co/TiO$_2$) (2) Addition of SiO$_2$ and/or Al$_2$O$_3$ to TiO$_2$ improves its attrition resistance; addition of TiO$_2$ or of La$_2$O$_3$ to Al$_2$O$_3$ improves its attrition resistance (3) Attrition resistance of Co/Al$_2$O$_3$ is improved when the γ-Al$_2$O$_3$ is formed from synthetic boehmite having a crystallite diameter of 4–5 nm and is pretreated in acidic solution having a pH of 1–3	[230–235]

[a] USY: ultrastable Y-zeolite. [b] REO-Y: rare-earth exchanged Y-zeolite.

3.2. Prevention of Chemical Degradation (by Vapor–Solid and Solid–Solid Reactions)

The most serious problems of oxidation of metal catalysts, overreduction of oxide catalysts, and reaction of the active catalytic phase with carrier or promoter, can be minimized or prevented by careful catalyst and process design (as enumerated in Table 18 and illustrated in Table 19). For example, the loss of Rh due to solid-state reaction with alumina in the automotive three-way catalyst can be prevented by supporting Rh on ZrO_2 in a separate layer from Pt and/or Pd on alumina [215–222] In Fischer–Tropsch synthesis, the oxidation of the active cobalt phase in supported cobalt catalysts to inactive oxides, aluminates, and silicates can be minimized by employing a two- or three-stage process in which product steam is moderated in the first stage by limiting conversion and in subsequent states by interstage removal of water [223] It can also be moderated by addition of noble metal promoters that facilitate and maintain high reducibility of the cobalt and by coating the alumina or silica support with materials such as ZrO_2 that are less likely to react with cobalt to form inactive phases.

3.3. Prevention of Fouling by Coke and Carbon

Rostrup-Nielsen and Trimm [57], Trimm [59], and Bartholomew [60] have discussed principles and methods for avoiding coke and carbon formation. General methods of preventing coke or carbon formation are summarized in Table 18. Most of these are based on one important fundamental principle: *carbon or coke results from a balance between the reactions that produce atomic carbon or coke precursors and the reactions of these species with H_2, H_2O, or O_2 that remove them from the surface.* If the conditions favor formation over gasification, these species accumulate on the surface and react further to form less active forms of carbon or coke, which either coat the surface with an inactive film or plug the pores, causing loss of catalyst effectiveness, pore plugging, or even destruction of the carrier matrix.

Methods to lower rates of formation of carbon or coke precursors relative to their rates of gasification vary with the mechanism of formation (*i.e.*, gas, surface, or bulk phase) and the nature of the active catalytic phase (e.g., metal or oxide). For example, gas phase formation can be minimized by choosing reaction conditions that minimize the formation of free radicals, by using free-radical traps, by introducing gasifying agents (e.g., H_2, H_2O) or gas diluents, and by minimizing the void space available for homogeneous reaction. Similarly, the formation and growth of carbon or coke species on metal surfaces is minimized by choosing reaction conditions that minimize the formation of atomic carbon or coke precursors and by introducing gasifying agents. Selective membranes or supercritical conditions can also be used to lower the gas-phase and surface concentrations of coke precursors. Since carbon or coke formation on metals apparently requires a critical ensemble of surface metal atoms and/or dissolution of carbon into the bulk metal, introduction of modifiers that change ensemble sizes (e.g., Cu or S in Ni or Ru) or that lower the solubility of carbon (e.g., Pt in Ni) can be effective in minimizing these forms of deactivation.

For example, in a detailed STM study of submonolayers of Au on Ni(111), Besenbacher and co-workers [71] found that the electron density of Ni atoms in the vicinity of Au atoms was increased; from density functional theory (DFT) calculations they concluded that the strength of carbon adsorption (and hence the tendency to form graphite) was decreased on next-nearest neighbor Ni atoms; from studies of the effects of S adsorption on methane activation and graphite formation on pure Ni, they were able to infer that the ensemble size needed for methane dissociation is smaller than that for graphite formation. These fundamental insights were used in the design of a 0.3% Au-promoted 16% $Ni/MgAl_2O_4$ catalyst that loses no activity over 4000 h during steam reforming of n-butane, while the corresponding unpromoted Ni catalyst loses about 5% of its initial activity (see Figure 32). In contrast to the moderating effects of noble metal additives, addition of 0.5% Sn to cobalt substantially increases the rate of carbon filament formation from ethylene [72], an effect desirable in the commercial production of carbon filament fibers.

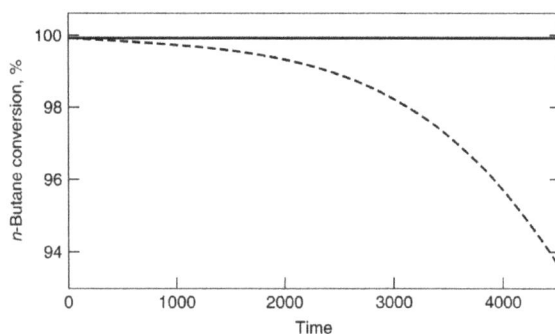

Figure 32. Conversion of n-butane as a function of time during steam reforming in a 3% n–butane–7% hydrogen–3% water in helium mixture at a space velocity of 1.2 h^{-1}. The dashed curve shows the n-butane conversion for the Ni catalyst (16.8% Ni) and the solid curve for the Au/Ni catalyst (16.4% Ni/0.3% Au). Reproduced from [71]. Copyright 1998, American Association for the Advancement of Science.

Coke deposition on oxide or sulfide catalysts occurs mainly on strongly acidic sites; accordingly the rate of coking can be lowered by decreasing the acidity of the support. For example, silanation of HY and HZSM-5 zeolites decreases their activities but improves catalyst life [236]. In steam reforming, certain catalyst additives, e.g., MgO, K_2O, or U_3O_8, facilitate H_2O or CO_2 adsorption and dissociation to oxygen atoms, which in turn gasify coke precursors [8,60,70].

Similarly, for steam reforming catalysts used for light alcohol and oxygenate conversion, the addition of partially reducible oxides, like ceria, in nickel perovskite ($La_{1-x}Ce_xNiO_3$) catalysts [237] or as a support for a cobalt catalyst [238], reduce the rate of carbon deposition. Alternatively, the reaction atmosphere may be modified to increase the gasification rate by adding oxidizing reactants (e.g., O_2 and/or CO_2) to reduce the rate of coke deposition [63]. This process is often described as autothermal reforming because it tends to balance the endothermic steam reforming reactions with exothermic reactions that make the process thermally neutral.

As in the case of poisoning (see below), there are certain reactor bed or catalyst geometries that minimize the effects of coking on the reaction. For example, specific film-mass transport or pore diffusion regimes favor coke or carbon deposition on either the outside or inside of the catalyst pellet [239,240]. Choosing supports with relatively large pores minimizes pore plugging; choice of large-diameter, mechanically-strong pellets avoids or delays reactor plugging. However, in view of the rapidity at which coke and carbon can deposit on, plug, and even destroy catalyst particles, the importance of preventing the onset of such formation cannot be overemphasized.

Reforming of naphtha provides an interesting case study of catalyst and process designs to avoid deactivation by coking [8,196–198,241]. The classical Pt/Al_2O_3 catalyst is bifunctional; that is, the metal catalyzes dehydrogenation, while the acid sites of the Al_2O_3 catalyze isomerization and hydrocracking. Together, the two functions catalyze dehydrocylization and aromatization. Addition of Re, Sn, or Ge, to Pt and sulfiding of the Pt–Re catalyst substantially reduce coke formation by diluting large Pt ensembles that would otherwise produce large amounts of coke, while addition of Sn and Ir improves selectivity for dehydrogenation relative to hydrogenolysis, the latter of which leads to coke formation. Naphtha reforming processes are designed for (1) high enough H_2 pressure to favor gasification of coke precursors while minimizing hydrocracking, (2) maintenance of Cl and S contents throughout the bed to ensure optimum acidity and coke levels, and (3) low enough overall pressure to thermodynamically and kinetically favor dehydrogenation and dehydrocylization. Accordingly, optimal process conditions are a compromise between case 1 and case 3. The above-mentioned improvements in catalyst technologies, especially resistance to coking, have enabled important process improvements, such as optimal operation at lower pressure; thus, processes have evolved over the past two to three decades from conventional fixed-bed reactors at high pressure (35 bar) using nonregenerative Pt catalysts to low pressure (3.5 bar), slowly moving-bed, continuously regenerated units with highly selective Pt/Sn catalysts, resulting in substantial economic benefits [198,241].

3.4. Prevention of Poisoning

Since poisoning is generally due to strong adsorption of feed impurities and since poisoned catalysts are generally difficult or impossible to regenerate, it is best prevented by removal of impurities from the feed to levels that will enable the catalyst to operate at its optimal lifetime. For example, it is necessary to lower the feed concentration of sulfur compounds in conventional methanation and Fischer–Tropsch processes involving base metal catalysts to less than 0.1 ppm in order to ensure a catalyst lifetime of 1–2 years. This is typically accomplished using a guard bed of porous ZnO at about 200 °C. In cracking or hydrocracking reactions on oxide catalysts, it is important to remove strongly basic compounds, such as ammonia, amines, and pyridines, from the feed; ammonia in some feedstocks, for example, can be removed by aqueous scrubbing. The poisoning of catalysts by metal impurities can be moderated by selective poisoning of the unwanted metal. For example, in catalytic cracking of nickel-containing petroleum feedstocks, nickel sites, which would otherwise produce copious amounts of coke, are selectively poisoned by antimony [242]. The poisoning of hydrotreating catalysts by nickel and vanadium metals can be minimized by (1) using a guard bed of inexpensive Mo catalyst or graded catalyst bed with inexpensive, low-activity Mo at the top (bed entrance) and expensive, high-activity catalyst at the bottom (see Figure 33) and (2) by depositing coke prior to the metals, since these metal deposits can be physically removed from the catalyst during regeneration [243].

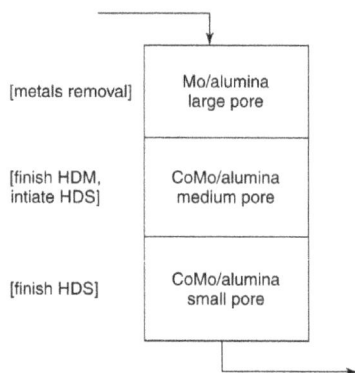

Figure 33. Staged reactor system with decreasing pore size strategy for hydrodemetalization (HDM)/hydrodesulfurization (HDS) of residuum. Reproduced from [214]. Copyright 1993, Marcel Dekker.

It may be possible to lower the rate of poisoning through careful choice of reaction conditions that lower the strength of poison adsorption [60] or by choosing mass-transfer-limiting regimes that limit deposits to the outer shell of the catalyst pellet, while the main reaction occurs uninterrupted on the interior of the pellet [239]. The manner in which the active catalytic material is deposited on a pellet (e.g., uniformly or in an eggshell or egg yolk pattern) can significantly influence the life of the catalyst [17,244].

An example of reducing catalyst poisoning (and oxidation) through process design has been reported in a process patent for staged hydrocarbon synthesis via the Fischer–Tropsch reaction [245]. While cobalt catalysts are favored because of their high activities and while it is desirable to achieve high conversions of CO in the process, the one-pass conversion for cobalt is limited by (1) its tendency to be oxidized at high partial pressures of product water observed at high CO conversions and (2) its tendency under these conditions to form the oxygenated products (e.g., alcohols and aldehydes) that poison or suppress its synthesis activity. One alternative is to separate products and recycle the unused CO and H_2, but this requires costly recompression and separation of the oxygenates. Costly separation and/or poisoning can be prevented by operating a first-stage reactor containing a cobalt catalyst to a moderately high conversion followed by reacting the remaining CO and H_2 in a second stage to above 95% conversion on an iron catalyst, which is not sensitive to the oxygenates and which shifts some of the product water to H_2 and CO_2, thus minimizing its hydrothermal degradation.

An example of reducing catalyst poisoning through catalyst design occurs in abatement of emissions for automotive and motorcycle engines [18,212]. Application of an alumina or zeolite coating, or alternatively preparing the active phase in a sublayer, provides a diffusion barrier that prevents or slows the access of poisons from the fuel or oil (e.g., phosphorus and/or zinc from lubricating oil or corrosion products) to the catalyst surface. The principle is to optimize the pore size distribution of the diffusion barrier to provide access to the catalytic phase of relatively small hydrocarbon, CO, NO, and O_2 molecules, while preventing access of larger molecules, such as from lubricating oil and/or particulates.

Finally, another strategy that has been employed to reduce the impact of poisoning, particularly for sulfur, is the inclusion of traps or "getters" as part of the catalyst. These species, including rare earth oxides of thulium (Tm) [50] or Ce [51] and simple zinc oxide, essentially act as sacrificial stoichiometric reactants to protect the active metal by preferentially adsorbing the poison. These traps can extend the catalyst life, but because they are not catalytic as they perform, they are necessarily temporary agents if the poison remains in the feed to the process.

3.5. Prevention of Sintering

Since most sintering processes are irreversible or are reversed only with great difficulty, it is important to choose reaction conditions and catalyst properties that avoid such problems. Metal growth is a highly activated process; thus, by choosing reaction temperatures lower than 0.3–0.5 times the melting point of the metal, rates of metal sintering can be greatly minimized. The same principle holds true in avoiding recrystallization of metal oxides, sulfides, and supports. Of course, one approach to lowering reaction temperature is to maximize activity and surface area of the active catalytic phase.

Although temperature is the most important variable in the sintering process, differences in reaction atmosphere can also influence the rate of sintering. Water vapor, in particular, accelerates the crystallization and structural modification of oxide supports. Accordingly, it is vital to minimize the concentration of water vapor in high temperature reactions on catalysts containing high surface area supports.

Besides lowering temperature and minimizing water vapor, it is possible to lower sintering rates through addition of thermal stabilizers to the catalyst. For example, the addition of higher melting noble metals (such as rhodium or ruthenium) to a base metal (such as nickel) increases the thermal stability of the base metal [106]. Addition of Ba, Zn, La, Si, and Mn oxide promoters improves the thermal stability of alumina [246]. These additives can affect product selectivity, but generally positively toward desired products, and always through extending the productive life of the catalysts [8].

Designing thermally stable catalysts is a particular challenge in high temperature reactions, such as automotive emissions control, ammonia oxidation, steam reforming, and catalytic combustion. The development of thermally stable automotive catalysts has received considerable attention, thus providing a wealth of scientific and technological information on catalyst design (e.g., Refs. [8,215–222]). The basic design principles are relatively simple: (1) utilize thermally and hydrothermally stable supports, e.g., high-temperature δ- or θ-aluminas or alkaline-earth or rare-earth oxides that form ultrastable spinels with γ-alumina; (2) use PdO rather than Pt or Pt–Rh for high temperature converters, since PdO is considerably more thermally stable in an oxidizing atmosphere because of its strong interaction with oxide supports; and (3) use multilayer strategies and/or diffusion barriers to prevent thermally induced solid-state reactions (e.g., formation of Rh aluminate) and to moderate the rate of highly exothermic CO and hydrocarbon oxidations. For example, a typical three-way automotive catalyst may contain alkaline-earth metal oxides (e.g., BaO) and rare-earth oxides (e.g., La_2O_3 and CeO_2), for stabilizing Pt and/or PdO on alumina, and ZrO_2 as a thermal stabilizer for the

CeO$_2$ (an oxygen storage material) and as a noninteracting support for Rh in a separate layer or in a separate phase in a composite layer (see Figure 34).

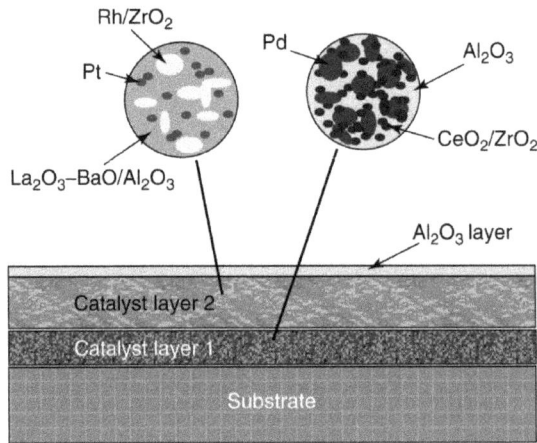

Figure 34. Conceptual design (by C. H. Bartholomew) of an advanced three-way catalyst for auto emissions control. Catalyst layer 1 is wash-coated first onto the monolithic substrate and consists of (**a**) well-dispersed Pd, which serves to oxidize CO/hydrocarbons and to reduce NO and (**b**) CeO$_2$/ZrO$_2$ crystallites (in intimate contact with Pd), which store/release oxygen respectively, thereby improving the performance of the Pd. Catalyst layer 2 (added as a second wash coat) is a particle composite of Rh/ZrO$_2$ (for NO reduction) and Pt/La$_2$O$_3$–BaO/Al$_2$O$_3$ (with high to moderately-high activity for oxidation of CO and hydrocarbons). A thin (50–80 μm) coat of Al$_2$O$_3$, deposited over catalyst layer 2, acts as a diffusion barrier to foulants and/or poisons. Both the Al$_2$O$_3$ layer and catalyst layer 2 protect the sulfur-sensitive components of catalyst layer 1 from poisoning by SO$_2$..

Often, ideal metal dispersions require metal nanoparticles to be distributed closely together, but these particles are thermodynamically unstable on the surface and undergo rapid sintering, as described in Section 2.3 above. Recently, in an attempt to reduce sintering rates, researchers have attempted to stabilize the metal nanoparticles by first dispersing them on a support, encapsulating them in the same or another metal oxide, and then opening porosity to the particles (e.g., [247,248]). These approaches have met with varying degrees of success, but point to promising new areas of synthesis techniques that have the potential to reduce or to eliminate deactivation by sintering.

3.6. Prevention of Mechanical Degradation

While relatively few studies have focused on this topic, there are nevertheless principles that guide the design of processes and catalysts in preventing or minimizing mechanical degradation (see Table 19). In terms of catalyst design, it is important to (1) choose supports, support additives, and coatings that have high fracture toughness, (2) use preparation methods that favor strong bonding of primary particles and agglomerates in pellets and monolith coatings, (3) minimize (or rather optimize) porosity (thus maximizing density), and (4) use binders, such as carbon, to facilitate plastic deformation and thus protect against brittle fracture. Processes (and to some extent preparation procedures) should be designed to minimize (1) highly turbulent shear flows or cavitation that lead to fracture of particles or separation of coatings, (2) large thermal gradients or thermal cycling leading to thermal stresses, and (3) formation of chemical phases of substantially different densities or formation of carbon filaments leading to fracture of primary particles and agglomerates. Nevertheless, thermal or chemical tempering can be used in a controlled fashion to strengthen catalyst particles or agglomerates.

Examples of catalyst design to minimize attrition can be found in the recent scientific [230,231] and patent [232–235] literature focusing on the Fischer–Tropsch synthesis in slurry reactors. These studies indicate that (1) spray drying of particles improves their density and attrition resistance; (2) addition of silica and/or alumina into titania improves its attrition resistance, while addition of only 2000–3000 ppm of titania to γ-alumina improves alumina's attrition resistance; and (3) preformed alumina spheres promoted with La_2O_3 provide greater attrition resistance relative to silica. Increasing attrition resistance is apparently correlated with increasing density [230,231,235]. According to Singleton and co-workers [235], attrition resistance of Co/Al_2O_3 is improved when the γ-alumina support is (1) formed from synthetic boehmite having a crystallite diameter of 4–5 nm and (2) is pretreated in acidic solution having a pH of 1–3 (see Figure 35); moreover, attrition resistance decreases in the order $Co/Al_2O_3 > Co/SiO_2 > Co/TiO_2$ and is greater for catalysts prepared by aqueous *versus* nonaqueous impregnation.

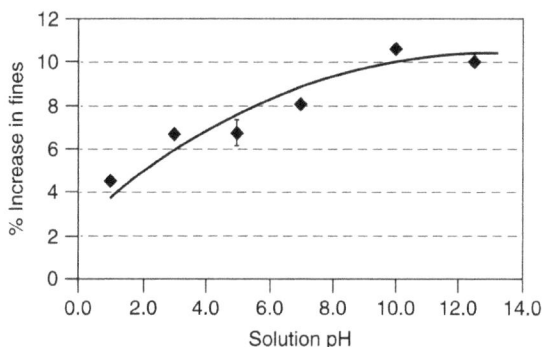

Figure 35. Effect of solution pH on the attrition resistance of 70-μm γ-Al_2O_3 particles measured in jet-cup tests [235]. The % increase in fines is defined at the % increase of particles of less than 11 μm.

4. Regeneration of Deactivated Catalysts

Despite our best efforts to prevent it, the loss of catalytic activity in most processes is inevitable. When the activity has declined to a critical level, a choice must be made among four alternatives: (1) restore the activity of the catalyst, (2) use it for another application, (3) reclaim and recycle the important and/or expensive catalytic components, or (4) discard the catalyst. The first alternative (regeneration and reuse) is almost always preferred; catalyst disposal is usually the last resort, especially in view of environmental considerations.

The ability to reactivate a catalyst depends upon the reversibility of the deactivation process. For example, carbon and coke formation is relatively easily reversed through gasification with hydrogen, water, or oxygen. Sintering on the other hand is generally irreversible, although metal redispersion is possible under certain conditions in selected noble metal systems. Some poisons or foulants can be selectively removed by chemical washing, mechanical treatments, heat treatments, or oxidation [249,250]; others cannot be removed without further deactivating or destroying the catalyst.

The decision to regenerate/recycle or discard the entire catalyst depends largely on the rate of deactivation. If deactivation is very rapid, as in the coking of cracking catalysts, repeated or continuous regeneration becomes an economic necessity. Precious metals are almost always reclaimed where regeneration is not possible. Disposal of catalysts containing nonnoble heavy metals (e.g., Cr, Pb, or Sn) is environmentally problematic and should be a last resort; if disposal is necessary, it must be done with great care, probably at great cost. Accordingly, a choice to discard depends upon a combination of economic and legal factors [250]. Indeed, because of the scarcity of landfill space and an explosion of environmental legislation, both of which combine to make waste-disposal prohibitively expensive, there is a growing trend to regenerate or recycle spent catalysts [251,252]. A sizeable catalyst regeneration industry benefits petroleum refiners by helping to control catalyst costs and to limit liabilities [253,254]; it provides for *ex situ* regeneration of catalyst and recovery/recycling of metals, e.g., of cobalt, molybdenum, nickel, and vanadium from hydroprocessing catalysts [251].

Consistent with its importance, the scientific literature treating catalyst regeneration is significant and growing (includes nearly 1000 journal articles since 1990). Regeneration of sulfur-poisoned catalysts has been reviewed by Bartholomew and co-workers [28]. Removal of coke and carbon from catalysts has received attention in reviews by Trimm [59,250], Bartholomew [60], and Figueiredo [1]. Redispersion of sintered catalysts has been discussed by Ruckenstein and Dadyburjor [101], Wanke [102], and Baker and co-workers [103]. Useful case studies of regeneration of hydrotreating [255] and hydrocarbon-reforming catalysts [256] have also been reported. The proceedings of the 9th International Symposium on Catalyst Deactivation (2001) contains 12 papers treating catalyst regeneration [257]. Regeneration, recycling, and disposal of deactivated heterogeneous catalysts have been reviewed briefly by Trimm [250].

The patent literature treating catalyst regeneration/reactivation is enormous (more than 17,000 patents); the largest fraction of this literature describes processes for regeneration of catalysts in three important petroleum refining processes, *i.e.*, FCC, catalytic hydrocarbon reforming, and alkylation. However, a significant number of patents also claim methods for regenerating absorbents and catalysts used in aromatization, oligomerization, catalytic combustion, SCR of NO, hydrocracking, hydrotreating, halogenation, hydrogenation, isomerization, partial oxidation of hydrocarbons, carbonylations, hydroformylation, dehydrogenation, dewaxing, Fisher–Tropsch synthesis, steam reforming, and polymerization.

Conventional methods for regenerating (largely *in situ*) coked, fouled, poisoned, and/or sintered catalysts in some of these processes and representative examples thereof [258–297] are summarized in Table 20, while the basic principles and limitations involved in regeneration of coked, poisoned, and sintered catalysts are briefly treated in the subsections that follow.

4.1. Regeneration of Catalyst Deactivated by Coke or Carbon

Carbonaceous deposits can be removed by gasification with O_2, H_2O, CO_2, and H_2. The temperature required to gasify these deposits at a reasonable rate varies with the type of gas, the structure and reactivity of the carbon or coke, and the activity of the catalyst. Walker and co-workers [302] reported the following order (and relative magnitudes) for rates of uncatalyzed gasification at 10 kN/m^3 and 800 °C: O_2 (105) > H_2O (3) > CO_2 (1) > H_2 (3×10^{-3}). However, this activity pattern does not apply in general for other conditions and for catalyzed reactions [1]. Nevertheless, the order of decreasing reaction rate of O_2 > H_2O > H_2 can be generalized.

Rates of gasification of coke or carbon are greatly accelerated by the same metal or metal oxide catalysts upon which carbon or coke deposits. For example, metal-catalyzed coke removal with H_2 or H_2O can occur at a temperature as low as 400 °C [1]; β-carbon deposited in methanation can be removed with H_2 over a period of a few hours at 400–450 °C and with oxygen over a period of 15–30 min at 300 °C [60]. However, gasification of more graphitic or less reactive carbons or coke species in H_2 or H_2O may require temperatures as high as 700–900 °C [1], conditions, of course, that result in catalyst sintering.

Because catalyzed removal of carbon with oxygen is generally very rapid at moderate temperatures (e.g., 400–600 °C), industrial processes typically regenerate catalysts deactivated by carbon or coke in air. Indeed, air regeneration is used to remove coke from catalysts in catalytic cracking [81], hydrotreating processes [255], and catalytic reforming [256].

One of the key problems in air regeneration is avoiding hot spots or overtemperatures which could further deactivate the catalyst. The combustion process is typically controlled by initially feeding low concentrations of air and by increasing oxygen concentration with increasing carbon conversion [255,303]; nitrogen gas can be used as a diluent in laboratory-scale tests, while steam is used as a diluent in full-scale plant operations [303]. For example, in the regeneration of hydrotreating catalysts, McCulloch [255] recommends keeping the temperature at less than 450 °C to avoid the γ- to α-alumina conversion, MoO_3 sublimation, and cobalt or nickel aluminate formation, which occur at 815, 700, and 500–600 °C respectively.

Because coke burn-off is a rapid, exothermic process, the reaction rate is controlled to a large extent by film heat and mass transfer. Accordingly, burn-off occurs initially at the exterior surface and then progresses inward, with the reaction occurring mainly in a shrinking shell consistent with a "shell-progressive" or "shrinking-core" model, as illustrated in Figure 36 [304]. As part of this same work, Richardson [304] showed how experimental burn-off rate data can be fitted to various coking transport models, e.g., parallel or series fouling. Burn-off rates for coke deposited on SiO_2/Al_2O_3 catalysts were reported by Weisz and Goodwin [305]; the burning rate was found to be independent of initial coke level, coke type, and source of catalyst.

Table 20. Conventional Methods for and Representative Examples of Catalyst Regeneration from Scientific and Patent Literatures.

Deactivation mechanism Process/Reaction Catalyst	Problem/cause	Method(s) of regeneration/phenomena studied/conclusions	Ref.
Deactivation by coke, carbon			
Alkene aromatization oligomerization Zeolites, esp. ZSM-5, -22, -23, beta-zeolite, ferrierite	Catalyst fouling by condensation of heavy oligomers to coke	(1) ZSM-5 catalyst for light olefin oligomerization containing 2–3% coke is treated in 8–10% steam/air mixture (1300 kPa, 93 °C inlet) in a fluidized bed (2) A coked crystalline alumogallosilicate is contacted with oxygen at a concentration of 0.05–10 vol%, 420–580 °C, and 300–4000 h^{-1}	[258,259]
Alkylation of isoparaffins on solid catalysts Sulfated zirconia, USY^{a}, Nafion, silicalite, ZSM-5	Rapid catalyst deactivation due to coke formation; unacceptable product quality, and thermal degradation of catalyst during regeneration	(1) Coked zeolite is regenerated in liquid phase ($P > 3500$ kPa) fluid bed with H$_2$ in two steps: (a) at reaction temperature (20–50 °C) and (b) at 25 °C above reaction temperature (2) Coked Pd- and Pt/Y-zeolite catalysts containing 10–13% coke are regenerated in either air or H$_2$; H$_2$ treatment enables removal of most of the coke at low to moderate temperatures; higher temperatures are required for air (3) USY and other zeolites are regenerated in supercritical isobutane	[260–263]

Table 20. *Cont.*

Deactivation mechanism Process/Reaction Catalyst	Problem/cause	Method(s) of regeneration/phenomena studied/conclusions	Ref.
		Deactivation by coke, carbon	
Catalytic reforming of naphtha Pt/Al$_2$O$_3$ promoted with Re, Sn, Ge, or Ir	Poisoning and fouling by coke produced by condensation of aromatics and olefins	(1) Coke on Pt bimetallic reforming catalyst is removed off-stream in a fixed or moving bed at 300–600 °C, followed by oxychlorination (350–550 °C) (2) Coke on Pt/zeolite is removed in halogen-free oxygen-containing gas at $T < 415$ °C (3) Sintering during oxidation of coke on Pt–Ir/Al$_2$O$_3$ catalyst can be minimized by low regeneration temperatures (4) Study of influence of heating rate, temperature, and time on structural properties of regenerated Pt–Sn/Al$_2$O$_3$ (5) Study of effects of Cl, Sn content, and regeneration sequence on dispersion and selectivity of Pt–Sn/Al$_2$O$_3$ (6) Regenerated Pt–Re/Al$_2$O$_3$ is more stable than the fresh catalyst in n-heptane conversion and more selective for toluene	[264–269]
Dehydrogenation of propane and butane Cr$_2$O$_3$/Al$_2$O$_3$, Cr$_2$O$_3$/ZrO$_2$, FeO/K/MgO, Pt/Al$_2$O$_3$, Pt–Sn/Al$_2$O$_3$, Pt–Sn/KL-zeolite	Catalyst activity is low due to equilibrium limitations and build-up of product H$_2$; rapid loss of activity occurs due to coke formation	(1) Temperatures gradients were measured during burn off of coke formed on a chromia–alumina catalyst during butene dehydrogenation; data were used in developing a mathematical model for predicting temperatures and coke profiles (2) Coked supported palladium catalyst used in the dehydrogenation of dimethyltertrahydronaphthalenes to dimethylnaphthalenes is reactivated with an organic polar solvent at a temperature below 200 °C	[270.271]

Table 20. *Cont.*

Deactivation mechanism Process/Reaction Catalyst	Problem/cause	Method(s) of regeneration/phenomena studied/conclusions	Ref.
Deactivation by coke, carbon			
Fischer–Tropsch synthesis Co/Al$_2$O$_3$	Loss of activity due to blocking of sites by carbon overlayers and heavy hydrocarbons	(1) Carbidic surface carbon deposited on cobalt can be largely removed in hydrogen at 170–200 °C and in steam at 300–400 °C (2) Slurry-phase cobalt catalysts may lose 50% activity during reaction over a period of a few days; the activity can be rejuvenated *in situ* by injecting H$_2$ gas into vertical draft tubes inside the reactor	[272–274]
Fluid catalytic cracking (FCC) of heavy hydrocarbons USY or REO-Y^b in silica matrix	Rapid loss of activity due to poisoning of acid sites and blocking of small zeolite pores by coke	(1) Process and apparatus for increasing the coke burning capacity of FCC regenerators; auxiliary regenerator partially burns off the coke at turbulent or fast fluidized-bed conditions (2) Multistage fluidized-bed regeneration of spent FCC catalyst in a single vessel by incorporating two relatively dense phase fluidized beds beneath a common dilute phase region	[275,276]
Hydrocracking of heavy naphtha CoMo, NiW, MoW on Al$_2$O$_3$ or SiO$_2$–Al$_2$O$_3$; Pt or Pd on Y-zeolite, mordenite, or ZSM-5	Loss of activity due to poisoning of acid sites and blocking of small zeolite pores by coke	(1) Regeneration of noble metal/zeolite via progressive partial removal of carbonaceous deposits under controlled oxidizing conditions to maximize sorption of a probe molecule while minimizing metal sintering (2) Regeneration of noble metal/zeolite in air at about 600 °C, followed by a mild treatment in aqueous ammonia to improve catalytic activity	[277,278]

Table 20. *Cont.*

Deactivation mechanism Process/Reaction Catalyst	Problem/cause	Method(s) of regeneration/phenomena studied/conclusions	Ref.
		Deactivation by coke, carbon	
		(1) TPO studies of oxidative regeneration of CoMo and NiW HDS catalysts; sulfur is removed at 225–325 °C, carbon at 375–575 °C. Redispersion of NiW was observed by EXAFS	
		(2) Physicochemical changes in CoMo and NiCoMo HDS catalysts during oxidative regeneration, including redispersion of Co, Ni, and Mo oxides and surface area loss, were examined	
Hydrotreating of gas oil Al_2O_3-supported Mo and CoMo, NiMo, NiCoMo, MoW, NiW	Loss of activity due to formation of types I, II, and III coke on metal sulfide and alumina surfaces and in pores	(3) Changes in NiMo catalyst structure and coke composition during reaction and regeneration were examined and correlated	[279,280, 294–297]
		(4) Properties of NiMo catalyst deactivated during shale oil hydrogenation and regenerated in O_2 or H_2 were examined. Regeneration in 1.6% O_2 was more effective than that in 5% H_2. Ni aluminate spinel was observed after burn off	
		(5) Hard and soft cokes formed on CoMo catalysts during HDS of gas oil were characterized. At low coke levels, hard coke was more easily removed in H_2 than in O_2	
		(6) Spent catalysts are washed with solvent and contacted with steam at about 600 °C	
Methanol to olefins or gasoline Silica–alumina, Y-zeolite, ZSM-5, other zeolites, and aluminophosphate molecular sieves	Severe coking and deactivation of silica–alumina and Y-zeolite catalysts observed during high conversions of MeOH; also substantial coking of ZSM-5, other zeolites, and aluminophosphate molecular sieves	(1) Kinetics of coke burnoff from a SAPO-34 used in converting methanol to olefins were studied; kinetics are strongly dependent on the nature of the coke. Kinetics are slowed by strong binding of coke to acid sites	[281,282]
		(2) ZSM-34 catalyst used in conversion of methanol to light olefins is effectively regenerated in H_2-containing gas; this approach avoids the formation of catalyst-damaging products such as steam that would be formed during burn off in air	

Table 20. *Cont.*

Deactivation mechanism Process/Reaction Catalyst	Problem/cause	Method(s) of regeneration/phenomena studied/conclusions	Ref.
		Poisoning	
FCC of residuum USY or REO-Y in silica matrix	(1) Poisoning of acid sites by N- containing compounds. (2) Deposition of Ni and V metals on acid sites which change selectivity and decrease activity	(1) Organometallic solutions of Sb and Bi are added to process stream to passivate Ni by forming inactive Ni–Sb and Ni–Bi species (2) V metal deposits are trapped by reaction with magnesium orthosilicate to form an unreactive magnesium vanadium silicate (3) Spent metal-contaminated catalyst is demetallized by chlorinating and washing, followed by contacting with NH_4F and one antimony compound (4) Metal-contaminated catalyst is contacted with an aqueous solution of a carboxylic acid (e.g., formic, acetic, citric, or lactic acid) (5) Metal-contaminated catalyst is contacted with HCl, HNO_3, or H_2SO_4 (6) Metal-contaminated catalyst is contacted with reducing CO gas to form gaseous metal carbonyls that are separated from the catalyst	[281,282,298–301]
Hydrogenation or dechlorination Ni/SiO$_2$, Pd/Al$_2$O$_3$	Poisoning of metal sites by arsenic, sulfur, and other poisons	(1) Regeneration of Ni/SiO$_2$ catalyst poisoned by thiophene using a sequence of oxidation–reduction treatments at low P_{O_2} and 1 atm H_2 respectively (2) Regeneration in dilute hypochlorite solution of a Pd/Al$_2$O$_3$ catalyst deactivated during the aqueous-phase dechlorination of trichloroethylene in the presence of sulfite or HS$^-$ ions present in ground water	[285,286]
Hydrotreating of residuum Al$_2$O$_3$-supported Mo and CoMo	Pore-mouth poisoning and blockage by Ni, V, and Fe sulfides present in feed as organometallics	(1) Regeneration of catalysts containing V, Ni, or Fe by contacting with H_2O_2 solution and organic acid (2) Following removal of coke by air or solvent wash, catalyst is acid leached to remove undesired metals	[287,288]

Table 20. *Cont.*

Deactivation mechanism Process/Reaction Catalyst	Problem/cause	Method(s) of regeneration/phenomena studied/conclusions	Ref.
Thermal degradation			
Catalytic reforming of naphtha Pt/Al$_2$O$_3$ promoted with Re, Sn, Ge, or Ir; Pt/KL-zeolite	Sintering of Pt causing formation of large metal crystallites and loss of active surface area	(1) Redispersion of Pt–Ir bimetallic catalysts using a wet HCl/air treatment, since the conventional oxychlorination is not effective (2) Redispersion of Pt/KL-zeolite using wet HCl/air treatment followed by brief calcination and reduction (3) Redispersion of Pt–Re/Al$_2$O$_3$ in Cl$_2$ and O$_2$ (4) Redispersion of supported Pt, other noble metals, and Ni in Cl$_2$ and O$_2$	[266,269,289,290]
Hydrocracking of heavy naphtha CoMo, NiW, MoW on Al$_2$O$_3$ or SiO$_2$–Al$_2$O$_3$; Pt or Pd on Y-zeolite, mordenite, or ZSM-5	Sintering of noble metal causing formation of large metal crystallites and loss of active surface area	Redispersion of noble metals on molecular sieves including silica-aluminates, ALPOS, SAPOS	[291]
Hydrotreating of gas oil and residuum Al$_2$O$_3$-supported Mo and CoMo	Sintering of Mo and Co sulfides causing formation of large sulfide crystals and loss of active surface area	(1) Oxidative regeneration of hydroprocessing catalyst at 600 °C optimizes surface area and Mo dispersion (2) Oxidative regeneration in several steps with a final oxidation at 500–600 °C to restore residual catalyst activity	[292,293]

[a] USY: ultrastable Y-zeolite. [b] REO-Y: rare-earth exchanged Y-zeolite.

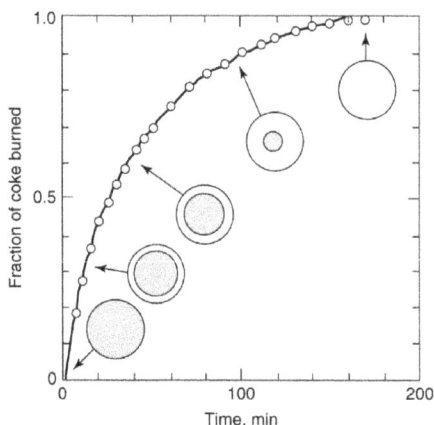

Figure 36. Shell-progressive regeneration of fouled pellet Reproduced from [304]. Copyright 1972, American Chemical Society.

4.2. Regeneration of Poisoned Catalysts

Much of the previous literature has focused on regeneration of sulfur-poisoned catalysts used in hydrogenations and steam reforming. Studies of regeneration of sulfur-poisoned Ni, Cu, Pt, and Mo with oxygen/air, steam, hydrogen, and inorganic oxidizing agents have been reported [28]. Rostrup-Nielsen [306] indicates that up to 80% removal of surface sulfur from Mg- and Ca-promoted Ni, steam reforming catalysts occurs at 700 °C in steam. The presence of both SO_2 and H_2S in the gaseous effluent suggests that the following reactions occur:

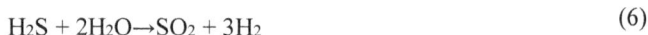

$$Ni\text{-}S + H_2O \rightarrow NiO + H_2S \tag{5}$$

$$H_2S + 2H_2O \rightarrow SO_2 + 3H_2 \tag{6}$$

Although this treatment is partially successful in the case of low-surface-area steam reforming catalysts, the high temperatures required for these reactions would cause sintering of most high-surface-area nickel catalysts.

Regeneration of sulfur-poisoned catalysts, particularly base metal catalysts, in air or oxygen has been largely unsuccessful. For example, the treatment of nickel steam-reforming catalysts in steam and air results in the formation of sulfates, which are subsequently reduced back to nickel sulfide upon contact with hydrogen. Nevertheless, sulfur can be removed as SO_2 at very low oxygen partial pressures, suggesting that regeneration is possible under carefully controlled oxygen atmospheres, including those provided by species such as CO_2 or NO that dissociate to oxygen. Apparently, at low oxygen pressures, the oxidation of sulfur to SO_2 occurs more rapidly than the formation of nickel oxide, while at atmospheric pressure the converse is true, *i.e.*, the sulfur or sulfate layer is rapidly buried in a nickel oxide layer. In the latter circumstance, the sulfur atoms diffuse to the nickel surface during reduction, thereby restoring the poisoned surface. Regeneration of sulfur-poisoned noble metals in air is more easily accomplished than with steam, although it is frequently attended by sintering. Regeneration of sulfur-poisoned nickel catalysts using hydrogen is impractical because (1) adsorption of sulfur is reversible only at high temperatures at which sintering rates are also high and (2) rates of removal of sulfur in H_2 as H_2S are slow even at high temperature.

Inorganic oxidizing agents such as $KMnO_4$ can be used to oxidize liquid phase or adsorbed sulfur to sulfites or sulfates [16]. These electronically shielded structures are less toxic than the unshielded sulfides. This approach has somewhat limited application, *i.e.*, in partial regeneration of metal catalysts used in low temperature liquid-phase hydrogenation reactions or in liquid-phase destruction of chlorinated organic compounds. For example, Lowrey and Reinhard [286] reported successful regeneration in dilute hypochlorite solution of a Pd/Al_2O_3 catalyst deactivated during the aqueous-phase dechlorination of trichloroethylene (TCE) in the presence of sulfite or HS^- ions. These poisons are formed by sulfate-reducing bacteria present in natural groundwater and are apparently adsorbed on the alumina or Pd surfaces more strongly than sulfate ions. Figure 37 illustrates how readily the poisoned catalyst is regenerated by dilute hypochlorite solutions; indeed, it is evident in Figure 37b that regeneration every 5–10 days successfully maintains the catalytic conversion of TCE around 25% (a value only slightly less than that observed for reaction in distilled water).

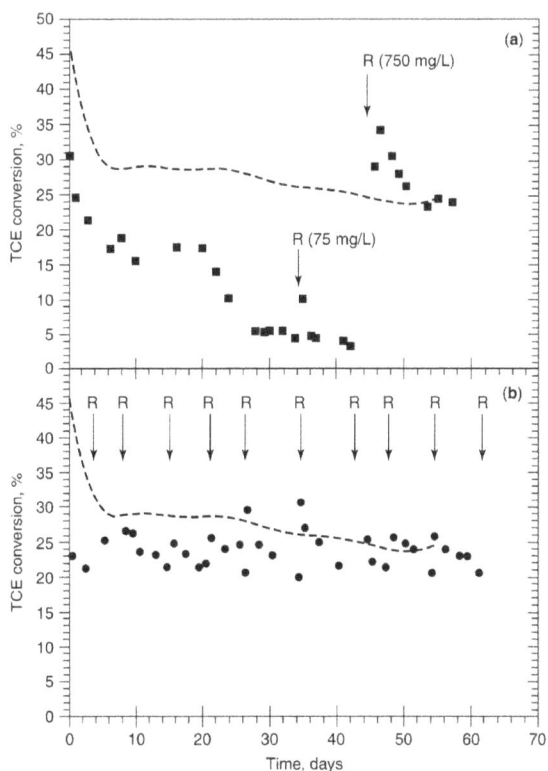

Figure 37. Effect of regeneration (R) with hypochlorite of Pd/Al_2O_3 catalysts used for aqueous phase dechlorination of trichloroethylene in the presence of HS^-/SO_3^{2-}. Reproduced from [286]. Copyright 1992, American Chemical Society.

4.3. Detailed Case Study on Regeneration of Selective Catalytic Reduction (SCR) Catalysts

4.3.1. Introduction to SCR: Key to Abatement of NO_x from Coal Utility Boilers

NO_x, generally defined as NO and NO_2, emissions from coal utility boilers (approximately 30% of total NO_x emissions in the U.S.) contribute substantially to the formation of acid rain and photochemical smog, which in turn damage human health, property, agriculture, lakes, and forests. Selective catalytic reduction (SCR) technology has been used in utility boilers since the 1980s in Japan and Europe in response to stringent NO_x removal regulations. By 2000, SCR systems had been installed in coal-fired boilers totaling roughly 25 and 55 GW in Japan and Europe respectively [307,308]. Equivalent stringent NO_x abatement regulations were enacted later in the U.S. by the EPA, including

(1)   the 1990 ARP and OTC mandates, requiring states to reduce NO_x emissions by 80%;

(2)   the 1995 OTC-Phase 1 requiring Reasonably Available Control Technology (RACT);

(3)   the 1998 NO_x SIP Call setting up a regional cap-and-trade program for 20 eastern states based on an equivalent NO_x emission rate of 0.15 $lb/10^6$-Btu; and

(4)   the 2005 Clean Air Interstate Rule (CAIR) requiring all states to meet Best Available Retrofit Technology (BART) for existing plants, equivalent to emission rates of less than 0.05–0.10 $lb/10^6$-Btu [309,310].

By 2006, about 100 GW of coal-fired steam boilers in the U.S. used SCR. Presently, the U.S. has about 140 GW [309] of coal-boiler SCR capacity; world-wide, an estimated 300 GW of coal-boiler SCR is in operation.

Prior to the more recent stringent U.S. emissions regulations, boiler and engine manufacturers successfully reduced NO_x emissions by 30–60% using modifications to combustion processes, including reducing excess air, adding two-stage combustion features, altering burner design, *etc.* However, meeting the new reduction targets of 80–90% is, in general, only possible through catalytic after-treatment (SCR). Given ever more restrictive NO_x emission standards and the fact that worldwide power production from coal could double or triple in the next decade to an estimated 1500 GW [311], total installed SCR unit capacity is expected to grow commensurately, providing continued investment and design challenges in this area.

4.3.2. Selective Catalytic Reduction of NO_x

4.3.2.1. Reaction Chemistry and Preferred Catalysts

Selective catalytic reduction (SCR) is a process in which a reducing agent, typically NH_3, reacts selectively with the NO_x to produce N_2 without consumption of the excess O_2 present in the flue gas. Desirable stoichiometric reactions for SCR of NO and NO_2 (Equations 7 and 8) occur with high activity and selectivity to N_2 within a narrow temperature window of 300–400 °C on preferred commercial catalysts.

$$4NH_3 + 4NO + O_2 \rightarrow 4N_2 + 6H_2O \tag{7}$$

$$4NH_3 + 2NO_2 + O_2 \rightarrow 3N_2 + 6H_2O \qquad (8)$$

Undesirable side reactions include oxidations of SO_2 (present in the flue gas) and the reducing agent NH_3. While only a small fraction of the SO_2 present in the flue gas is catalytically oxidized to SO_3, this acid precursor either corrodes downstream heat-exchange surfaces or reacts with NH_3 to form ammonium sulfates, which in turn can foul catalyst and/or heat exchange surfaces. Oxidation of NH_3 to either NO or N_2 may also occur at temperatures above 400 °C.

A typical commercial vanadia catalyst consists of 1 wt% V_2O_5 and 10 wt% WO_3 (alternatively 6 wt% MoO_3) supported on high-surface-area TiO_2 (mostly anatase, 60–80 m^2/g). TiO_2 has the decided advantage over Al_2O_3 as a support, since the former stabilizes the active vanadia species and does not form a bulk sulfate in the presence of SO_2-containing flue gases; thus TiO_2 promotes activity and extends catalyst life. WO_3 and MoO_3 prevent the transformation of anatase to rutile; they reside on basic sites of TiO_2, blocking adsorption of SO_3, thereby preventing sulfation of the support. Additionally, WO_3 and MoO_3 increase Brønsted acidity, promoting NO_x reduction while lowering SO_2 oxidation rate. Commercial vanadia-titania catalysts are typically supplied in the form of extruded monoliths or plates (see Figure 38), forms which minimize pressure drop [8].

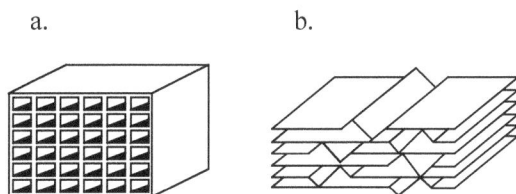

Figure 38. SCR catalyst support geometries: (**a**) extruded ceramic monolith; and (**b**) plate. Reproduced from [8]. Copyright 2006, Wiley-Interscience.

4.3.2.2. SCR Process Options

Two process options in terms of SCR reactant placement have found broad use for SCR units installed in coal-fired plants:

(1)   the *high dust unit (HDU)* involving placement of the SCR unit after the economizer and prior to the air heater, particulate collector, and SO_2 scrubber; and

(2)   the *tail end unit (TEU)* involving placement of the SCR unit following the SO_2 scrubber.

The HDU is used more widely in the U.S. and the TEU more frequently in Europe and Japan.

The HDU has the advantage of providing flue gas to the SCR unit at its ideal temperature range of 300–400 °C and disadvantages of

(1)   deactivation of the catalyst due to erosion, fouling, and poisoning by fly ash thereby limiting its useful life to about 3–4 years;

(2)   large monolith channel design to limit plugging by fly ash, but which also limits the amount of active catalyst per reactor volume; and

(3)   requirement for a low activity catalyst to limit oxidation of SO_2 to SO_3 and the attendant formation of ammonium sulfates which foul and corrode downstream heat exchangers.

The TEU enables use of a smaller volume of high activity catalyst with small diameter channels, since particulates and SO_2 have already been removed upstream; moreover, since deactivation rate is much lower due to the absence of fly ash and other poisons, catalyst life is substantially extended (*i.e.*, to 15–20 years). A significant disadvantage is that the outlet scrubber gas, which is only about 120 °C, must be reheated to at least 200–250 °C for the SCR to occur at reasonable rates. The energy cost of reheating only 100 °C can be as much as 4–6% of the boiler capacity, unless a regenerative heat exchanger is used. In addition, the SCR catalyst must be designed to operate at significantly lower temperatures (200–290 °C relative to a typical 300–400 °C for an HDU).

Given the long life of the TEU catalyst, no regeneration is necessary. However, regeneration of the HDU catalyst is highly desirable, since the regeneration cost is significantly lower than the cost of a new catalyst. With this background, further discussion focuses on the deactivation and regeneration of the HDU catalyst.

4.3.3. Catalyst Deactivation, Rejuvenation, and Regeneration

4.3.3.1. Catalyst Deactivation

SCR catalysts have typical process lifetimes around 2–7 years, depending upon their application and placement in a power plant or other such facility. The principal causes of SCR catalyst deactivation [8,312] are fourfold:

(1)   fouling/masking of (deposition of solids on) catalyst surfaces, pores, and channels by fly ash components (e.g., sulfates and phosphates of Ca, K, and Na) or ammonium bisulfate;

(2)   chemical poisoning of active sites by elements present in upstream lubricants or originating in the fuel such as As, Se, and P and alkali and alkaline earth metals;

(3)   hydrothermal sintering of the titania, especially as a result of high-temperature excursions; and

(4)   abrasion or erosion by fly ash.

Erosion, fouling, and masking from fly ash and poisoning by As and alkali metals are specific to SCR catalysts installed near the hot, high-particulate side of a coal-fired boiler, accounting for the significantly lower catalyst life of 2–4 years for this configuration.

Formation of ammonium bisulfate depends on flue gas temperature, SO_3 concentration and NH_3 concentration [313]. Deposition of ammonium bisulfate is more likely to occur in catalyst pores at lower reactor temperatures in low-dust or tail-end (TEU) SCR units and on cooler surfaces of heat exchangers. Figure 39a shows typical activity loss *versus* time performance for a set of commercial V/Ti catalysts tested in a DOE pilot SCR unit installed in a slip-stream near the exit of a coal-fired boiler (HDU location) using high sulfur, Eastern U.S. coals; 20% of the initial catalyst activity is lost in about 14,000 h (1.6 years); however, the plant will not shut down until 50–60% of the initial activity has been lost (around 3–4 years). Activity and NH_3 slip are plotted against NH_3/NO ratio for the same catalysts in Figure 39b. To maintain NH_3 slip (exit NH_3 concentration) below a target maximum of 2–5 ppm (2 is highly preferred), the NH_3/NO ratio must be maintained near 0.8; under these conditions NO conversion is about 88%.

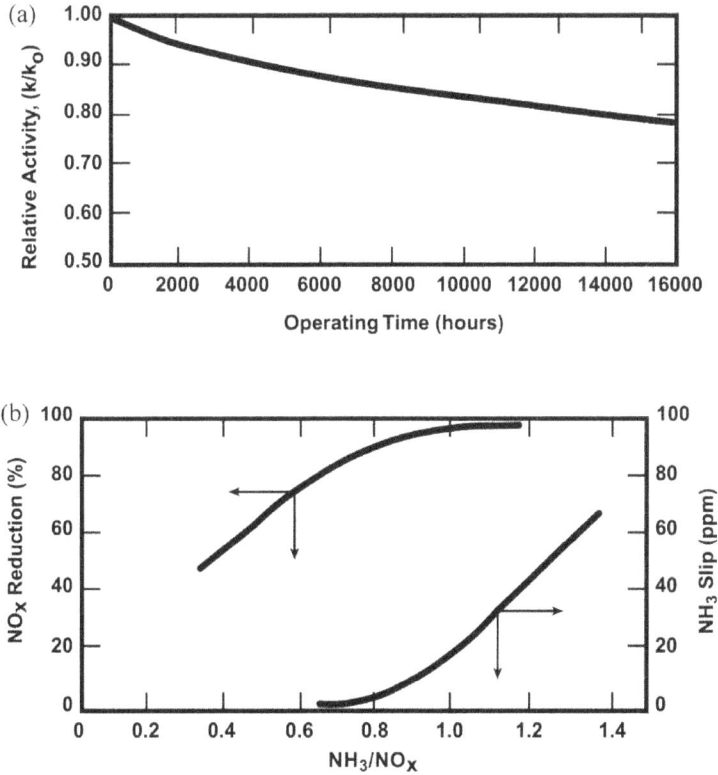

Figure 39. (a) Catalyst activity (k/k_o) *vs.* time; (b) Typical SCR performance. Reproduced from [313].

Prevention of deactivation requires optimal choices of catalyst design and process conditions. Abrasion, fouling, and/or poisoning by fly ash can be prevented by installation of a hot-side electrostatic precipitator or installing an active, low-temperature catalyst at the tail end of the process. Sintering is minimized by using catalyst promoters that enhance thermal stability and by maintaining reaction temperatures below critical values. The MoO_3 promoter extends catalyst life (in coal boilers) by preferentially adsorbing vapor-phase As which would otherwise adsorb on active V^{4+} sites. Free CaO in the fly ash (up to 3%) also scavenges As to low levels, forming calcium arsenide particles which are collected with the fly ash. Many U.S. coals contain adequate CaO; however, if the CaO content of the coal is too low, it can be added to the boiler or fuel. However, CaO levels above 3% of the fly ash are undesirable, since CaO reacts with SO_2 to form $CaSO_4$ which masks the exterior surface of the catalyst. Fouling by ammonium bisulfate is minimized by keeping exit SO_3 and NH_3 concentrations low and maintaining reaction temperatures above about 230 °C; SO_3 formation is minimized by keeping reaction temperatures below 350 °C or by using lower activity V_2O_5/TiO_2 or zeolite catalysts that have low selectivities for SO_3. Ultimately, however, extra catalyst volume is typically added to SCR reactors to extend periods between catalyst replacements.

For plants fueled by coal, substantial carry-over of inorganic ash occurs to HDU SCR units, a small, but significant fraction of which deposits on monolith walls, masks or blocks catalyst macropores, and plugs flow channels [314]. Extensive fouling necessitates the use of air lancing to purge the ash out of the catalyst channels. Figure 40 reveals the extent of serious channel plugging and erosion of an SCR catalyst in a pilot plant following several thousand hours of operation in flue gas containing coal fly ash. Plugging and excessive pressure drop are avoided by keeping monolith cell width at or above 7 mm.

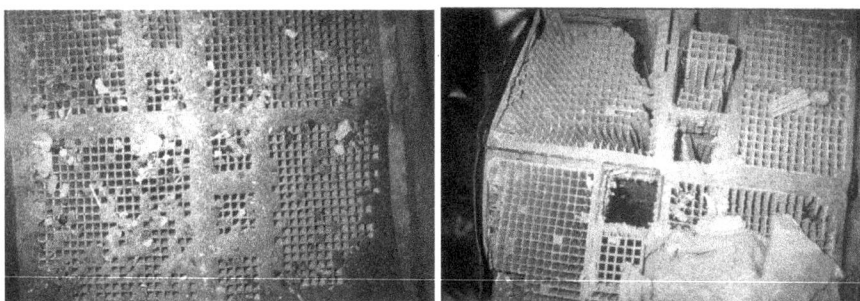

Figure 40. Catalyst channel plugging (**left**) and damage due to erosion (**right**) during operation in an SCR facility. Reproduced from [313].

The type and extent of chemical deactivation depends on operating conditions, fuel type, catalyst geometry, shut-downs for boiler maintenance, *etc.* Mini-pilot tests and subsequent full-scale SCR operating experience have provided little evidence of poisoning by basic minerals from Western United States coals; rather they indicate that deactivation occurs principally by masking of catalyst layers and plugging of catalyst pores by $CaSO_4$ and other fly ash minerals. Moreover, laboratory analysis of catalysts exposed to power plant slip streams indicates that mineral poisons do not penetrate deep into catalyst pores [315,316] nor do they adsorb on Brønsted acid sites unless plant conditions cause moisture to condense on the catalyst.

4.3.3.2. Plant Operating Strategy to Maximize Catalyst Life

A typical SCR unit consists of a series of two to four catalyst layers (three is most common for coal boiler cleanup) through which the flue gas usually flows downward (see Figure 41). A layer of fresh catalyst can be added as catalyst performance declines over time [317]. Two general schemes are followed for replacing the spent catalyst, both of which take into consideration the *relative activity* or *design activity level*, a parameter that is usually defined as the ratio of NO_x conversion at any time divided by that produced by the fresh catalyst. Once the NO_x reduction performance declines to the minimum design activity level (typically 65–75% of fresh activity), the catalyst can either be replaced entirely (*simultaneous* replacement scheme) or one layer can be replaced at a time (*sequential* replacement scheme), usually beginning at the top and working down [313,318].

The sequential method results in increased overall catalyst life (on a per-volume-replaced basis), while annual replacement cost would be 60% lower for the simultaneous scheme (see Figure 42 [319]). Thus, optimal, cost-effective design of an SCR unit requires considering both the initial capital and annual costs.

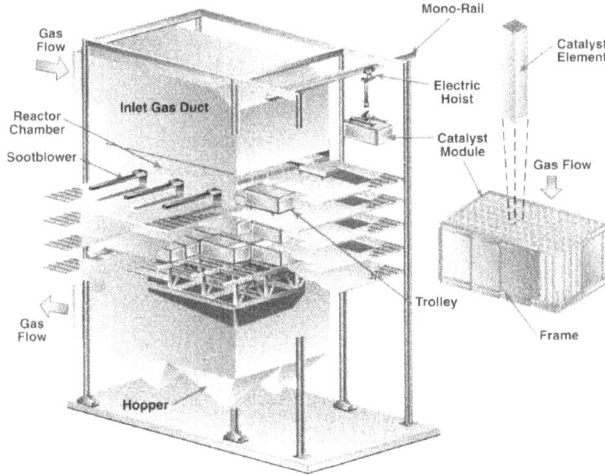

Figure 41. Vertical-flow fixed-bed SCR reactor. DOE SCR demonstration facility at Gulf Power Company's Plant Crist. Reproduced from [313].

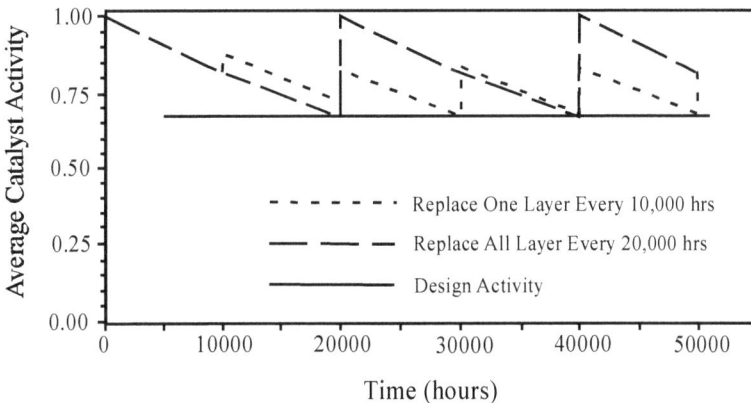

Figure 42. SCR replacement strategies: comparison of total replacement on a 20,000 h cycle relative to sequential replacement on a 10,000 h cycle while maintaining constant catalyst volume. Reproduced from [319].

Operating experience for commercial SCR installations has been better than anticipated. Catalyst lifetimes of 3–4 years at overall efficiencies of 75–90% for HDU's have been observed for electric boiler installations [312]. The principal contributors to operating cost include catalyst replacement cost, shutdown cost for catalyst replacement, and plant derating cost associated with catalyst pressure drop. Catalyst replacement or regeneration was typically required within 2–3 years and catalyst replacement times varied from 2–7 days. Pressure drop ranged from 0.8–15 cm of water for the various catalyst configurations and volumes. Pressure drops for plate type catalysts were significantly lower than for monolithic catalysts.

4.3.3.3. Catalyst Rejuvenation and Regeneration

While high-dust-catalyst life of 2–3 years is acceptable, advances in SCR catalyst regeneration technologies make it possible to extend life by several additional years. Recent experience indicates that even after long-term exposure to fly ash, foulants, and poisons, SCR catalysts may be successfully regenerated to the original performance or better [307,308,320–322].

4.3.3.4. Methods of Renewing Catalysts

Deactivated catalysts may be *cleaned, rejuvenated, and/or regenerated. Cleaning* commonly refers to removal of physical restrictions such as monolith channels plugged with fly ash or channel surfaces covered with a loose dust layer; these restrictions are easily removed *in situ* using compressed air, although cleaning will also be done as a first step in the other methods. *Rejuvenation* refers to relatively mild treatments that remove catalyst poisons or foulants inside the catalyst pores and restore part of the catalytic activity; these treatments are often done *in situ* or *on-site. Rejuvenation* involves removal of blinding layers and partial removal of some poisons; thus, activity is partly recovered, but none is added. *Regeneration* involves the *off-site*, complete restoration of catalytic activity through a series of relatively sophisticated treatments, some of which remove not only poisons and foulants, but also a part or much of the active catalytic materials from the support; hence, regeneration also involves restoration of the catalytically active materials bringing the catalyst to its original state or one of even higher activity. SCR catalysts are routinely and regularly cleaned or "blown out" during operation, while rejuvenation or regeneration is typically done after approximate 50–60% of the initial activity of the catalyst has been lost. *In situ rejuvenation* (ISR) *treatments* were practiced early (e.g., 1990s and early 2000s), while *off-site regeneration* (OSR) is now the *predominant practice* because of its greater effectiveness.

4.3.3.5. Rejuvenation or Regeneration?

According to McMahon [322], rejuvenating SCR catalyst may be more cost-effective than regenerating, if the catalyst is fairly new or the SCR system does not operate year around (as in the case of plants operating only during high pollutant levels, known as the "ozone season"). Otherwise, the choice between rejuvenation and regeneration depends largely on economics, *i.e.*,

(1)   the plant's dispatch economics, including transportation costs;
(2)   length of catalyst service;
(3)   costs of removing and replacing the catalyst;
(4)   the impact of the fuels combusted, *i.e.*, coal, oil, or gas; and
(5)   the location of the catalyst in the plant, *i.e.*, HDU or TGU.

Examples of rejuvenation treatments are found in the scientific and patent literature. For example, work by Zheng and Johnsson [323] and others (e.g., [324,325]) indicates that activity of poisoned catalysts might be partially regenerated by washing with water, sulfuric acid, NH_4Cl, and/or catalyst precursor solutions (e.g., ammonium paratungstate and vanadyl sulfate), as well as a combination of washing and treatment with gaseous SO_2. The extent to which these rejuvenation methods are effective in restoring a significant fraction of the original catalyst activity varies significantly.

4.3.3.5.1. Rejuvenation

On-site rejuvenation methods generally include the following procedural types: (i) removal of dust in the monolith channels with compressed air followed by (ii) washing catalyst in a tank containing agitated, deionized water to remove the $CaSO_4$ coating and alkali metal salts deposited by fly ash (the solution is generally mildly acidic due to impurities on the catalyst) or acidic aqueous solution (pH = 1–2 in either case) in a tank; (iii) rinsing vigorously with deionized water (usually in the same tank) to remove the dissolved and suspended deposits; and (iv) drying slowly in clean air at room temperature followed by drying gently in hot air. Examples of on-site regeneration methods include those developed and practiced in the time frame of 1995–2002 by SCR-Tech, SBW, Saar Energie, Steag, EnBW, HEW, BHK, and Integral [326–328]. The method described by Schneider and Bastuck [327] provided for adding catalytic materials, *i.e.*, vanadium and tungsten oxides (via impregnation of the V and W salts) to the cleaned catalyst.

The patent of Budin *et al.* [328] provides for more sophisticated treatments, including use of (i) nonionic surfactants and complex-forming or ion-exchange additives, (ii) washing with an acid or base, (iii) using acoustic radiation to remove fly-ash components, and (iv) addition of catalytic materials (oxides of V, W, Mo free of alkali and alkaline-earth metals, halogen, and sulfur) to restore activity, although few details or conditions of use are provided. In fact, no examples are provided in any of the patents cited directly above; accordingly, it is unclear to what extent and under what conditions the more sophisticated methods were used for on-site regeneration. The methods claimed by Budin *et al.* [328] are clearly more readily applied in off-site regeneration, as will be clear from the discussion below.

4.3.3.5.2. Regeneration

Bullock & Hartenstein [320], Cooper *et al.* [329], and McMahon [322] build a strong case for off-site regeneration and a comprehensive catalyst management program.

4.3.3.6. A Comprehensive Approach to Catalyst Management

The approach [320,322] includes

(1)   strategies for extending catalyst life and reusability and planning for catalyst removal/rotation to coincide with power plant outages;

(2)   catalyst inspection and testing before and following regeneration with replacement of badly damaged catalyst which is unregenerable;

(3)   off-site regeneration using a series of robust washing and chemical treatments to remove channel blockages, deactivated catalyst metals, and poisons, followed by chemical treatments to restore active catalytic materials; and

(4)   gentle drying/calcination in air to high temperatures to produce catalytically active oxides.

4.3.3.7. Common Regeneration Practices

Normal regeneration procedures [307,308,320,322,330–333] are designed to enhance removal of blockages, deactivated catalyst, and poisons and restore active catalytic material. These typically include the following steps:

(1)   pressurized wet and dry treatments to remove channel blockages and outer dust layers;

(2)   washing of catalyst units in tanks containing agitated water augmented with surfactants, dispersants, ion-exchange materials, emulsifiers, acid, base, and/or acoustic radiation to remove the outer $CaSO_4$ coating, alkali metal salts deposited in the catalyst pores, and deactivated (e.g., As-poisoned) catalyst;

(3)   rinsing repeatedly in deionized water and repeating ultrasonic treatments between or in concert with chemical treatments, with a final rinse to finish removal of any catalyst or fouling residue;

(4)   reimpregnation of the clean support with salts of the active catalytic materials (V, Mo, and W); and

(5)   drying (calcining) at low heating rates to decompose the salts of the active catalytic materials to active metal oxides of V, Mo, and W.

4.3.3.8. Regeneration Process Profile: SCR-Tech Regeneration Process

SCR-Tech is the most prominent and experienced off-site regeneration company with 13 years of experience in the regeneration business and a documented record of research and development, going back to their German parent company ENVICA, who in 1997 began developing an offsite regeneration process. SCR-Tech was the first and until 2008 the only company in the U.S. to perform off-site regeneration. In September 2007, Evonik Energy Services (formerly Steag) opened an SCR catalyst regeneration facility in the U.S.

The SCR-Tech regeneration process involves a number of different process steps illustrated in Figure 43. Upon receipt of a shipment of catalyst, catalyst elements from several modules are inspected and analyzed; results of the analysis provide a basis for determining the precise protocol for treatment, *i.e.*, the number and order of processing steps [334,335]. A large catalyst module is then led through a protocol of soaking, washing, ultrasonic treatment, arsenic and/or phosphorus removal (as needed), replenishment of V and Mo, neutralization, and rinsing in various soaking pits, as shown in Figure 43; all of these wet chemical steps are performed at controlled pH and temperature. Finally, the catalyst is dried, inspected, and packaged for shipment. Performance guarantees are provided for complete removal of blinding layers, catalyst activity (typically higher after regeneration), SO_2 to SO_3 conversion rate (typically lower), mechanical stability (the same), and deactivation rate (the same) such that all properties of the regenerated catalyst are as good or better than the new catalyst.

A comparison of the physical appearances of SCR monoliths and plates before and after regeneration in Figure 44 reveals the rigor of the SCR-Tech cleaning process. The nearly complete removal of poisons originally in high concentrations by the regeneration process is demonstrated in Figure 45. Surface concentrations of CaO, P_2O_5, SiO_2, and SO_4 were also substantially reduced.

Table 21 compares the costs of regenerating *versus* buying a new catalyst [322]. This case is for a typical 500 MW unit with 650 m^3 of catalyst contained in 450 modules (150 modules in each of 3 layers). The purchase cost of new catalyst in 2006 was $3500 to 4500 per m^3. The cost to regenerate the catalyst is approximately 60% of this price. Thus, the purchase cost of one layer is $758,000 to $975,000 as compared to a regeneration cost of $455,000 to $585,000 resulting in savings per layer of $303,000 to $390,000 or $910,000 to $1.2 million for three layers. Assuming the SCR unit runs year around (as most do now) and catalyst life is three years, the annual savings due to regeneration is in the range of $300,000 to $600,000. The disposal cost for an SCR catalyst can range from $50 to $2,000/ton, the upper figure based on the cost of treating the vanadium as hazardous waste. Hence the disposal cost could be as high as $500,000 for a layer of catalyst. According to McMahon, SCR catalysts can be regenerated from 3 to 7 times.

4.3.4. SCR Catalyst Case Study Summary Observations and Conclusions

1. Off-site regeneration processes are more sophisticated and demanding than on-site rejuvenation processes; the off-site regeneration processes provide significantly more efficient cleaning and reconstitution of the catalyst with full recovery of activity—sometimes greater than the fresh catalyst activity. Rejuvenation provides only partial (up to 85%) recovery of the original activity.
2. The development of offsite processes for regeneration of SCR catalysts is relatively new, having occurred largely over the past 10–15 years. SCR-Tech was the first and until 2008 the only company to operate an off-site regeneration facility in the U.S.
3. Because surface deposits are a primary deactivation mechanism, especially in HDU catalysts, extensive multi-step treatments are required, but rejuventation or regeneration appear to be a cost-effective method of catalyst management for SCR catalysts.

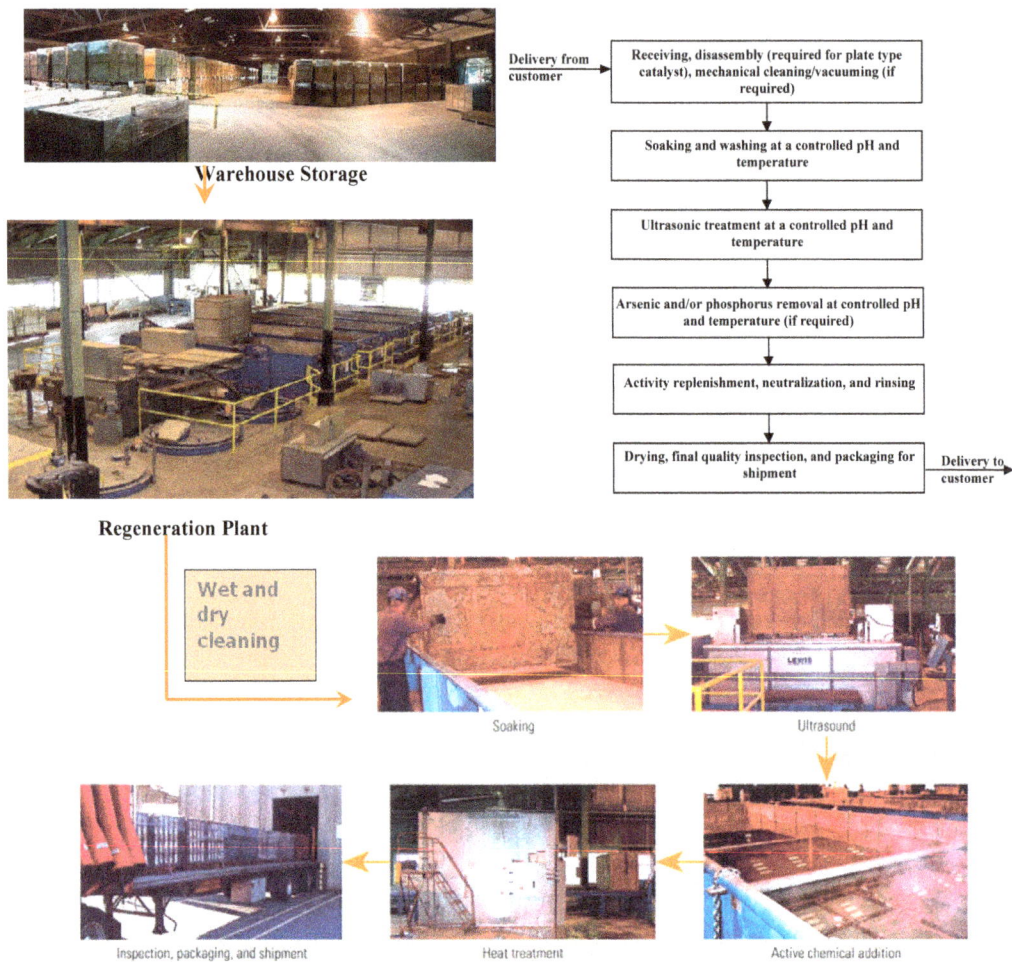

Figure 43. SCR-Tech catalyst regeneration process. Reproduced from [322,335–337]. Reproduced with permission of Electric Power and CoaLogix, Inc.

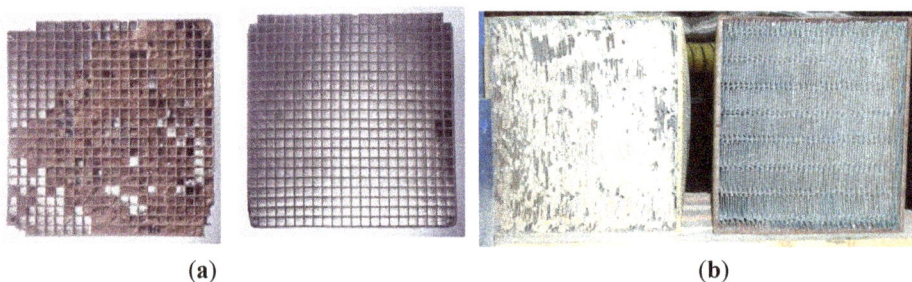

(a)   (b)

Figure 44. (a) Monolith and (b) plate SCR catalysts before and after SCR-Tech regenerative treatment. Reproduced from [334]. Courtesy CoaLogix, Inc.

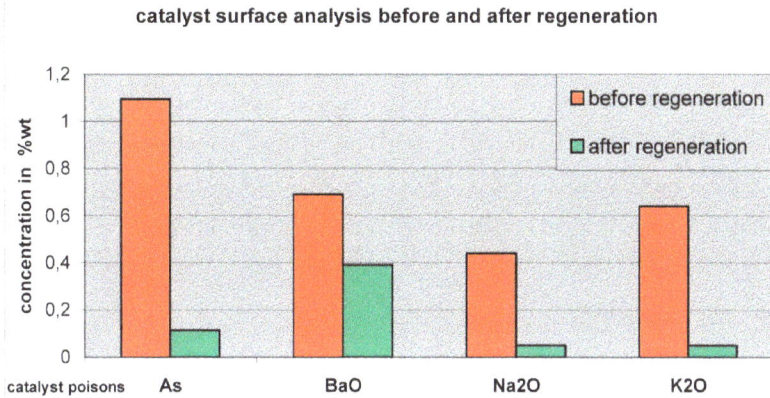

catalyst surface analysis before and after regeneration

Figure 45. Concentration of principle poisons before and after regeneration. Reproduced from [320]. Courtesy CoaLogix, Inc.

Table 21. Cost per layer (217 m^3 or 150 modules) of new *versus* regenerated SCR catalyst. Adapted from [322]. Copyright 2006, Electric Power.

Catalyst Handling Step	New	Regenerated
Removal from SCR system	Comparable	Comparable
Transport out	Comparable	Comparable
Purchase price	$758,000–$975,000	$455,000–$585,000
Shipping	Comparable	Comparable
Installation	Comparable	Comparable
Net savings from regeneration	$303,000–$390,000 pls disposal cost	
Disposal cost	$20,000–$500,000	0

4.4. Redispersion of Sintered Catalysts

During catalytic reforming of hydrocarbons on platinum-containing catalysts, growth of 1-nm platinum metal clusters to 5–20-nm crystallites occurs. An important part of the catalyst regeneration procedure is the redispersion of the platinum phase by a high temperature treatment in oxygen and chlorine, generally referred to as "oxychlorination." A typical oxychlorination treatment involves exposure of the catalyst to HCl or CCl$_4$ at 450–550 °C in 2–10% oxygen for a period of 1–4 h (see details in Table 22). During coke burning, some redispersion occurs, e.g., dispersion (D) increases from 0.25 to 0.51, while during oxychlorination the dispersion is further increased, e.g., from 0.51 to 0.81 [256]. A mechanism for platinum redispersion by oxygen and chlorine is shown in Figure 46 [256]. It involves the adsorption of oxygen and chlorine on the surface of a platinum crystallite and formation of AlCl$_3$, followed by the formation of PtCl$_2$(AlCl$_3$)$_2$ complexes that dissociatively adsorb on alumina to oxychloro-platinum complexes. These latter complexes form monodisperse platinum clusters upon subsequent reduction.

Table 22. Typical Regeneration Procedure for Reforming Catalysts [a].

(1)   *Preliminary operations:* cool the catalyst to about 200 °C and strip hydrocarbons and H_2 with N_2
(2)   *Elimination of coke by combustion:* inject dilute air (0.5% O_2) at 380 °C and gradually increase oxygen content to about 2% by volume while maintaining temperature below 450–500 °C to prevent further sintering of the catalyst. To prevent excessive leaching of Cl_2, HCl or CCl_4 may be injected during the combustion step
(3)   *Restoration of catalyst acidity:* restoration of acidity occurs at 500 °C by injection of a chlorinated compound in the presence of 100–200 ppm water in air
(4)   *Redispersion of the metallic phase:* expose the catalyst to a few Torr of HCl or CCl_4 in 2–10% O_2 in N_2 at 510–530 °C for a period of about 4 h. After redispersion, O_2 is purged from the unit and the catalyst is reduced in H_2

[a] Ref. [255,256].

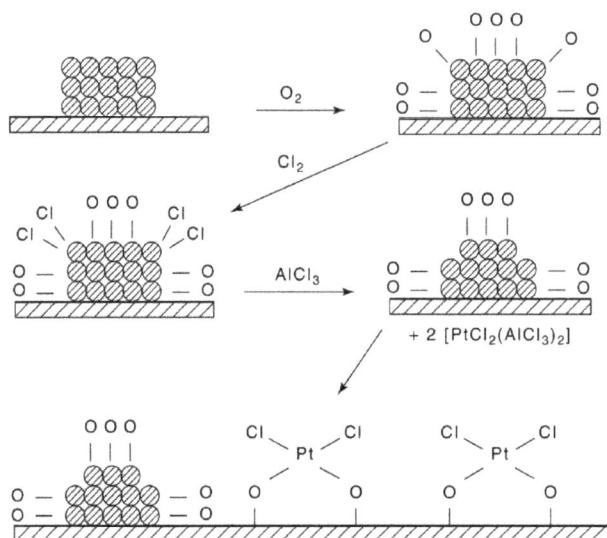

Figure 46. Proposed mechanism for redispersion by oxychlorination of alumina-supported platinum. Reproduced from [256]. Copyright 1982, Brill Nijhoff Publishers.

Some guidelines and principles regarding the redispersion process are worth enumerating:

(1)   In cases involving a high degree of Pt sintering or poisoning, special regeneration procedures may be required. If large crystallites have been formed, several successive oxychlorinations are performed [256].

(2)   Introducing oxygen into reactors in parallel rather than in series results in a significant decrease in regeneration time [101].

(3)   Introduction of hydrocarbons present in the reactor recycle after regeneration is said to stabilize the catalyst; solvents such as ammonium acetate, dilute nitric acid containing lead nitrate, and EDTA and its diammonium salt are reported to dissolve out metal aggregates without leaching out the dispersed metal [101].

(4)   The procedures for redispersion of Pt/alumina are not necessarily applicable to Pt on other supports or to other metals. For example, Pt/silica is redispersed at lower temperature and higher Cl_2 concentration (150–200 °C and 25% Cl_2). Pd/alumina can be redispersed in pure O_2 at 500 °C. While Pt–Re/alumina is readily redispersed by oxychlorination at 500 °C, Pt–Ir/alumina is not redispersed in the presence of O_2, unless the catalyst is pretreated with HCl [266].

An extensive scientific and patent literature of redisperson describes the use of chlorine, oxygen, nitric oxide, and hydrogen as agents for redispersion of sintered catalysts (summarized in Table 23). Most of the early literature shows positive effects for chlorine compounds in the presence of oxygen in redispersing alumina-supported platinum and other noble metals. Recent literature demonstrates the need for understanding the detailed surface chemistry in order to successfully develop and improve redispersion processes, especially in more complex catalyst systems such as alumina-supported bimetallics. For example, on the basis of a fundamental study of the redispersion surface chemistry, Fung [266] developed a redispersion procedure for Pt–Ir bimetallic catalysts using a wet HCl/air treatment, since the conventional oxychlorination is not effective for this catalyst.

Redispersion of alumina-supported platinum and iridium crystallites is also possible in a chlorine-free oxygen atmosphere, if chlorine is present on the catalyst. The extent of redispersion depends on the properties of the Pt/Al_2O_3 catalyst and temperature; for example, the data in Figure 47 [102] for two different catalysts [catalyst 1 is a commercial Pt/Al_2O_3 (Engelhard); catalyst 2 is Pt/Al_2O_3 (Kaiser KA-201) impregnated with chloroplatinic acid] show that the maximum increases in dispersion occur at about 550 °C. The data also show that redispersion does not occur in a hydrogen environment. The question whether redispersion of platinum occurs only in oxygen without chlorine present on the catalyst remains controversial.

Two models, "the thermodynamic redispersion model" and "the crystallite splitting model," have been advanced to explain the redispersion in oxygen [101,102,361]. The "thermodynamic" redispersion model hypothesizes the formation of metal oxide molecules that detach from the crystallite, migrate to active sites on the support, and form surface complexes with the support. Upon subsequent reduction, the metal oxide complexes form monodisperse metal clusters. In the "crystallite splitting" model, exposure of a platinum crystallite to oxygen at 500 °C leads to formation of a platinum oxide scale on the outer surface of the crystallite, which stresses and ultimately leads to splitting of the particle [361]. Dadyburjor hypothesizes that the crystallite splitting model is most applicable to the behavior of large crystallites and to all particles at relatively small regeneration times, while the thermodynamic migration model is useful for small particles and most particles after longer regeneration times.

Table 23. Representative Patents Prior to 1990 Treating Catalyst Redispersion.

Dispersing agent class	Dispersing agent	Metals/support	Patent No.	Ref.
Chlorine-Containing				
	Cl_2, Cl + halogen	Pt/zeolite	U.S. 4,645,751	[338]
	Cl, H_2O, O_2	Pt/zeolite	U.S. 4,657,874	[339]
	HCl, Cl–O	Ir	U.S. 4,491,636	[340]
	Cl, O_2	Pt–Ir, Ir	U.S. 4,467,045	[341]
	HCl, Cl	Pt–Ir–Re, Pt–Ir/zeolites	U.S. 4,359,400	[342]
	Cl, halogen	Ir, Pt–Ir/Al_2O_3	U.S. 4,480,046	[343]
	Cl–H_2O	Pt–Ir–Se/Al_2O_3	U.S. 4,492,767	[344]
	HCl–O–He	Pt–Ir–Se/Al_2O_3	U.S. 4,491,635	[345]
	Cl, O_2	Pt/zeolite	U.S. 4,855,269	[346]
	HCl, Cl, H_2O, O	Pt/zeolite	U.S. 4,925,819	[347]
	HCl, O	Ir, Pt–Ir/Al_2O_3	U.S. 4,444,896	[348]
	Cl, halogen	Ir, Pt–Ir/Al_2O_3	U.S. 4,444,895	[349]
	HCl	Ir, Pt–Ir/Al_2O_3	U.S. 4,517,076	[350]
Oxygen				
	O_2	Pt, Re/Al_2O_3	U.S. 4,482,637	[351]
Oxygen/N_2				
	O_2, N_2	Cu/Cr, Mn, Ru, Pd, Zn, Si, Mg, Ca, Sr, Ba	U.S. 4,855,267	[352]
Other				
	NO, NO + halogen	Pt, Pd/zeolite	Eu 0,306,170	[353]
	Halogen	Ru, Os, Rh, Pd/Al_2O_3	U.S. 4,891,346	[354]
	Halide	Ir, Pt–Ir/Al_2O_3	U.S. 4,447,551	[355]
	Halide, halogen/H_2O	Ir, Pt–Ir/Al_2O_3	U.S. 4,472,514	[356]
	Halogen	Ir, Pt–Ir/Al_2O_3	U.S. 4,473,656	[357]
	NO, NO + halogen, Cl	Group VIII metals/Al_2O_3, SiO_2, zeolites	U.S. 4,952,543	[358]
	H_2-halides, O_2	Ir, Pt–Ir/Al_2O_3	U.S. 4,444,897	[359]
	Halogen, H_2O	Ir, Pt–Ir/Al_2O_3	U.S. 4,472,515	[360]

4.4.1. Case Study: Cobalt based Fischer-Tropsch (FT) Catalyst Regeneration

Fischer-Tropsch (FT) synthesis is a catalytic process used to produce long chain hydrocarbons from synthesis gas consisting of carbon monoxide and hydrogen. Cobalt catalysts were initially developed by Franz Fischer and Hans Tropsch in the 1920s and similar cobalt-based catalysts are still in use today [8]. Although more expensive than iron based catalysts that are also used for FTS, supported cobalt FT catalysts are more active and selective for the desired liquid and wax products.

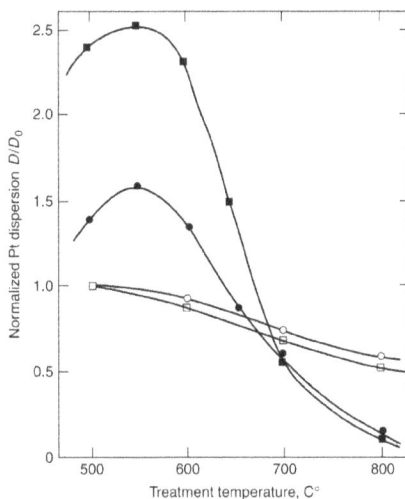

Figure 47. Effects of 1-h treatments in O_2 (closed symbols) and H_2 (open symbols) on the dispersion of Pt/Al$_2$O$_3$ catalysts: $\circ$,$\bullet$ Pt/Al$_2$O$_3$ (Engelhard), $\square$,$\blacksquare$ Pt/KA-201 alumina (Kaiser). Reproduced from [102]. Copyright 1982, Brill Nijhoff Publishers.

A recent review by the Davis group at the Center for Applied Energy Research at the University of Kentucky with Bukur at the University of Texas A&M in Qatar [362] focused on the results of studies using synchrotron radiation to characterize Co FT catalysts. The review includes a detailed consideration and analysis of the mechanisms and processes of sintering, oxidation, aluminate formation, and coking and carbide formation and under what operating conditions each is important. They summarize their and others' previous findings that oxidation primarily occurs on small (<2 nm) cobalt crystallites and at high partial pressures of water [362–366]. Further, they highlight the potentially complicated transformations between CoO and aluminates [362,364,367]. These complications highlight a complex mechanism that may be related to chemical-assisted sintering of Co FTS catalysts through a combination of the effect of CoO reduction during the initial activation of the catalysts and water exposure during operation. First, CoO, present either due to incomplete reduction of the catalysts [368] or oxidation of the small (<2 nm) crystallites as suggested by Davis' group [369,370] can apparently increase the sintering rate due to mobility that allows them to aggregate into larger CoO clusters that are subsequently reduced to metallic Co, as inferred from evidence presented in a number of studies [79,362,368–371]. Primarily, X-ray absorption near edge (XANES) analysis shows simultaneous increasing extent of reduction and increasing Co-Co coordination, due both to removal of oxygen and increases in particle size. Second, water is believed to cause chemical-assisted sintering [80,367,372–374], especially at high partial pressures that occur at CO conversions above about 65% [223], although the exact mechanisms are debated. Minor surface oxidation [373,374] and surface wetting [375] have been proposed, although Saib et al. have shown that cobalt oxidation is not an important deactivation route [79] in catalysts with Co particles >~8 nm, which are typical in commercial FTS catalysts.

A number of articles by researchers at Sasol, Eindhoven University of Technology, and the University of South Africa detailed the causes of deactivation and demonstrated the regenerability of alumina-supported cobalt FT catalysts [79,368,371,376–382]. Through a combination of studies on single crystal [377] and actual catalysts from pilot plants operated under industrial FT conditions [368,371], they concluded that contrary to prior hypotheses, neither formation of cobalt aluminates nor oxidation of the cobalt were significant deactivation mechanisms. In fact, extent of Co oxidation actually decreased with time on stream [371]. However, Co sintering and carbon deposition were identified as the primary means of deactivation. In unpublished presentations by these authors, the relative contributions of carbon deposition and sintering to the deactivation were reported as roughly equal. More interestingly, both of these deactivation mechanisms could be largely reversed through high pressure oxidation treatment [376,378], which removes both inactive carbon and redisperses the cobalt. Through high resolution transmission electron micrographs (HRTEM), the mechanism of redispersion of the cobalt was identified as the Kirkendall effect, which results in the formation of spherical shells of cobalt oxide that during subsequent reduction disperse into smaller crystallites of cobalt (see Figure 48). Bezemer *et al.* have previously shown that unpromoted Co FT catalysts require Co crystallites of at least 6 nm in diameter to achieve maximum turnover frequency, but this is the optimum size because larger crystallites display the same surface activity as the 6 nm particles [383]. The oxidative regeneration and reduction process described by Hauman *et al.* [376] and Weststrate *et al.* [377] recovers ~95% of the fresh catalyst activity by removing the carbon deposits and returning the sintered cobalt particles to near the optimum 6 nm size. While the rate per mass of catalyst is nearly constant following regeneration, some smaller particles are produced on model catalysts because the rate on a turnover frequency basis decreases by roughly 1/3 compared to the fresh catalysts [376].

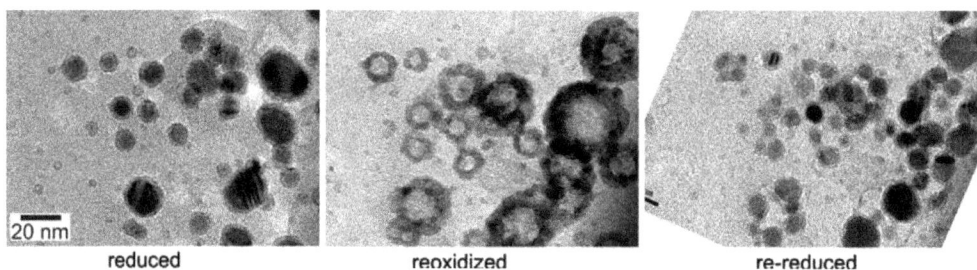

reduced   reoxidized   re-reduced

Figure 48. Bright field TEM images showing redispersion of cobalt particles supported on a flat model silica by oxidative treatment. The center image shows hollow spheres created by the Kirkendall effect, which form dispersed smaller particles upon re-reduction in the right hand image. Reproduced from [378]. Copyright 2011, Springer.

These results are significant because they show the power of careful evaluation of the root causes of deactivation in an important catalytic system and then show how proper choice of regeneration conditions can extend the life of the catalysts by redispersion of the active metal. However, promoters may not be redispersed as completely as the cobalt during repeated regeneration. Although traditional promoters, like Pt and Ru, appear to remain with the Co and maintain their effect, some promoters like Au tend to segregate and lose their promotion effect, as indicated by TPR peaks shifting to higher temperatures [384].

5. Summary

This article focuses on the causes, mechanisms, prevention, modeling, and treatment (experimental and theoretical) of deactivation. Several general, fundamental principles are evident:

(1)   The causes of deactivation are basically of three kinds: chemical, mechanical, and thermal. The five intrinsic mechanisms of catalyst decay, (a) poisoning, (b) fouling, (c) thermal degradation, (d) chemical degradation, and (e) mechanical failure, vary in their reversibility and rates of occurrence. Poisoning and thermal degradation are generally slow, irreversible processes, while fouling with coke and carbon is generally rapid and reversible by regeneration with O_2 or H_2.

(2)   Catalyst deactivation is more easily prevented than cured. Poisoning by impurities can be prevented through careful purification of reactants or mitigated to some extent by adding traps or "getters" as components of the catalyst. Carbon deposition and coking can be prevented by minimizing the formation of carbon or coke precursors through gasification, careful design of catalysts and process conditions, and by controlling reaction rate regimes, e.g., mass transfer regimes, to minimize effects of carbon and coke formation on activity. Sintering is best avoided by minimizing and controlling the temperature of reaction, although recent developments have focused on encapsulating metal crystallites to eliminate mobility, while still allowing access for reactants and products.

(3)   Catalyst regeneration is feasible in some circumstances, especially to recover activity loss due to rapid coking or longer term deactivation associated with loss of active metal dispersion. Typically, regeneration or rejuvenation strategies are dictated by process or economic necessity to obtain desired process run lengths. Life cycle operating strategies are important considerations when evaluating catalyst regeneration/rejuvenation *versus* replacement decisions. Rejuvenation treatments can extend the useful life of catalysts. Selective catalytic reduction catalysts provide an example of rejuvenation practiced in a commercial process.

Acknowledgments

The authors wish to acknowledge Brigham Young University for its support and the employees of MDPI, for their untiring assistance in publishing this review.

Author Contributions

Calvin H. Bartholomew was the primary author of this review. Morris D. Argyle provided assistance in writing and revising the case studies and updating the article in response to the reviewers' comments.

104

Conflicts of Interest

The authors declare no conflict of interest.

References

1. Figueiredo, J.L. Carbon formation and gasification on nickel. In *Progress in Catalyst Deactivation; (NATO Advanced Study Institute Series E, No. 54)*; Martinus Nijhoff Publishers: Boston, MA, USA, 1982; pp. 45–63.
2. Hughes, R. *Deactivation of Catalysts*; Academic Press, London, UK, 1984.
3. Oudar, J.; Wise, H. *Deactivation and Poisoning of Catalysts*; Marcel Dekker: New York, NY, USA, 1985.
4. Butt, J.B.; Petersen, E.E. *Activation, Deactivation, and Poisoning of Catalysts*; Academic Press: San Diego, CA, USA, 1988.
5. Denny, P.J.; Twigg, M.V. Factors determining the life of industrial heterogeneous catalysts. In *Catalyst Deactivation 1980 (Studies in Surface Science and Catalysis)*; Delmon, B., Froment, G.F., Eds.; Elsevier: Amsterdam, The Nerthelands, 1980; Volume 6; pp. 577–599.
6. Bartholomew, C.H. Catalyst Deactivation. *Chem. Eng.* **1984**, *91*, 96–112.
7. Butt, J.B. Catalyst deactivation and regeneration. In *Catalysis—Science and Technology*; Anderson, J.R., Boudart, M., Eds.; Springer-Verlag: New York, NY, USA, 1984; Volume 6, pp. 1–63.
8. Bartholomew, C.H.; Farrauto, R.J. *Fundamentals of Industrial Catalytic Processes*, 2nd ed.; Wiley-Interscience: Hoboken, NJ, USA, 2006.
9. Delmon, B.; Froment, G.F. *Catalyst Deactivation 1980 (Studies in Surface Science and Catalysis)*; Elsevier: Amsterdam, The Nerthelands, 1980; Volume 6.
10. Delmon, B.; Froment, G.F. *Catalyst Deactivation 1987 (Studies in Surface Science and Catalysis)*; Elsevier: Amsterdam, The Nerthelands, 1987; Volume 34.
11. Bartholomew, C.H.; Butt, J.B. *Catalyst Deactivation 1991 (Studies in Surface Science and Catalysis)*; Elsevier: Amsterdam, The Nerthelands, 1991; Volume 68.
12. Delmon, B.; Froment, G.F. *Catalyst Deactivation 1994 (Studies in Surface Science and Catalysis)*; Elsevier: Amsterdam, The Nerthelands, 1994; Volume 88.
13. Bartholomew, C.H.; Fuentes, G.A. *Catalyst Deactivation 1997 (Studies in Surface Science and Catalysis)*; Elsevier: Amsterdam, The Nethelands, 1997; Volume 111.
14. Delmon, B.; Froment, G.F. *Catalyst Deactivation 1999 (Studies in Surface Science and Catalysis)*; Elsevier: Amsterdam, The Nethelands, 1999; Volume 126.
15. Moulijn, J.A. Catalyst Deactivation. *Appl. Catal. A* **2001**, *212*, 1–255.
16. Maxted, E.B. The poisoning of metallic catalysts. *Adv. Catal.* **1951**, *3*, 129–177.
17. Hegedus, L.L.; McCabe, R.W. Catalyst poisoning. In *Catalyst Deactivation 1980 (Studies in Surface Science and Catalysis)*; Delmon, B., Froment, G.F., Eds.; Elsevier: Amsterdam, The Nethelands, 1980; Volume 6, pp. 471–505.
18. Hegedus, L.L.; McCabe, R.W. *Catalyst Poisoning*; Marcel Dekker: New York, NY, USA, 1984.

19. Butt, J.B. Catalyst poisoning and chemical process dynamics. In *Progress in Catalyst Deactivation (NATO Advanced Study Institute Series E, No. 54)*; Figueiredo, J.L., Ed.; Martinus Nijhoff Publishers: Boston, MA, USA, 1982; pp. 153–208.

20. Barbier, J. Effect of poisons on the activity and selectivity of metallic catalysts. In *Deactivation and Poisoning of Catalysts*; Oudar, J., Wise, H., Eds.; Marcel Dekker: New York, NY, USA, 1985; pp. 109–150.

21. Bartholomew, C.H. Mechanisms of nickel catalyst poisoning. In *Catalyst Deactivation 1987 (Studies in Surface Science and Catalysis)*; Delmon, B., Froment, G.F., Eds.; Elsevier: Amsterdam, The Nethelands, 1987; Volume 34, pp. 81–104.

22. Rostrup-Nielsen, J.R. Promotion by poisoning. In *Catalyst Deactivation 1991 (Studies in Surface Science and Catalysis)*; Bartholomew, C.H., Butt, J.B., Eds.; Elsevier: Amsterdam, The Nethelands, 1991; Volume 68, pp. 85–101.

23. Inga, J.; Kennedy, P.; Leviness, S. A1 Fischer-tropsch process in the presence of nitrogen contaminants. WIPO Patent WO 2005/071044, 4 August 2005.

24. Völiter, V.J.; Hermann, M. Katalytische Wirksamkeit van reinem und von CO-vergiftetem Platin bei der p-H_2-Umwandlung. *Z. Anorg. Allg. Chem.* **1974**, *405*, 315.

25. Baron, K. Carbon monoxide oxidation on platinum-lead films. *Thin Solid Films* **1978**, *55*, 449–462.

26. Clay, R.D.; Petersen, E.E. Catalytic activity of an evaporated platinum film progressively poisoned with arsine. *J. Catal.* **1970**, *16*, 32–43.

27. Madon, R.J.; Seaw, H. Effect of Sulfur on the Fischer-Tropsch Synthesis. *Catal. Rev.—Sci. Eng.* **1977**, *15*, 69–106.

28. Bartholomew, C.H.; Agrawal, P.K.; Katzer, J.R. Sulfur poisoning of metals. *Adv. Catal.* **1982**, *31*, 135–242.

29. Rostrup-Nielsen, J.R. Sulfur Poisoning. In *Progress in Catalyst Deactivation (NATO Advanced Study Institute Series E, No. 54)*; Figueiredo, J.L., Ed.; Martinus Nijhoff Publishers: Boston, MA, USA, 1982; pp. 209–227.

30. Wise, H.; McCarty, J.; Oudar, J. Sulfur and carbon interactions with metal surfaces. In *Deactivation and Poisoning of Catalysts*; Oudar, J., Wise, H., Eds.; Marcel Dekker: New York, NY, USA, 1985; pp. 1–50.

31. Rostrup-Nielsen, J.R.; Nielsen, P.E.H. Catalyst deactivation in synthesis gas production, and important syntheses. In *Deactivation and Poisoning of Catalysts*; Oudar, J., Wise, H., Eds.; Marcel Dekker: New York, NY, USA, 1985; pp. 259–323.

32. Grossmann, A.; Erley, W.; Ibach, H. Adsorbate-induced surface stress and surface reconstruction: oxygen, sulfur and carbon on Ni(111). *Surf. Sci.* **1995**, *337*, 183–189.

33. Ruan, L.; Stensgaard, I.; Besenbacher, F.; Lægsgaard, E. Observation of a missing-row structure on an fcc (111) surface: The (5 $\sqrt{3}$ ×2)S phase on Ni(111) studied by scanning tunneling microscopy. *Phys. Rev. Lett.* **1993**, *71*, 2963–2966.

34. Kitajima, Y.; Yokoyama, T.; Ohta, T.; Funabashi, M.; Kosugi, N.; Kuroda, H. Surface EXAFS and XANES studies of (5v3× 2)S/Ni(111). *Surf. Sci.* **1989**, *214*, L261–L269.

35. Perdereau, M.; Oudar, J. Structure, mécanisme de formation et stabilité de la couche d'adsorption du soufre sur le nickel. *Surf. Sci.* **1970**, *20*, 80–98.

36. Oudar, J. Sulfur adsorption and poisoning of metallic catalysts. *Catal. Rev.—Sci. Eng.* **1980**, *22*, 171–195.

37. McCarroll, J.J.; Edmonds, T.; Pitkethly, R.C. Interpretation of a complex low energy electron diffraction pattern: Carbonaceous and sulphur-containing structures on Ni(111). *Nature* **1969**, *223*, 1260–1262.

38. Edmonds, T.; McCarroll, J.J.; Pitkethly, R.C. Surface structures formed during the interaction of sulphur compounds with the (111) face of nickel. *J. Vac. Sci. Technol.* **1971**, *8*, 68–74.

39. Bartholomew, C.H. Mechanisms of catalyst deactivation. *Appl. Catal. A* **2001**, *212*, 17–60.

40. Ruan, L.; Besenbacher, F.; Stensgaard, I.; Lægsgaard, E. Atom-resolved studies of the reaction between H_2S and O on Ni(110). *Phys. Rev. Lett.* **1992**, *69*, 3523–3526.

41. Hepola, J.; McCarty, J.; Krishnan, G.; Wong, V. Elucidation of behavior of sulfur on nickel-based hot gas cleaning catalysts. *Appl. Catal. B* **1999**, *20*, 191–203.

42. Erley, W.; Wagner, H. Sulfur poisoning of carbon monoxide adsorption on Ni(111). *J. Catal.* **1978**, *53*, 287–294.

43. Rendulic, K.D.; Winkler, A. The initial sticking coefficient of hydrogen on sulfur- and oxygen-covered polycrystalline nickel surfaces. *Surf. Sci.* **1978**, *74*, 318–320.

44. Goodman, D.W.; Kiskinova, M. Chemisorption and reactivity studies of H_2 and CO on sulfided Ni(100). *Surf. Sci.* **1981**, *105*, L265–L270.

45. Kiskinova, M.; Goodman, D.W. Modification of chemisorption properties by electronegative adatoms: H_2 and CO on chlorided, sulfided, and phosphided Ni(100). *Surf. Sci.* **1981**, *108*, 64–76.

46. Johnson, S.; Madix, R.S. Desorption of hydrogen and carbon monoxide from Ni(100), Ni(100)p(2 × 2)S, and Ni(100)c(2 × 2)S surfaces. *Surf. Sci.* **1981**, *108*, 77–98.

47. Madix, R.J.; Thornberg, M.; Lee, S.B. CO-sulfur interaction on Ni(110); evidence for local interactions, not long range electronic effects. *Surf. Sci.* **1983**, *133*, L447–L451.

48. Hardegree, E.L.; Ho, P.; White, J.M. Sulfur adsorption on Ni(100) and its effect on CO chemisorption: I. TDS, AES and work function results. *Surf. Sci.* **1986**, *165*, 488–506.

49. Erekson, E.J.; Bartholomew, C.H. Sulfur poisoning of nickel methanation catalysts: II. Effects of H_2S concentration, CO and H_2O partial pressures and temperature on reactivation rates. *Appl. Catal.* **1983**, *5*, 323–336.

50. Jacobs, G.; Ghadiali, F.; Pisanu, A.; Padro, C.L.; Borgna, A.; Alvarez, W.E.; Resasco, D.E. Increased Sulfur Tolerance of Pt/KL Catalysts Prepared by Vapor-Phase Impregnation and Containing a Tm Promoter. *J. Catal.* **2000**, *191*, 116–127.

51. Jongpatiwut, S.; Sackamduang, P.; Rirksomboon, T.; Osuwan, S.; Alvarez, W.E.; Resasco, D.E. Sulfur- and water-tolerance of Pt/KL aromatization catalysts promoted with Ce and Yb. *Appl. Catal. A* **2002**, *230*, 177–193.

52. Jacobs, G.; Ghadiali, F.; Pisanu, A.; Borgna, A.; Alvarez, W.E.; Resasco, D.E. Characterization of the morphology of Pt clusters incorporated in a KL zeolite by vapor phase and incipient wetness impregnation. Influence of Pt particle morphology on aromatization activity and deactivation. *Appl. Catal. A* **1999**, *188*, 79–98.

53. McVicker, G.B.; Kao, J.L.; Ziemiak, T.J.J.; Gates, W.E.; Robbins, J.L.; Treacy, M.M.J.; Rice, S.B.; Vanderspurt, T.H.; Cross, V.R.; Ghosh, A.K. Effect of Sulfur on the Performance and on the Particle Size and Location of Platinum in Pt/KL Hexane Aromatization Catalysts. *J. Catal.* **1993**, *139*, 48–61.

54. Farbenindustrie, I.G. Improvements in the Manufacture and Production of Unsaturated Hydrocarbons of Low Boiling Point. British Patent 322,284, 5 December 1929.

55. Kritzinger, J.A. The role of sulfur in commercial iron-based Fischer–Tropsch catalysis with focus on C_2-product selectivity and yield. *Catal. Today* **2002**, *71*, 307–318.

56. Sparks, D.E.; Jacobs, G.; Gnanamani, M.K.; Pendyala, V.R.R.; Ma, W.; Kang, J.; Shafer, W.D.; Keogh, R.A.; Graham, U.M.; Gao, P.; *et al*. Poisoning of cobalt catalyst used for Fischer–Tropsch synthesis. *Catal. Today* **2013**, *215*, 67–72.

57. Rostrup-Nielsen, J.R.; Trimm, D.L. Mechanisms of carbon formation on nickel-containing catalysts. *J. Catal.* **1977**, *48*, 155–165.

58. Trimm, D.L. The Formation and Removal of Coke from Nickel Catalyst. *Catal. Rev.—Sci. Eng.* **1977**, *16*, 155–189.

59. Trimm, D.L. Catalyst design for reduced coking (review). *Appl. Catal.* **1983**, *5*, 263–290.

60. Bartholomew, C.H. Carbon deposition in steam reforming and methanation. *Catal. Rev.—Sci. Eng.* **1982**, *24*, 67–112.

61. Albright, L.F.; Baker, R.T.K. *Coke Formation on Metal Surfaces (ACS Symposium Series 202)*; American Chemical Society: Washington, DC, USA, 1982.

62. Menon, P.G. Coke on catalysts—harmful, harmless, invisible and beneficial types. *J. Mol. Catal.* **1990**, *59*, 207–220.

63. Rostrup-Nielsen, J.R. Conversion of hydrocarbons and alcohols for fuel cells. *Phys. Chem. Chem. Phys.* **2001**, *3*, 283–288.

64. Trane-Restrup, R.; Resasco, D.E.; Jensen, A.D. Steam reforming of light oxygenates. *Catal. Sci. Technol.* **2013**, *3*, 3292–3302.

65. De Lima, S.M.; da Silva, A.M.; da Costa, L.O.O.; Assaf, J.M.; Jacobs, G.; Davis, B.H.; Mattos, L.V.; Noronha, F.B. Evaluation of the performance of Ni/La_2O_3 catalyst prepared from $LaNiO_3$ perovskite-type oxides for the production of hydrogen through steam reforming and oxidative steam reforming of ethanol. *Appl. Catal. A* **2010**, *377*, 181–190.

66. Deken, J.D.; Menon, P.G.; Froment, G.F.; Haemers, G.; On the nature of carbon in $Ni\alpha$-Al_2O_3 catalyst deactivated by the methane-steam reforming reaction. *J. Catal.* **1981**, *70*, 225–229.

67. Durer, W.G.; Craig, J.H., Jr.; Lozano, J. Surface carbon and its effects on hydrogen adsorption on Rh(100). *Appl. Surf. Sci.* **1990**, *45*, 275–277.

68. Moeller, A.D.; Bartholomew, C.H. Deactivation by carbon of nickel and nickel-molybdenum methanation catalysts. *Prepr.—Am. Chem. Soc., Div. Fuel Chem.* **1980**, *25*, 54–70.

69. Marschall, K.-J.; Mleczko, L. Short-contact-time reactor for catalytic partial oxidation of methane. *Ind. Eng. Chem. Res.* **1999**, *38*, 1813–1821.

70. Rostrup-Nielsen, J.R. Catalytic steam reforming. In *Catalysis—Science and Technology*; Anderson, J.R., Boudart, M., Eds.; Springer-Verlag: New York, NY, USA, 1984; Volume 5, pp. 1–117.

71. Besenbacher, F.; Chorkendorff, I.; Clausen, B.S.; Hammer, B.; Molenbroek, A.M.; Nørskov, J.K.; Stensgaard, I. Design of a surface alloy catalyst for steam reforming. *Science* **1998**, *279*, 1913–1915.

72. Nemes, T.; Chambers, A.; Baker, R.T.K. Characteristics of carbon filament formation from the interaction of Cobalt–Tin particles with ethylene. *J. Phys. Chem. B* **1998**, *102*, 6323-6330.

73. Bartholomew, C.H.; Strasburg, M.V.; Hsieh, H. Effects of support on carbon formation and gasification on nickel during carbon monoxide hydrogenation. *Appl. Catal.* **1988**, *36*, 147–162.

74. Vance, C.K.; Bartholomew, C.H. Hydrogenation of carbon dioxide on group viii metals: III, Effects of support on activity/selectivity and adsorption properties of nickel. *Appl. Catal.* **1983**, *7*, 169–177.

75. Baker, R.T.K.; Chludzinski, J.J., Jr. Filamentous carbon growth on nickel-iron surfaces: The effect of various oxide additives. *J. Catal.* **1980**, *64*, 464–468.

76. Brown, D.E.; Clark, J.T.K.; Foster, A.I.; McCarroll, J.J.; Sims, M.L. Inhibition of coke formation in ethylene steam cracking. In *Coke Formation on Metal Surfaces (ACS Symposium Series 202)*; Albright, L.F., Baker, R.T.K., Eds.; American Chemical Society: Washington, DC, USA, 1982; pp. 23–43.

77. Bitter, J.H.; Seshan, K.; Lercher, J.A. Deactivation and coke accumulation during CO_2/CH_4 reforming over Pt catalysts. *J. Catal.* **1999**, *183*, 336–343.

78. Rostrup-Nielsen, J.R. Coking on nickel catalysts for steam reforming of hydrocarbons. *J. Catal.* **1974**, *33*, 184–201.

79. Saib, A.M.; Moodley, D.J.; Ciobica, I.M.; Hauman, M.M.; Sigwebela, B.H.; Weststrate, C.J.; Niemanstsverdriet, J.W.; van de Loosdrecht, J. Fundamental understanding of deactivation and regeneration of cobalt Fischer–Tropsch synthesis catalysts. *Catal. Today* **2010**, *154*, 271–282.

80. Tsakoumis, N.E.; Rønning, M.; Borg, O.; Rytter, E.; Holmen, A. Deactivation of cobalt based Fischer–Tropsch catalysts: A review. *Catal. Today* **2010**, *154*, 162–182.

81. Gates, B.C.; Katzer, J.R.; Schuit, G.C.A. *Chemistry of Catalytic Processes*; McGraw-Hill: New York, NY, USA, 1979.

82. Naccache, C. Deactivation of acid catalysts. In *Deactivation and Poisoning of Catalysts*; Oudar, J., Wise, H., Eds.; Marcel Dekker: New York, NY, USA, 1985; pp. 185–203.

83. Appleby, W.G.; Gibson, J.W.; Good, G.M. Coke Formation in Catalytic Cracking. *Ind. Eng. Chem. Process Des. Dev.* **1962**, *1*, 102–110.

84. Beuther, H.; Larson, O.H.; Perrotta, A.J. The mechanism of coke formation on catalysts. In *Catalyst Deactivation 1980 (Studies in Surface Science and Catalysis)*; Delmon, B., Froment, G.F., Eds.; Elsevier: Amsterdam, The Nerthelands, 1980; Volume 6, pp. 271–282.

85. Gayubo, A.G.; Arandes, J.M.; Aguayo, A.T.; Olazar, M.; Bilbao, J. Deactivation and acidity deterioration of a silica/alumina catalyst in the isomerization of cis-butene. *Ind. Eng. Chem. Res.* **1993**, *32*, 588–593.

86. Augustine, S.M.; Alameddin, G.N.; Sachtler, W.M.H. The effect of Re, S, and Cl on the deactivation of Pt γ-Al$_2$O$_3$ reforming catalysts. *J. Catal.* **1989**, *115*, 217–232.

87. Guisnet, M.; Magnoux, P. Coking and deactivation of zeolites: Influence of the Pore Structure. *Appl. Catal.* **1989**, *54*, 1–27.

88. Bauer, F.; Kanazirev, V.; Vlaev, C.; Hanisch, R.; Weiss, W. Koksbildung in ZSM-5-Katalysatoren. *Chem. Tech.* **1989**, *41*, 297–301.

89. Grotten, W.A.; Wojciechowski, B.W.; Hunter, B.K. On the relationship between coke formation chemistry and catalyst deactivation. *J. Catal.* **1992**, *138*, 343–350.

90. Bellare, A.; Dadyburjor, D.B. Evaluation of modes of catalyst deactivation by coking for cumene cracking over zeolites. *J. Catal.* **1993**, *140*, 510–525.

91. Uguina, M.A.; Serrano, D.P.; Grieken, R.V.; Vènes, S. Adsorption, acid and catalytic changes induced in ZSM-5 by coking with different hydrocarbons. *Appl. Catal. A* **1993**, *99*, 97–113.

92. Li, C.; Chen, Y.-W.; Yang, S.-J.; Yen, R.-B. *In-situ* FTIR investigation of coke formation on USY zeolite. *Appl. Surf. Sci.* **1994**, *81*, 465–468.

93. Buglass, J.G.; de Jong, K.P.; Mooiweer, H.H. Analytical studies of the coking of the zeolite ferrierite. *J. Chem. Soc. Abstr.* **1995**, *210*, 105-PETR.

94. Chen, D.; Rebo, H.P.; Moljord, K.; Holmen, A. Effect of coke deposition on transport and adsorption in zeolites studied by a new microbalance reactor. *Chem. Eng. Sci.* **1996**, *51*, 2687–2692.

95. Guisnet, M.; Magnoux, P.; Martin, D. Roles of acidity and pore structure in the deactivation of zeolites by carbonaceous deposits. In *Catalyst Deactivation 1997 (Studies in Surface Science and Catalysis)*; Bartholomew, C.H., Fuentes, G.A., Eds.; Elsevier: Amsterdam, The Nethelands, 1997; Volume 111, pp. 1–19.

96. Masuda, T.; Tomita, P.; Fujikata, Y.; Hashimoto, K. Deactivation of HY-type zeolite catalyst due to coke deposition during gas-oil cracking. In *Catalyst Deactivation 1999 (Studies in Surface Science and Catalysis)*; Delmon, B., Froment, G.F. Eds.; Elsevier: Amsterdam, The Nethelands, 1999; Volume 126, pp. 89–96.

97. Cerqueira, H.S.; Magnoux, P.; Martin, D.; Guisnet, M. Effect of contact time on the nature and location of coke during methylcyclohexane transformation over a USHY zeolite. In *Catalyst Deactivation 1999 (Studies in Surface Science and Catalysis)*; Delmon, B., Froment, G.F., Eds.; Elsevier: Amsterdam, The Nethelands, 1999; Volume 126, pp. 105–112.

98. Wanke, S.E.; Flynn, P.C. The sintering of supported metal catalysts. *Catal. Rev.—Sci. Eng.* **1975**, *12*, 93–135.

99. Wynblatt, P.; Gjostein, N.A. Supported metal crystallites. *Prog. Solid State Chem.* **1975**, *9*, 21–58.

100. Ruckenstein, E.; Pulvermacher, B. Kinetics of crystallite sintering during heat treatment of supported metal catalysts. *AIChE J.* **1973**, *19*, 356–364.

101. Ruckenstein, E.; Dadyburjor, D.B. Sintering and redispersion in supported metal catalysts. *Rev. Chem. Eng.* **1983**, *1*, 251–356.

102. Wanke, S.E. Sintering of commercial supported platinum group metal catalysts. In *Progress in Catalyst Deactivation (NATO Advanced Study Institute Series E, No. 54)*; Figueiredo, J.L., Ed.; Martin Nijhoff Publishers: Boston, MA, USA, 1982; pp. 315–328.

103. Baker, R.T.; Bartholomew, C.H.; Dadyburjor, D.B. Sintering and Redispersion: Mechanisms and Kinetics. In *Stability of Supported Catalysts: Sintering and Redispersion*; Horsley, J.A., Ed.; Catalytica: Mountain View, CA, USA, 1991; pp. 169-225.

104. Bartholomew, C.H. Model catalyst studies of supported metal sintering and redispersion kinetics. In *Catalysis (Specialist Periodical Report)*; Spivey, J.J., Agarwal, S.K., Eds.; Royal Society of Chemistry, Cambridge, UK, 1993; Volume 10, pp. 41–82.

105. Bartholomew, C.H. Sintering kinetics of supported metals: new perspectives from a unifying GPLE treatment. *Appl. Catal. A* **1993**, *107*, 1–57.

106. Bartholomew, C.H. Sintering kinetics of supported metals: perspectives from a generalized power law approach. In *Catalyst Deactivation 1994 (Studies in Surface Science and Catalysis)*; Delmon, B., Froment, G.F., Eds.; Elsevier: Amsterdam, The Netherlands, 1994; Volume 88, pp. 1–18.

107. Bartholomew, C.H. Sintering and redispersion of supported metals: Perspectives from the literature of the past decade. In *Catalyst Deactivation 1997 (Studies in Surface Science and Catalysis)*; Bartholomew, C.H., Fuentes, G.A., Eds.; Elsevier: Amsterdam, The Netherlands, 1997; Volume 111, pp. 585–592.

108. Bartholomew, C.H.; Sorenson, W. Sintering kinetics of silica- and alumina-supported nickel in hydrogen atmosphere. *J. Catal.* **1983**, *81*, 131–141.

109. Moulijn, J.A.; van Diepen, A.E.; Kapteijn, F. Catalyst deactivation: is it predictable?: What to do? *Appl. Catal. A* **2001**, *212*, 3–16.

110. Bridger, G.W.; Spencer, M.S. Methanol synthesis. In *Catalyst Handbook*, 2nd ed.; Twigg, M.V., Ed.; Manson Publishing: London, UK, 1996; pp. 441–468.

111. Fuentes, G.A. Catalyst deactivation and steady-state activity: A generalized power-law equation model. *Appl. Catal.* **1985**, *15*, 33–40.

112. Fuentes, G.A.; Ruiz-Trevino, F.A. Towards a better understanding of sintering phenomena in catalysis. In *Catalyst Deactivation 1991 (Studies in Surface Science and Catalysis)*; Bartholomew, C.H., Butt, J.B., Eds., Elsevier: Amsterdam, The Netherlands, 1991; Volume 68, pp. 637–644.

113. Bournonville, J.P.; Martino, G. Sintering of Alumina Supported Platinum. In *Catalyst Deactivation 1980 (Studies in Surface Science and Catalysis)*; Delmon, B., Froment, G.F., Eds., Elsevier: Amsterdam, The Netherlands, 1980; Volume 6, pp. 159–166.

114. Somorjai, G.A. Small-Angle X-Ray Scattering and Low Energy Electron Diffraction Studies on Catalyst Surfaces, In *X-ray and Electron Methods of Analysis (Progress in Analytical Chemistry)*; Van Olphen, H., Parrish, W., Eds., Plenum Press; New York, NY, USA, 1968; Volume 1, pp. 101–126.

115. Seyedmonir, S.R.; Strohmayer, D.E.; Guskey, G.J.; Geoffroy, G.L.; Vannice, M.A. Characterization of supported silver catalysts: III. Effects of support, pretreatment, and gaseous environment on the dispersion of Ag. *J. Catal.* **1985**, *93*, 288–302.

116. Trimm, D.L. Thermal stability of catalyst supports. In *Catalyst Deactivation 1991 (Studies in Surface Science and Catalysis)*; Bartholomew, C.H., Butt, J.B. Eds.; Elsevier: Amsterdam, The Nethelands, 1991; Volume 68, pp. 29–51.

117. Shastri, A.G.; Datye, A.K.; Schwank, J. Influence of chlorine on the surface area and morphology of TiO_2. *Appl. Catal.* **1985**, *14*, 119–131.

118. Oberlander, R.K. Aluminas for catalysts: Their preparations and properties. In *Applied Industrial Catalysis*; Leach, B.E., Ed.; Academic Press: Orlando, FL, USA, 1984; Volume 3, pp. 64–112.

119. Wefers, K.; Misra, C. *Oxides and Hydroxides of Aluminum*; Alcoa Technical Paper No. 19; Alcoa Laboratories: Pittsburg, PA, USA, 1987.

120. Hegedus, L.L.; Baron, K. Phosphorus accumulation in automotive catalysts. *J. Catal.* **1978**, *54*, 115–119.

121. Summers, J.; Hegedus, L.L. Modes of catalyst deactivation in stoichiometric automobile exhaust. *Ind. Eng. Chem. Prod. Res. Dev.* **1979**, *18*, 318–324.

122. Peter-Hoblyn, J.D.; Valentine, J.M.; Sprague, B.N.; Epperly, W.R. Methods for reducing harmful emissions from a diesel engine. U.S. Patent 6,003,303, 21 December 1999.

123. Manson, I. Self-regenerating diesel exhaust particulate filter and material. U.S. Patent 6,013,599, 11 January 2000.

124. Deeba, M.; Lui, Y.K.; Dettling, J.C. Four-way diesel exhaust catalyst and method of use. U.S. Patent 6,093,378, 25 July 2000.

125. Dry, M.E. The fischer-tropsch synthesis. In *Catalysis—Science and Technology*; Anderson, J., Boudart, M., Eds.; Springer-Verlag: New York, NY, USA, 1981; pp. 159–218.

126. Huber, G.W.; Guymon, C.G.; Stephenson, B.C.; Bartholomew, C.H. Hydrothermal stability of Co/SiO_2 Fischer-Tropsch synthesis catalysts. In *Catalyst Deactivation 2001 (Studies in Surface Science and Catalysis)*; Spivey, J.J., Roberts, G.W., Davis, B.H., Eds.; Elsevier: Amsterdam, The Netheland, 2001; Volume 139, pp. 423–430.

127. Busca, G.; Lietti, L.; Ramis, G.; Berti, F. Chemical and mechanistic aspects of the selective catalytic reduction of NO_x by ammonia over oxide catalysts: A review. *Appl. Catal. B* **1998**, *18*, 1–36.

128. Kobylinski, T.P.; Taylor, B.W.; Yong, J.E. Stabilized ruthenium catalysts For NO_x reduction. *Proc. SAE*, **1974**, Paper 740250.

129. Shelef, M.; Gandhi, H.S. The reduction of nitric oxide in automobile emissions: Stabilisation of catalysts containing ruthenium. *Platinum Met. Rev.* **1974**, *18*, 2–14.

130. Gandhi, H.S.; Stepien, H.K.; Shelef, M. Optimization of ruthenium-containing, stabilized, nitric oxide reduction catalysts. *Mat. Res. Bull.* **1975**, *10*, 837–845.

131. Bartholomew, C.H. Reduction of nitric oxide by monolithic-supported palladium-nickel and palladium-ruthenium alloys. *Ind. Eng. Chem. Prod. Res. Dev.* **1975**, *14*, 29–33.

132. Clark, R.W.; Tien, J.K.; Wynblatt, P. Loss of palladium from model platinum-palladium supported catalysts during annealing. *J. Catal.* **1980**, *61*, 15–18.

133. Shen, W.M.; Dumesic, J.A.; Hill, C.G. Criteria for stable Ni particle size under methanation reaction conditions: Nickel transport and particle size growth via nickel carbonyl. *J. Catal.* **1981**, *68*, 152–165.

134. Pannell, R.B.; Chung, K.S.; Bartholomew, C.H. The stoichiometry and poisoning by sulfur of hydrogen, oxygen and carbon monoxide chemisorption on unsupported nickel. *J. Catal.* **1977**, *46*, 340–347.

135. Lohrengel, G.; Baerns, M. Determination of the metallic surface area of nickel and its dispersion on a silica support by means of a microbalance. *Appl. Catal.* **1981**, *1*, 3–7.

136. Qamar, I.; Goodwin, J.G. Fischer-Tropsch Synthesis over Composite Ru Catalysts In Proceedings of 8th North American Catalysis Society Meeting, Philadelphia, PA, USA, 1983, Paper C-22.

137. Goodwin, J.G.; Goa, D.O.; Erdal, S.; Rogan, F.H. Reactive metal volatilization from Ru/Al2O3 as a result of ruthenium carbonyl formation. *Appl. Catal.* **1986**, *24*, 199–209.

138. Watzenberger, O.; Haeberle, T.; Lynch, D.T.; Emig, G. Deactivation of Heteropolyacid Catalysts. In *Catalyst Deactivation 1991 (Studies in Surface Science and Catalysis)*; Bartholomew, C.H., Butt, J.B., Eds.; Elsevier: Amsterdam, The Nethelands, 1991; Volume 68, pp. 441–448.

139. Agnelli, M.; Kolb, M.; Mirodatos, C. Co hydrogenation on a nickel catalyst: 1. Kinetics and modeling of a low-temperature sintering process. *J. Catal.* **1994**, *148*, 9–21.

140. Lee, H.C.; Farrauto, R.J. Catalyst deactivation due to transient behavior in nitric acid production. *Ind. Eng. Chem. Res.* **1989**, *28*, 1–5.

141. Farrauto, R.J.; Lee, H.C. Ammonia oxidation catalysts with enhanced activity. *Ind. Eng. Chem. Res.* **1990**, *29*, 1125–1129.

142. Sperner, F.; Hohmann, W. Rhodium-platinum gauzes for ammonia oxidation. *Platinum Met. Rev.* **1976**, *20*, 12–20.

143. Hess, J.M.; Phillips, J. Catalytic etching of Pt/Rh gauzes. *J. Catal.* **1992**, *136*, 149–160.

144. Kuo, C.L.; Hwang, K.C. Does morphology of a metal nanoparticle play a role in ostwald ripening processes? *Chem. Mater.* **2013**, *25*, 365–371.

145. Bartholomew, C.H. Hydrogen adsorption on supported cobalt, iron, and nickel. *Catal. Lett.* **1990**, *7*, 27–51.

146. Wu, N.L.; Phillips, J. Catalytic etching of platinum during ethylene oxidation. *J. Phys. Chem.* **1985**, *89*, 591–600.

147. Wu, N.L.; Phillips, J. Reaction-enhanced sintering of platinum thin films during ethylene oxidation. *J. Appl. Phys.* **1986**, *59*, 769–779.

148. Wu, N.L.; Phillips, J. Sintering of silica-supported platinum catalysts during ethylene oxidation. *J. Catal.* **1988**, *113*, 129–143.

149. Bielanski, A.; Najbar, M.; Chrzaoszcz, J.; Wal, W. Deactivation of the V_2O_5-MoO_3, catalysts in the selective oxidation of eenzene to maleic anhydride and the changes in its morphology and chemical composition. In *Catalyst Deactivation 1980 (Studies in Surface Science and Catalysis)*; Delmon, B., Froment, G.F., Eds.; Elsevier: Amsterdam, The Nethelands, 1980; Volume 6, pp. 127–140.

150. Burriesci, N.; Garbassi, F.; Petrera, M.; Petrini, G.; Pernicone, N. Solid state reactions in Fe-Mo oxide catalysts for methanol oxidation during aging in industrial plants. In *Catalyst Deactivation 1980 (Studies in Surface Science and Catalysis)*; Delmon, B., Froment, G.F., Eds.; Elsevier: Amsterdam, The Nethelands, 1980; Volume 6, pp. 115–126.

151. Xiong, Y.L.; Castillo, R.; Papadopoulou, C.; Dada, L.; Ladriere, J.; Ruiz, P.; Delmon, B. The protecting role of antimony oxide against deactivation of iron molybdate in oxidation catalysts. In *Catalyst Deactivation 1991 (Studies in Surface Science and Catalysis)*; Bartholomew, C.H., Butt, J.B., Eds.; Elsevier: Amsterdam, The Nethelands, 1991; Volume 68, pp. 425–432.

152. Farrauto, R.J.; Hobson, M.; Kennelly, T.; Waterman, E. Catalytic chemistry of supported palladium for combustion of methane. *Appl. Catal. A* **1992**, *81*, 227–237.

153. Gai-Boyes, P.L. Defects in Oxide Catalysts: Fundamental Studies of Catalysis in Action. *Catal. Rev.—Sci. Eng.* **1992**, *34*, 1–54.

154. Delmon, B. Solid-state reactions in catalysts during ageing: Beneficial role of spillover. In *Catalyst Deactivation 1994 (Studies in Surface Science and Catalysis, Volume 88)*; Delmon, B., Froment, G.F.; Eds., Elsevier: Amsterdam, The Nethelands, 1994, pp. 113–128.

155. Erickson, K.M.; Karydis, D.A.; Boghosian, S.; Fehrmann, R. Deactivation and compound formation in sulfuric-acid catalysts and model systems. *J. Catal.* **1995**, *155*, 32–42.

156. Delmon, B. Solid state reactions in catalysts: An approach to real active systems and their deactivation. In *Catalyst Deactivation 1997 (Studies in Surface Science and Catalysis)*; Bartholomew, C.H., Fuentes, G.A., Eds.; Elsevier: Amsterdam, The Nethelands, 1997; Volume 111, pp. 39–51.

157. Jackson, N.B.; Datye, A.K.; Mansker, L.; O'Brien, R.J.; Davis, B.H. Deactivation and attrition of iron catalysts in synthesis gas. In *Catalyst Deactivation 1997 (Studies in Surface Science and Catalysis)*; Bartholomew, C.H., Fuentes, G.A., Eds.; Elsevier: Amsterdam, The Nethelands, 1997; Volume 111, pp. 501–516.

158. Eliason, S.A.; Bartholomew, C.H. Temperature-programmed reaction study of carbon transformations on iron fischer-tropsch catalysts during steady-state synthesis. In *Catalyst Deactivation 1997 (Studies in Surface Science and Catalysis)*; Bartholomew, C.H., Fuentes, G.A., Eds.; Elsevier: Amsterdam, The Nethelands, 1997; Volume 111, pp. 517–526.

159. Baranski, A.; Dziembaj, R.; Kotarba, A.; Golebiowski, A.; Janecki, Z.; Pettersson, J.B.C. Deactivation of iron catalyst by water-potassium thermal desorption studies. In *Catalyst Deactivation 1999 (Studies in Surface Science and Catalysis)*; Delmon, B., Froment, G.F., Eds.; Elsevier: Amsterdam, The Nethelands, 1999; Volume 126, pp. 229–236.

160. Querini, C.A.; Ravelli, F.; Ulla, M.; Cornaglia, L.; Miró, E. Deactivation of Co, K catalysts during catalytic combustion of diesel soot: Influence of the support. In *Catalyst Deactivation 1999 (Studies in Surface Science and Catalysis)*; Delmon, B., Froment, G.F., Eds.; Elsevier: Amsterdam, The Nethelands, 1999; Volume 126, pp. 257–264.

161. Wilson, J.; de Groot, C. Atomic-scale restructuring in high-pressure catalysis. *J. Phys. Chem.* **1995**, *99*, 7860–7866.

162. Parkinson, G.S.; Novotny, Z.; Argentero, G.; Schmid, M.; Pavelec, J.; Kosak, R.; Blaha, P.; Diebold, U. Carbon monoxide-induced adatom sintering in a Pd–Fe$_3$O$_4$ model catalyst. *Nat. Mater.* **2013**, *12*, 724–728.

163. Pham, H.N.; Reardon, J.; Datye, A.K. Measuring the strength of slurry phase heterogeneous catalysts. *Powder Technol.* **1999**, *103*, 95–102.

164. Kalakkad, D.S.; Shroff, M.D.; Kohler, S.; Jackson, N.; Datye, A.K. Attrition of precipitated iron Fischer-Tropsch catalysts. *Appl. Catal. A* **1995**, *133*, 335–350.

165. Callister, W.D. *Materials Science and Engineering: An Introduction*; John Wiley, & Sons, Inc.: New York, NY, USA, 2000.

166. Coble, R.L.; Kingery, W.D. Effect of porosity on physical properties of sintered alumina. *J. Am. Ceram. Soc.* **1956**, *39*, 377–385.

167. Deng, S.G.; Lin, Y.S. Granulation of sol-gel-derived nanostructured alumina. *AIChE J.* **1997**, *43*, 505–514.

168. Thoma, S.G.; Ciftcioglu, M.; Smith, D.M. Determination of agglomerate strength distributions: Part 1. Calibration via ultrasonic forces. *Powder Technol.* **1991**, *68*, 53–61.

169. Bankmann, M.; Brand, R.; Engler, B.H.; Ohmer, J. Forming of high surface area TiO$_2$ to catalyst supports. *Catal. Today* **1992**, *14*, 225–242.

170. Kenkre, V.M.; Endicott, M.R. A theoretical model for compaction of granular materials. *J. Am. Ceram. Soc.* **1996**, *79*, 3045–3054.

171. Song, H.; Evans, J.R.G. A die pressing test for the estimation of agglomerate strength. *J. Am. Ceram. Soc.* **1994**, *77*, 806–814.

172. Werther, J.; Xi, W. Jet attrition of catalyst particles in gas fluidized beds. *Powder Technol.* **1993**, *76*, 39–46.

173. Bhatt, B.L.; Schaub, E.S.; Hedorn, E.C.; Herron, D.M.; Studer, D.W.; Brown, D.M. Liquid phase Fischer-Tropsch synthesis in a bubble column. In *Proceedings of Liquefaction Contractors Review Conference*; Stiegel, G.J., Srivastava, R.D., Eds.; U.S. Department of Energy: Pittsburgh, PA, USA, 1992; pp. 402–423.

174. Pham, H.N.; Datye, A.K. The synthesis of attrition resistant slurry phase iron Fischer-Tropsch catalysts. *Catal. Today* **2000**, *58*, 233–240.

175. Bemrose, C.R.; Bridgewater, J. A review of attrition and attrition test methods. *Powder Technol.* **1987**, *49*, 97–126.

176. Ghadiri, M.; Cleaver, J.A.S.; Tuponogov, V.G.; Werther, J. Attrition of FCC powder in the jetting region of a fluidized bed. *Powder Technol.* **1994**, *80*, 175–178.

177. Weeks, S.A.; Dumbill, P. Method speeds FCC catalyst attrition resistance determinations. *Oil Gas J.* **1990**, *88*, 38–40.

178. Zhao, R.; Goodwin, J.G.; Jothimurugesan, K.; Spivey, J.J.; Gangwal, S.K. Comparison of attrition test methods: ASTM standard fluidized bed vs jet cup. *Ind. Eng. Chem. Res.* **2000**, *39*, 1155–1158.

179. Doolin, P.K.; Gainer, D.M.; Hoffman, J.F. Laboratory testing procedure for evaluation of moving bed catalyst attrition. *J. Test. Eval.* **1993**, *21*, 481.

180. Oukaci, R.; Singleton, A.H.; Wei, D.; Goodwin, J.G. Attrition resistance of Al_2O_3-supported cobalt F-T catalysts. In Proceedings of the 217th National Meeting, ACS Division of Petroleum Chemistry, Anaheim, CA, USA, 21–25 March 1999; p. 91.

181. Adams, M.J.; Mullier, M.A.; Seville, J.P.K. Agglomerate strength measurement using a uniaxial confined compression test. *Powder Technol.* **1994**, *78*, 5–13.

182. Emig, G.; Martin, F.G. Development of a fluidized bed catalyst for the oxidation of *n*-butane to maleic anhydride. *Ind. Eng. Chem. Res.* **1991**, *30*, 1110–1116.

183. Silver, R.G.; Summers, J.C.; Williamson, W.B. *Catalysis and Automotive Pollution Control II*, Elsevier: Amsterdam, The Nethelands, 1991; p. 167.

184. Fisher, G.B.; Zammit, M.G.; LaBarge, J. *Investigation of Catalytic Alternatives to Rhodium in Emissions Control*; SAE Report 920846; SAE International: Warrendale, PA, USA, 1992.

185. Farrauto, R.J.; Heck, R.M. Catalytic converters: state of the art and perspectives. *Catal. Today* **1999**, *51*, 351–360.

186. Beer, G.L. Extended catalyst life two stage hydrocarbon synthesis process. U.S. Patent 6,169,120, 2 January 2001.

187. Bartholomew, C.H.; Stoker, M.W.; Mansker, L.; Datye, A. Effects of pretreatment, reaction, and promoter on microphase structure and Fischer-Tropsch activity of precipitated iron catalysts. In *Catalyst Deactivation 1999 (Studies in Surface Science and Catalysis)*; Delmon, B., Froment, G.F., Eds.; Elsevier: Amsterdam, The Nethelands, 1999; Volume 126, pp. 265–272.

188. Nagano, O.; Watanabe, T. Method for producing methacrolein. U.S. Patent 5,728,894, 17 March 1998.

189. Maillet, T.; Barbier, J.; Duprez, D. Reactivity of steam in exhaust gas catalysis III. Steam and oxygen/steam conversions of propane on a Pd/Al_2O_3 catalyst. *Appl. Catal. B* **1996**, *9*, 251–266.

190. Mathys, G.M.K.; Martens, L.R.M.; Baes, M.A.; Verduijn, J.P.; Huybrechts, D.R.C. Alkene oligomerization. WIPO Patent WO 1993/016020 (A3), 16 September 1993.

191. Mathys, G.M.K.; Martens, L.R.M.; Baes, M.A.; Verduijn, J.P.; Huybrechts, D.R.C. Alkene oligomerization. U.S. Patent 5,672,800, 30 September 1997.

192. Stine, L.O.; Muldoon, B.S.; Gimre, S.C.; Frame, R.R. Process for oligomer production and saturation. U.S. Patent 6,080,903, 26 June 2000.

193. Subramaniam, B.; Arunajatesan, V.; Lyon, C.J. Coking of solid acid catalysts and strategies for enhancing their activity. In *Catalyst Deactivation 1999 (Studies in Surface Science and Catalysis)*; Delmon, B., Froment, G.F., Eds.; Elsevier: Amsterdam, The Nethelands, 1999; Volume 126, pp. 63–77.

194. Ginosar, D.M.; Fox, R.V.; Kong, P.C. Solid catalyzed isoparaffin alkylation at supercritical fluid and near-supercritical fluid conditions. WIPO Patent WO 1999/033769 (A1), 8 July 1999.

195. Ribeiro, F.H.; Bonivardi, A.L.; Kim, C. Transformation of platinum into a stable, high-temperature, dehydrogenation-hydrogenation catalyst by ensemble size reduction with rhenium and sulfur. *J. Catal.* **1994**, *150*, 186–198.

196. Ginosar, D.; Subramaniam, B. Coking and activity of a reforming catalyst in near-critical and dense supercritical reaction mixtures. In *Catalyst Deactivation 1994 (Studies in Surface Science and Catalysis)*; Delmon, B., Froment, G.F., Eds.; Elsevier: Amsterdam, The Nethelands, 1994; Volume 88, pp. 327–334.

197. Petersen, E.E. Catalyst deactivation: Opportunity amidst woe. In *Catalyst Deactivation 1997 (Studies in Surface Science and Catalysis)*; Bartholomew, C.H., Fuentes, G.A., Eds.; Elsevier: Amsterdam, The Nethelands, 1997; Volume 111, pp. 87–98.

198. Gosselink, J.W.; Veen, J.A.R.V. Coping with catalyst deactivation in hydrocarbon processing. In *Catalyst Deactivation 1999 (Studies in Surface Science and Catalysis)*; Delmon, B., Froment, G.F., Eds.; Elsevier: Amsterdam, The Nethelands, 1999; Volume 126, pp. 3–16.

199. Lin, L.; Zao, T.; Zang, J.; Xu, Z. Dynamic process of carbon deposition on Pt and Pt–Sn catalysts for alkane dehydrogenation. *Appl. Catal.* **1990**, *67*, 11–23.

200. Resasco, D.E.; Haller, G.L. Catalytic dehydrogenation of lower alkanes. In *Catalysis (Specialist Periodiodical Report*; Spivey, J.J., Agarwal, S.K., Eds.; Royal Society of Chemistry, Cambridge, UK, 1994; Volume 11, pp. 379–411.

201. Cortright, R.D.; Dumesic, J.A. Microcalorimetric, Spectroscopic, and Kinetic Studies of Silica Supported Pt and Pt/Sn Catalysts for Isobutane Dehydrogenation. *J. Catal.* **1994**, *148*, 771–778.

202. Weyten, H.; Keizer, K.; Kinoo, A.; Luyten, J.; Leysen, R. Dehydrogenation of propane using a packed-bed catalytic membrane reactor. *AIChE J.* **1997**, *43*, 1819–1827.

203. Praserthdam, P.; Mongkhonsi, T.; Kunatippapong, S.; Jaikaew, B.; Lim, N. Determination of coke deposition on metal active sites of propane dehydrogenation catalysts. In *Catalyst Deactivation 1997 (Studies in Surface Science and Catalysis)*; Bartholomew, C.H., Fuentes, G.A., Eds.; Elsevier: Amsterdam, The Nethelands, 1997; Volume 111, pp. 153–158.

204. Rose, B.H.; Kiliany, T.R. Dual catalyst system. WIPO Pat. App. WO 2000/069993 (A1), 12 May 2000.

205. Guerrero-Ruiz, A.; Sepulveda-Escribano, A.; Rodriguez-Ramos, I. Cooperative action of cobalt and MgO for the catalysed reforming of CH_4 with CO_2. *Catal. Today* **1994**, *21*, 545–550.

206. Qin, D.; Lapszewicz, J. Study of mixed steam and CO_2 reforming of CH_4 to syngas on MgO-supported metals. *Catal. Today* **1994**, *21*, 551–560.

207. Stagg, S.; Resasco, D. Effects of promoters and supports on coke formation on Pt catalysts during CH_4 reforming with CO_2. In *Catalyst Deactivation 1997* (Studies in Surface Science and Catalysis); Bartholomew, C.H., Fuentes, G.A., Eds.; Elsevier: Amsterdam, The Nethelands, 1997; Volume 111, pp. 543–550.

208. Fujimoto, K.; Tomishige, K.; Yamazaki, O.; Chen, Y.; Li, X. Development of catalysts for natural gas reforming: Nickel-magnesia solid solution catalyst. *Res. Chem. Intermed.* **1998**, *24*, 259–271.

209. Marshall, C.L.; Miller, J.T. Process for converting methanol to olefins or gasoline. U.S. Patent 5,191,142, 2 March 1993.

210. Gayubo, A.G.; Aguayo, A.T.; Campo, A.E.S.D.; Benito, P.L.; Bilbao, J. The role of water on the attenuation of coke deactivation of a SAPO-34 catalyst in the transformation of methanol into olefins. In *Catalyst Deactivation 1999 (Studies in Surface Science and Catalysis)*; Delmon, B., Froment, G.F., Eds.; Elsevier: Amsterdam, The Nethelands, 1999; Volume 126, pp. 129–136.

211. Barger, P.T. SAPO catalysts and use thereof in methanol conversion processes. U.S. Patent 5,248,647, 28 Sepetember 1993.

212. Cox, J.P.; Evans, J.M. Exhaust gas catalyst for two-stroke engines. WIPO Patent WO 1999/042202 (A1), 20 February 1998.

213. Leviness, S.C.; Mart, C.J.; Behrmann, W.C.; Hsia, S.J.; Neskora, D.R. Slurry hydrocarbon synthesis process with increased catalyst life. WIPO Patent WO 1998/050487 (A1), 2 May 1997.

214. Bartholomew, C.H. Catalyst deactivation in hydrotreating of residua: A review. In *Catalytic Hydroprocessing of Petroleum and Distillates*; Oballa, M., Shih, S. Eds.; Marcel Dekker: New York, NY, USA, 1993; pp. 1–32.

215. Summers, J.; Williamson, W.B. Palladium-only catalysts for closed-loop control. In *Environmental Catalysis 1993*; Armor, J.N., Ed.; American Chemical Society: Washington, DC, USA, 1993; Volume 552, pp. 94–113.

216. Dettling, J.; Hu, Z.; Lui, Y.K.; Smaling, R.; Wan, C.Z.; Punke, A. Smart Pd TWC technology to meet stringent standards. In *Studies in Surface Science and Catalysis*; Elsevier: Amsterdam, The Nethelands, 1995; Volume 96, pp. 461–472.

217. Berndt, M.; Ksinsik, D. Catalyst and process for its preparation. U.S. Patent 4,910,180, 20 March 1990.

218. Prigent, M.; Blanchard, G.; Phillippe, P. Catalyst supports/catalysts for the treatment of vehicular exhaust gases. U.S. Patent 4,985,387, 15 January 1991.

219. Williamson, W.B.; Linden, D.G.; Summers, J.C., II. High-temperature three-way catalyst for treating automotive exhaust gases. U.S. Patent 5,041,407, 20 August 1991.

220. Williamson, W.B.; Linden, D.G.; Summers, J.C., II. High durability and exhuast catalyst with low hydrogen sulfide emissions. U.S. Patent 5,116,800, 26 May 1992.

221. Narula, C.K.; Watkins, W.L.H.; Chattha, M.S. Binary La-Pd oxide catalyst and method of making the catalyst. U.S. Patent 5,234,881, 10 August 1993.

222. Wan, C.-Z.; Tauster, S.J.; Rabinowitz, H.N. Catalyst composition containing platinum and rhodium components. U.S. Patent 5,254,519, 19 October 1993.

223. Argyle, M.D.; Frost, T.S.; Bartholomew, C.H. Modeling cobalt fischer tropsch catalyst deactivation with generalized power law expressions. *Top. Catal.* **2014**, *57*, 415–429.

224. Furuya, T.; Yamanaka, S.; Hayata, T.; Koezuka, J.; Yoshine, T.; Ohkoshi, A. Hybrid catalytic combustion for stationary gas turbine—Concept and small scale test results. In Proceedings of Gas Turbine Conference and Exhibition, Anaheim, CA, USA, 31 May–4 June, 1987; ASME Paper 87-GT-99.

225. Kawakami, T.; Furuya, T.; Sasaki, Y.; Yoshine, T.; Furuse, Y.; Hoshino, M. Feasibility study on honeycomb ceramics for catalytic combustor. In Proceedings of Gas Turbine and Aeroengine Congress and Exposition, Toronto, ON, Canada, 4–8 June 1989; ASME Paper 89-GT-41.

226. Betta, R.D.; Ribeiro, F.; Shoji, T.; Tsurumi, K.; Ezawa, N.; Nickolas, S. Catalyst structure having integral heat exchange. U.S. Patent 5,250,489, 5 October 1993.

227. Fujii, T.; Ozawa, Y.; Kikumoto, S. High pressure test results of a catalytic combustor for gas turbine. *J. Eng. Gas Turbines Power* **1998**, *120*, 509–513.

228. Borio, D.O.; Schbib, N.S. Simulation and optimization of a set of catalytic reactors used for dehydrogenation of butene into butadiene. *Comput. Chem. Eng.***1995**, *19*, S345–S350.

229. Fiato, R.A. Two stage process for hydrocarbon synthesis. U.S. Patent 5,028,634, 2 July 1991.

230. Zhao, R.; Goodwin, J.G., Jr.; Jothimurugesan, K.; Gangwal, S.K.; Spivey, J.J. Spray-dried Iron Fischer-Tropsch catalysts. 1. Effect of structure on the attrition resistance of the catalysts in the calcined state. *Ind. Eng. Chem. Res.* **2001**, *40*, 1065–1075.

231. Zhao, R.; Goodwin, J.G., Jr.; Jothimurugesan, K.; Gangwal, S.K.; Spivey, J.J. Spray-dried iron Fischer-Tropsch catalysts. 2. Effect of carburization on catalyst attrition resistance. *Ind. Eng. Chem. Res.* **2001**, *40*, 1320–1328.

232. Singleton, A.H.; Oukaci, R.; Goodwin, J.G. Processes and catalysts for conducting fischer-tropsch synthesis in a slurry bubble column reactor. U.S. Patent 5,939,350, 17 August 1999.

233. Plecha, S.; Mauldin, C.H.; Pedrick, L.E. Titania catalysts, their preparation and use in Fischer-Tropsch synthesis. U.S. Patent 6,087,405, 11 July 2000.

234. Plecha, S.; Mauldin, C.H.; Pedrick, L.E. Titania catalysts, their preparation and use in fischer-tropsch synthesis. U.S. Patent 6,124,367, 26 September 2000.

235. Singleton, A.H.; Oukaci, R.; Goodwin, J.G. Reducing fischer-tropsch catalyst attrition losses in high agitation reaction systems. WIPO Patent WO 2000/071253 (A3), 30 November 2000.

236. Seitz, M.; Klemm, E.; Emig, G. Poisoning and regeneration of NOx adsorbing catalysts for automotive applications. In *Catalyst Deactivation 1999 (Studies in Surface Science and Catalysis)*; Delmon, B., Froment, G.F. Eds.; Elsevier: Amsterdam, The Nethelands, 1999; Volume 126, pp. 211–218.

237. de Lima, S.M.; da Silva, A.M.; da Costa, L.O.O.; Assaf, J.M.; Mattos, L.V.; Sarkari, R.; Venugopal, A.; Noronha, F.B. Hydrogen production through oxidative steam reforming of ethanol over Ni-based catalysts derived from $La_{1-x}Ce_xNiO_3$ perovskite-type oxides. *Appl. Catal. B* **2012**, *121–122*, 1–9.

238. Song, H.; Ozkan, U.S. Ethanol steam reforming over Co-based catalysts: Role of oxygen mobility. *J. Catal.* **2009**, *261*, 66–74.

239. Masamune, S.; Smith, J.M. Performance of fouled catalyst pellets. *AIChE J.* **1966**, *12*, 384–394.

240. Murakami, Y.; Kobayashi, T.; Hattori, T.; Masuda, M. Effect of intraparticle diffusion on catalyst fouling. *Ind. Eng. Chem. Fundam.* **1968**, *7*, 599–605.

241. Stork, W.H.J. Molecules, catalysts and reactors in hydroprocessing of oil fractions. In *Hydrotreatment and Hydrocracking of Oil Fractions (Studies in Surface Science and Catalysis)*; Froment, G.F., Delmon, B., Grange, P., Eds.; Elsevier: New York, NY, USA, 1997; Volume 106, pp. 41–67.

242. Parks, G.D.; Schaffer, A.M.; Dreiling, M.J.; Shiblom, C.M. Surface Studies on the Interaction of Nickel and Antimony on Cracking Catalysts. *Prepr.—Am. Chem. Soc., Div. Petr. Chem.* **1980**, *25*, 335.

243. Trimm, D.L. Introduction to Catalyst Deactivation. In *Progress in Catalyst Deactivation (NATO Advanced Study Institute Series E, No. 54)*; Figueiredo, J.L., Ed.; Martinus Nijhoff Publishers: Boston, MA, USA,1982; pp. 3–22.

244. Becker, E.R.; Wei, J.J. Nonuniform distribution of catalysts on supports: I. Bimolecular Langmuir reactions. *J. Catal.* **1977**, *46*, 365–381.

245. Long, D.C. Staged hydrocarbon synthesis process. U.S. Patent 5,498,638, 12 March 1996.

246. Powell, B.R., Jr. Stabilitzation of high surface area aluminas. In Proceedings of the Materials Research Society Annual Meeting, Boston, MA, USA, 16–21 November 1980; Paper H9.

247. Dai, Y.; Lim, B.; Yang, Y.; Cobley, C.M.; Li, W.; Cho, E.C.; Grayson, B.; Fanson, P.T.; Campbell, C.T.; Sun, Y.; *et al.* A sinter-resistant catalytic system based on platinum nanoparticles supported on TiO$_2$ nanofibers and covered by porous silica. *Angew. Chem. Int. Ed.* **2010**, *49*, 8165–8168.

248. Shang, L.; Bian, T.; Zhang, B.; Zhang, D.; Wu, L.-Z.; Tung, C.-H.; Yin, Y.; Zhang, T. Graphene-supported ultrafine metal nanoparticles encapsulated by mesoporous silica: Robust catalysts for oxidation and reduction reactions. *Angew. Chem. Int. Ed.* **2014**, *53*, 250–254.

249. Heck, R.; Farrauto, R. *Catalytic Air Pollution Control: Commercial Technology*; Van Nostrand Reinhold: New York, NY, USA, 1995.

250. Trimm, D.L. The regeneration or disposal of deactivated heterogeneous catalysts. *Appl. Catal. A* **2001**, *212*, 153–160.

251. Berrebi, G.; Dufresne, P.; Jacquier, Y. Recycling of spent hydroprocessing catalysts: EURECAT technology. *Environ. Prog.* **1993**, *12*, 97–100.

252. D'Aquino, R.L. For catalyst regeneration, incentives anew. *Chem. Eng.* **2000**, *107*, 32–33, 35.

253. Chang, T. Regeneration industry helps refiners control costs, limit liabilities. *Oil Gas J.* **1998**, *96*, 49.

254. Blashka, S.R.; Duhon, W. Catalyst regeneration. *Int. J. Hydrocarbon Eng.* **1998**, *4*, 60–64.

255. McCulloch, D.C. Catalytic hydrotreating in petroleum refining. In *Applied Industrial Catalysis*; Leach, B.E., Ed.; Academic Press: New York, NY, USA, 1983; pp. 69–121.

256. Franck, J.P.; Martino, G. Deactivation and Regeneration of Catalytic-Reforming Catalysts. In *Progress in Catalyst Deactivation* (NATO Advanced Study Institute Series E., No. 54); Figueiredo, J.L., Ed.; Martinus Nijhoff Publishers: Boston, MA, USA, 1982; pp. 355–397.

257. Spivey, J.J.; Roberts, G.W.; Davis, B.H. *Catalyst Deactivation 2001 (Studies in Surface Science and Catalysis*; Elsevier: Amsterdam, The Nethelands, 2001; Volume 139.

258. Haddad, J.H.; Harandi, M.N.; Owen, H. Upgrading light olefin fuel gas in a fluidized bed catalyst reactor and regeneration of the catalyst. U.S. Patent 5,043,517, 27 August 1991.

259. Ueda, M.; Murakami, T.; Shibata, S.; Hirabayashi, K.; Kondoh, T.; Adachi, K.; Hoshino, N.; Inoue, S. Process for the regeneration of coke-deposited, crystalline silicate catalyst. U.S. Patent 5,306,682, 26 April 1994.

260. Zhang, S.Y.-F.; Gosling, C.D.; Sechrist, P.A.; Funk, G.A. Dual regeneration zone solid catalyst alkylation process. U.S. Patent 5,675,048, 7 October 1997.

261. Panattoni, G.; Querini, C.A. Isobutane alkylation with C$_4$ olefins: regeneration of metal-containing catalysts. In *Catalyst Deactivation 2001 (Studies in Surface Science and Catalysis)*; Spivey, J.J., Roberts, G.W., Davis, B.H., Eds.; Elsevier: Amsterdam, The Nethelands, 2001; Volume 139, pp. 181.

262. Thompson, D.N.; Ginosar, D.M.; Burch, K.C. Regeneration of a deactivated USY alkylation catalyst using supercritical isobutene. *Appl. Catal. A* **2005**, *279*, 109–116.

263. Petkovic, L.M.; Ginosar, D.M.; Burch, K.C. Supercritical fluid removal of hydrocarbons adsorbed on wide-pore zeolite catalysts. *J. Catal.* **2005**, *234*, 328–339.

264. Dufresne, P.; Brahma, N. Offsite regeneration process for a catalyst containing at least one precious metal. U.S. Patent 5,854,162, 29 December 1998.

265. Innes, R.A.; Holtermann, D.L.; Mulaskey, B.F. Low temperature regeneration of coke deactivated reforming catalysts. U.S. Patent 5,883,031, 16 March 1999.

266. Fung, S.C. Regenerating a reforming catalyst—this bimetallic noble-metal catalyst is used in a refining process to convert saturated-hydrocarbons to aromatics - simply burning off the carbon was not enough—a procedure had to be found for redispersing the metals on the alumina support as well. *Chemtech* **1994**, *24*, 40–44.

267. Alfonso, J.C.; Aranda, D.A.G.; Schmal, M.; Frety, R. Regeneration of a Pt-Sn/Al$_2$O$_3$ catalyst: Influence of heating rate, temperature and time of regeneration. *Fuel Proc. Technol.* **1997**, *50*, 35–48.

268. Arteaga, G.J.; Anderson, J.A.; Rochester, C.H. Effects of Catalyst Regeneration with and without Chlorine on Heptane Reforming Reactions over Pt/Al$_2$O$_3$ and Pt–Sn/Al$_2$O$_3$. *J. Catal.* **1999**, *187*, 219–229.

269. Pieck, C.L.; Vera, C.R.; Parera, J.M. Study of industrial and laboratory regeneration of Pt-Re/Al$_2$O$_3$ catalysts. In *Catalyst Deactivation 2001 (Studies in Surface Science and Catalysis)*; Spivey, J.J., Roberts, G.W., Davis, B.H., Eds.; Elsevier: Amsterdam, The Nethelands, 2001; Volume 139, pp. 279–286.

270. Acharya, D.R.; Hughes, R.; Kennard, M.A.; Liu, Y.P. Regeneration of fixed beds of coked chromia—alumina catalyst. *Chem. Eng. Sci.* **1992**, *47*, 1687–1693.

271. Sikkenga, D.L.; Zaenger, I.C.; Williams, G.S. Palladium catalyst reactivation. U.S. Patent 4,999,326, 12 March 1991.

272. Ekstrom, A.; Lapszewicz, J. On the role of metal carbides in the mechanism of the Fischer-Tropsch reaction. *J. Phys. Chem.* **1984**, *88*, 4577–4580.

273. Ekstrom, A.; Lapszewicz, J. The reactions of cobalt surface carbides with water and their implications for the mechanism of the Fischer-Tropsch reaction. *J. Phys. Chem.* **1987**, *91*, 4514–4619.

274. Pedrick, L.E.; Mauldin, C.H.; Behrmann, W.C. Draft tube for catalyst rejuvenation and distribution. U.S. Patent 5,268,344, 7 December 1993.

275. Owen, H.; Schipper, P.H. Process and apparatus for regeneration of FCC catalyst with reduced NOx and or dust emissions. U.S. Patent 5,338,439,16 August 1994.

276. Raterman, M.F. Two-stage fluid bed regeneration of catalyst with shared dilute phase. U.S. Patent 5,198,397, 30 March 1993.

277. Apelian, M.R.; Fung, A.S.; Hatzikos, G.H.; Kennedy, C.R.; Lee, C.-H.; Kiliany, T.R.; Ng, P.K.; Pappal, D.A. Regeneration of noble metal containing zeolite catalysts via partial removal of carbonaceous deposits. U.S. Patent 5,393,717, 28 February 1995.

278. Clark, D.E. Hydrocracking process using a reactivated catalyst. U.S. Patent 5,340,957, 23 August 1994.

279. Yoshimura, Y.; Sato, T.; Shimada, H.; Matsubayashi, N.; Imamura, M.; Nishijima, A.; Yoshitomi, S.; Kameoka, T.; Yanase, H. Oxidative regeneration of spent molybdate and tungstate hydrotreating catalysts. *Energy Fuels* **1994**, *8*, 435–445.

280. Oh, E.-S.; Park, Y.-C.; Lee, I.-C.; Rhee, H.-K. Physicochemical changes in hydrodesulfurization catalysts during oxidative regeneration. *J. Catal.* **1997**, *172*, 314–321.

281. Aquavo, A.T.; Gayubo, A.G.; Atutxa, A.; Olazar, M.; Bilbao, J. Regeneration of a catalyst based on a SAPO-34 used in the transformation of methanol into olefins. *J. Chem. Tech. Biotech.* **1999**, *74*, 1082–1088.

282. Forbus, N.P.; Wu, M.M.-S. Regeneration of methanol/methyl ether conversion catalysts. U.S. Patent 4,777,156, 11 October 1988.

283. Krishna, A.; Hsieh, C.; English, A.E.; Pecoraro, T.; Kuehler, C. Additives improve FCC process; increase catalyst life, lower regeneration temperatures and improve yield and quality of products cost effectively. *Hydrocarbon Processing* 1991, 59–66.

284. Altomare, C.; Koermer, G.; Schubert, P.; Suib, S.; Willis, W. A designed fluid cracking catalyst with vanadium tolerance. *Chem. Mater.* **1989**, *1*, 459–463.

285. Aguinaga, A.; Montes, M. Regeneration of a nickel/silica catalyst poisoned by thiophene. *Appl. Catal. A* **1992**, *90*, 131–144.

286. Lowry, G.V.; Reinhard, M. Pd-Catalyzed TCE Dechlorination in Groundwater: Solute Effects, Biological Control, and Oxidative Catalyst Regeneration. *Environ. Sci. Technol.* **2000**, *34*, 3217–3223.

287. Trinh, D.C.; Desvard, A.; Martino, G. Process for regenerating used catalysts by means of hydrogen peroxide aqueous solution stabilized with an organic compound. U.S. Patent 4,830,997, 16 May 1989.

288. Sherwood, D.E. Process for the reactivation of spent alumina-supported hydrotreating catalysts. U.S. Patent 5,230,791, 27 July 1993.

289. Fung, S.C. Deactivation and regeneration/redispersion chemistry of Pt/KL-zeolite. In *Catalyst Deactivation 2001 (Studies in Surface Science and Catalysis)*; Spivey, J.J., Roberts, G.W., Davis, B.H., Eds.; Elsevier: Amsterdam, The Nethelands, 2001; Volume 139, pp. 399–406.

290. Didillon, B. Catalyst regeneration process and use of the catalyst in hydrocarbon conversion processes. U.S. Patent 5,672,801, 30 September 1997.

291. Tsao, Y.-Y.P.; von Ballmoos, R. Reactivating catalysts containing noble metals on molecular sieves containing oxides of aluminum and phosphorus. U.S. Patent 4,929,576, 29 May 1990.

292. Dufresne, P.; Brahma, N.; Girardier, F. Off-site regeneration of hydroprocessing catalysts. *Rev. Inst. Fr. Pet.* **1995**, *50*, 283–293.

293. Clark, F.T.; Hensley, A.L., Jr. Process for regenerating a spent resid hydroprocessing catalyst using a group IIA metal. U.S. Patent 5,275,990, 4 January 1994.

294. Brito, A.; Arvelo, R.; Gonzalez, A.R. Variation in structural characteristics of a hydrotreatment catalyst with deactivation/regeneration cycles. *Ind. Eng. Chem. Res.* **1998**, *37*, 374–379.

295. Teixeira-da-Silva, V.L.S.; Lima, F.P.; Dieguez, L.C. Regeneration of a deactivated hydrotreating catalyst. *Ind. Eng. Chem. Res.* **1998**, *37*, 882–886.

296. Snape, C.E.; Diaz, M.C.; Tyagi, Y.R.; Martin, S.C.; Hughes, R. Characterisation of coke on deactivated hydrodesulfurisation catalysts and a novel approach to catalyst regeneration. In *Catalyst Deactivation 2001 (Studies in Surface Science and Catalysis)*; Spivey, J.J., Roberts, G.W., Davis, B.H., Eds.; Elsevier: Amsterdam, The Nethelands, 2001; Volume 139, pp. 359–365.

297. Sherwood, D.E., Jr.; Hardee, J.R., Jr. Method for the reactivation of spent alumina-supported hydrotreating catalysts. U.S. Patent 5,445,728, 29 August 1995.

298. Maholland, M.K.; Fu, C.-M.; Lowery, R.E.; Kubicek, D.H.; Bertus, B.J. Demetallization and passivation of spent cracking catalysts. U.S. Patent 5,021,377, 4 June 1991.

299. Kubicek, D.H.; Fu, C.-M.; Lowery, R.E.; Maholland, K.M. Reactivation of spent cracking catalysts. U.S. Patent 5,141,904, 25 August 1992.

300. Fu, C.-M.; Maholland, M.; Lowery, R.E. Reactivation of spent, metal-containing cracking catalysts. U.S. Patent 5,151,391, 29 September 1992.

301. Hu, Y.; Luo, B.; Sun, K.; Yang, Q.; Gong, M.; Hu, J.; Fang, G.; Li, Y. Dry regeneration-demetalization technique for catalyst for residuum and/or heavy oil catalytic cracking. U.S. Patent 6,063,721, 16 May 2000.

302. Walker, P.L., Jr.; Rusinko, F., Jr.; Austin, L.G. Gas Reactions of Carbon. *Adv. Catal.* **1959**, *11*, 133–221.

303. Fulton, J.W. Regenerating Spent Catalyst. *Chem. Eng.* **1988**, *96*, 111–114.

304. Richardson, J.T. Experimental determination of catalyst fouling parameters. Carbon profiles. *Ind. Eng. Chem. Process Des. Dev.* **1972**, *11*, 8–11.

305. Weisz, P.B.; Goodwin, R.B. Combustion of carbonaceous deposits within porous catalyst particles: II. Intrinsic burning rate. *J. Catal.* **1966**, *6*, 227–236.

306. Rostrup-Nielsen, J.R. Some principles relating to the regeneration of sulfur-poisoned nickel catalyst. *J. Catal.* **1971**, *21*, 171–178.

307. Hartenstein, H. *Feasibility of SCR Technology for NOx Control Technology for the Milton R. Young Station, Center, North Dakota*; Expert Report on behalf of the United States Department of Justice: Washington, DC, USA, July 2008.

308. Hartenstein, H.; Ehrnschwender, M.; Sibley, A.F. SCR Regeneration-10 Years of R&D and Commercial Application. In Proceedings of Power Plant Air Pollution Control Mega Symposium, Baltimore, MD, USA, 24–28 August 2008; Paper No. 104.

309. EPA; Office of Air and Radiation. Update on Cap and Trade Programs for SO_2 and NO_x. Presented at the Enviornmental Markets Association, 12 Annual Spring Conference, Miami, FL, USA, 29April–2 May 2008.

310. EPA. NO_x Budget Trading Program—Basic Information. Available online: http://www.epa. gov/airmarkt/progsregs/nox/docs/NBPbasicinfo.pdf (accessed on 31 October 2014).

311. McIlvaine Company. News Release: Coal to Play Bigger Role in Future Power Generation With World Capacity of 1450 GW by 2012. Available online: http://www.mcilvainecompany. com/brochures/Alerts_for_Internet/s9%20moni/1104coal%20to%20play%20bigger%20role %20in%20future%20power%20generation.htm (accessed on 31 October 2014).

312. Hjalmarsson, A.-K. *NOx Control Technologies for Coal Combustion*; IEACR/24; IEA Coal Research: Lexington, KY, USA, 1990.

313. U.S. Department of Energy; Southern Company Services. Control of Nitrogen Oxide Emissions: Selective Catalytic Reduction. *Clean Coal Technology*; Topical Report 9; U.S. Department of Energy: Pittsburgh PA, USA; Southern Company Services: Birmingham AL, USA; 1997.

314. Janssen, F.; Meijer, R. Quality control of DeNOx catalysts: Performance testing, surface analysis and characterization of DeNOx catalysts. *Catal. Today* **1993**, *16*, 157–185.

315. Ashton, J.; Nackos, A.; Bartholomew, C.H.; Hecker, W.C.; Baxter, L. Poisoning/Deactivation of Vanadia/Titanium Dioxide SCR Catalysts in Coal Systems. Presented at the 19th Annual Western States Catalysis Club Symposium, Albuquerque, NM, USA, 25 February 2005.

316. Guo, X.; Nackos, A.; Ashton, J.; Bartholomew, C.H.; Hecker, W.C.; Baxter, L. Poisoning Study of V_2O_5-WO_3/TiO_2 Catalysts by Na, K, and Ca. Presented at 19th Annual Western States Catalysis Club Symposium, Albuquerque, NM, USA, 25 February 2005.

317. Forzatti, P.; Lietti, L.; Tronconi, E. Nitrogen Oxides Removal–Industrial. In *Encyclopedia of Catalysis*; Horváth, T.I., Ed.; Wiley, Hoboken, NJ, USA, 2003.

318. Rubin, E.S., Kalagnanam, J.R., Frey, C.H., Berkenpas, M.B. Integrated Environmental Control Modeling of Coal-Fired Power Systems. *Journal of the Air and Waste Management Association* **1997**, 47, 1180-1188.

319. Cochran, J.R.; Ferguson, A.W. Selective Catalytic Reduction for NOx Emission Control. In Proceedings of the First International Conference on Air Pollution Zannetti, P., Brebbia, C.A., Garcia Gardea, J.E., Ayala Milian, G., Eds., Monterrey, Mexico, 16-18 Feb. 1993, Volume 1, pp. 703–718.

320. Bullock, P.E.; Hartenstein, H.U. O&M Cost Reduction of a Coal-Fired US Merchant Plant Through an Optimized SCR Management Strategy Involving Catalyst Regeneration. In Proceedings of the Conference on Selective Catalytic Reduction and Selective Non-Catalytic Reduction for NO_x Control, Pittsburgh, PA, USA, 15–17 May 2002.

321. Babcock Hitachi's Rejuvenation of Mehrum SCR Catalyst. *FGD and DeNOx Newsletter*; No. 315; The McIlvaine Company: Northfield, IL, USA; 2004.

322. McMahon, B. Catalyst regeneration: The business case. *Power* **2006**, *150*, 36-39.

323. Zheng, Y.; Jensen, A.; Johnsson, J. Laboratory investigation of selective catalytic reduction catalysts: Deactivation by potassium compounds and catalyst regeneration. *Ind. Eng. Chem. Res.* **2004**, *43*, 941–947.

324. Khodayari, R.; Odenbrand, C. Regeneration of commercial SCR catalysts by washing and sulphation: effect of sulphate groups on the activity. *Appl. Catal. B* **2001**, *33*, 277–291.

325. Khodayari, R.; Odenbrand, C. Regeneration of commercial TiO_2-V_2O_5-WO_3 SCR catalysts used in bio fuel plants. *Appl. Catal. B* **2001**, *30*, 87–89.

326. Dörr, H.-K.; Koch, G.; Bastuck, W. Method for renewed activation of honeycomb-shaped catalyst elements for denitrating flue gases. U.S. Patent, 6,387,836 (B1), 14 May 2002.

327. Schneider, P.; Bastuck, W. Method to re-enable honeycomb catalyst elements constructed for the denitrification of flue gases. DE Patent 10222915 (B4), 15 January 2004.

328. Budin, R.; Krotla, K.; Rabitsch, H. Process for regenerating used deNOx or dedioxin catalytic converters. U.S. Patent, 6,484,733 (B2), 26 November 2002.

329. Cooper, M.; Harrison, K.; Lin, C. An Experimental Program to Optimize SCR Catalyst Regeneration for Lower Oxidation of SO_2 to SO_3. Proceedings of the 2006 Environmental Controls Conference, Pittsburgh, PA, USA, 16–18 May 2006.

330. Patel, N.; Hartenstein, H.-U.; Wenz, F. Process for decoating a washcoat catalyst substrate. U.S. Patent 7,559,993 (B1), 14 July 2009.

331. Patel, N.; Hartenstein, H.-U.; Wenz, F. Process for decoating a washcoat catalyst substrate. U.S. Patent 6,929,701, 16 August 2005.

332. Tate, A.; Skipper, J.; Wenz, F. Environmentally sound handling of deactivated SCR catalyst. *Coal Power*, 31 July 2008.

333. Hartenstein, H.U.; Hoffman, T. Method of regeneration of SCR catalyst. U.S. Patent 7,723,251, 25 May 2010.

334. Servatius, P.; Schluttig, A. Lifetime extension of SCR-DeNOx catalysts using SCR-Tech's high efficiency ultrasonic regeneration process. In Proceedings of 2001 Conference of Selective Catalytic Reduction and Selective Non-Catalytic Reduction for NO_x Control, Pittsburgh, PA, USA, 16–18 May 2001.

335. Hartenstein, H.; Gutberlet, H. Catalyst Regeneration—An Integral Part of Proper Catalyst Management. Proceedings of the 2001 Workshop on Selective Catalytic Reduction, Baltimore, MD, USA, 13–15 November 2001.

336. Wenz, F.; Deneault, R.; Franklin, H.N. The Goals, Challenges and Successes of Regenerating Selective Catalytic Reduction Catalyst. In Proceedings of the Electric Power 2006 Conference–NO$_x$ Control III–SCR Technologies, Atlanta, GA, USA, 2–4 May 2006.

337. CoaLogix. Available online: http://www.coalogix.com (accessed on 31 October 2014).

338. McCullen, S.B.; Wong, S.S.; Huang, T.J. Regeneration of noble metal-highly siliceous zeolite with sequential hydrogen halide and halogen or organic-halogen compound treatment. U.S. Patent 4,645,751, 24 February 1987.

339. Borghard, W.S.; Huang, T.J.; McCullen, S.B.; Schoennagel, H.J.; Tsao, Y.-.P.; Wong, S.S. Redispersion of agglomerated noble metals on zeolite catalysts. U.S. Patent 4,657,874, 14 April 1987.

340. Fung, S.C.; Rice, R.W. Process using halogen/oxygen for reactivating iridium and selenium containing catalysts. U.S. Patent 4,491,636, 1 January 1985.

341. Fung, S.C. Redispersion of Ir catalysts by low temperature reduction step. U.S. Patent 4,467,045, 21 August 1984.

342. Landolt, G.R.; McHale, W.D.; Schoennagel, H.J. Catalyst regeneration procedure. U.S. Patent 4,359,400, 16 November 1982.

343. Fung, S.C.; Weissman, W.; Carter, J.L. Reactivation process for iridium-containing catalysts using low halogen flow rates. U.S. Patent 4,480,046, 30 October 1984.

344. Fung, S.C. Low temperature decoking process for reactivating iridium and selenium containing catalysts. U.S. Patent 4,492,767, 8 January 1985.

345. Fung, S.C.; Weissman, W.; Carter, J.L.; Kmak, W.S. Reactivating iridium and selenium containing catalysts with hydrogen halide and oxygen. U.S. Patent 4,491,635, 1 January 1985.

346. Mohr, D.H. Process for regenerating a monofunctional large-pore zeolite catalyst having high selectivity for paraffin dehydrocyclization. U.S. Patent 4,855,269, 8 August 1989.

347. Fung, S.C.; Tauster, S.J.; Koo, J.Y. Method of regenerating a deactivated catalyst. U.S. Patent 4,925,819, 15 May 1990.

348. Fung, S.C.; Kmak, W.S. Reactivation of iridium-containing catalysts by halide pretreat and oxygen redispersion. U.S. Patent 4,444,896, 24 April 1984.

349. Fung, S.C.; Weissman, W.; Carter, J.L. Reactivation process for iridium-containing catalysts using low halogen flow rates. U.S. Patent 4,444,895, 24 April 1984.

350. Boyle, J.P.; Gilbert, J.B. Reactivation of iridium-containing catalysts. U.S. Patent 4,517,076, 14 May 1985.

351. Buss, W.D.; Hughes, T.R. *In situ* hydrocarbon conversion catalyst regeneration and sulfur decontamination of vessels communicating with catalyst reactor. U.S. Patent 4,482,637, 13 November 1984.

352. Cheng, W.-H. Regeneration of methanol dissociation catalysts. U.S. Patent 4,855,267, 8 August 1989.

353. Huang, Y.-Y.; LaPierre, R.B.; McHale, W.D. Process for dispersing or redispersing a group VIII noble metal species on a porous inorganic support. European Patent 0,306,170 (B1), 8 March 1989.

354. Hucul, D.A. Redispersal of Ru, Os, Rh and Pd catalysts and processes therewith. U.S. Patent 4,891,346, 2 January, 1990.

355. Fung, S.C.; Rice, R.W. Process for reactivating iridium-containing catalysts. U.S. Patent 4,447,551, 8 May 1984.

356. Fung, S.C. Process for reactivating iridium-containing catalysts in series. U.S. Patent 4,472,514, 18 September 1984.

357. S.C. Fung; R.W. Rice Process for reactivating iridium-containing catalysts. U.S. Patent 4,473,656, 25 September 1984.

358. Huang, Y.-Y.; LaPierre, R.B.; McHale, W.D. Process for dispersing or redispersing a Group VIII noble metal species on a porous inorganic support. U.S. Patent 4,952,543, 28 August 1990.

359. Fung, S.C.; Weissman, W.; Carter, J.L.; Kmak, W.S. Reactivating iridium-containing catalysts with hydrogen halide and oxygen. U.S. Patent 4,444,897, 24 April 1984.

360. Fung, S.C. Low temperature decoking process for reactivating iridium containing catalysts. U.S. Patent 4,472,515 18 September 1984.

361. Dadyburjor, D.B. Regions of validity for models of regeneration of sintered supported metal catalysts. In *Catalyst Deactivation (Studies in Surface Science and Catalysis)*; Delmon, B., Froment, G.F., Eds.; Elsevier: Amsterdam, The Nethelands, 1980; Volume 6, pp. 341–351.

362. Jacobs, G.; Ma, W.; Gao, P.; Todic, B.; Bhatelia, T.; Bukur, D.B.; Davis, B.H. The application of synchrotron methods in characterizing iron and cobalt Fischer–Tropsch synthesis catalysts. *Catal. Today* **2013**, *214*, 100–139.

363. Van Steen, E.; Claeys, M.; Dry, M.E.; van de Loosdrecht, J.; Viljoen, E.L.; Visagie, J.L. Stability of nanocrystals: Thermodynamic analysis of oxidation and re-reduction of cobalt in water/hydrogen mixtures. *J. Phys. Chem. B* **2005**, *109*, 3575–3577.

364. Ma, W.; Jacobs, G.; Ji, Y.; Bhatelia, T.; Bukur, D.B.; Khalid, S.; Davis, B.H. Fischer–Tropsch synthesis: Influence of CO conversion on selectivities, H_2/CO usage ratios, and catalyst stability for a Ru promoted Co/Al_2O_3 catalyst using a slurry phase reactor. *Top. Catal.* **2011**, *54*, 757–767.

365. Jacobs, G.; Sarkar, A.; Ji, Y.; Patterson, P.M.; Das, T.K.; Luo, M.; Davis, B.H. Fischer-Tropsch synthesis: Characterization of interactions between reduction promoters and Co for Co/Al_2O_3–based GTL catalysts. In Proceedings of *AIChE Annual Meeting*, San Francisco, CA, USA, 12–17 November 2006.

366. Das, T.K.; Jacobs, G.; Patterson, P.M.; Conner, W.A.; Davis, B.H. Fischer–Tropsch synthesis: characterization and catalytic properties of rhenium promoted cobalt alumina catalysts. *Fuel* **2003**, *82*, 805–815.

367. Jacobs, G.; Das, T.K.; Patterson, P.M.; Luo, M.; Conner, W.A.; Davis, B.H. Fischer–Tropsch synthesis: Effect of water on Co/Al_2O_3 catalysts and XAFS characterization of reoxidation phenomena. *Appl. Catal. A* **2004**, *270*, 65–76.

368. Moodley, D.J.; Saib, A.M.; van de Loosdrecht, J.; Welker-Nieuwoudt, C.A.; Sigwebela, B.H.; Niemanstsverdriet, J.W. The impact of cobalt aluminate formation on the deactivation of cobalt-based Fischer–Tropsch synthesis catalysts. *Catal. Today* **2011**, *171*, 192–200.

369. Jermwongratanachai, T.; Jacobs, G.; Shafer, W.D.; Ma, W.; Pendyala, V.R.R.; Davis, B.H.; Kitiyanan, B.; Khalid, S.; Cronauer, D.C.; Kropf, A.J.; Marshall, C.L. Fischer–Tropsch synthesis: Oxidation of a fraction of cobalt crystallites in research catalysts at the onset of FT at partial pressures mimicking 50% CO conversion. *Top. Catal.* **2014**, *57*, 479–490.

370. Azzam, K.; Jacobs, G.; Ma, W.; Davis, B.H. Effect of cobalt particle size on the catalyst intrinsic activity for Fischer–Tropsch synthesis. *Catal. Lett.* **2014**, *144*, 389–394.

371. Saib, A.M.; Borgna, A.; van de Loosdrecht, J.; van Berge, P.J.; Niemanstsverdriet, J.W. XANES study of the susceptibility of nano-sized cobalt crystallites to oxidation during realistic Fischer–Tropsch synthesis. *Appl. Catal. A* **2006**, *312*, 12–19.

372. Dalai, A.K.; Davis, B.H. Fischer–Tropsch synthesis: A review of water effects on the performances of unsupported and supported Co catalysts. *Appl. Catal. A* **2008**, *348*, 1–15.

373. Sadeqzadeh, M.; Chambrey, S.; Piché, S.; Fongarland, P.; Luck, F.; Curulla-Ferré, D.; Schweich, D.; Bousquet, J.; Khodakov, A.Y. Deactivation of a Co/Al$_2$O$_3$ Fischer–Tropsch catalyst by water-induced sintering in slurry reactor: Modeling and experimental investigations. *Catal. Today* **2013**, *215*, 52–59.

374. Sadeqzadeh, M.; Hong, J.; Fongarland, P.; Curulla-Ferre, D.; Luck, F.; Bousquet, J.; Schweich, D.; Khodakov, A.Y. Mechanistic modeling of cobalt based catalyst sintering in a fixed bed reactor under different conditions of Fischer–Tropsch synthesis. *Ind. Eng. Chem. Res.* **2012**, *51*, 11955–11964.

375. Soled, S.L.; Kiss, G.; Kliewer, C.; Baumgartner, J.; El-Malki, E.-M. Learnings from Co Fischer-Tropsch catalyst studies. Present at 245th ACS National Meeting & Exposition, New Orleans, LA, USA, 7–11 April 2013.

376. Hauman, M.M.; Saib, A.M.; Moodley, D.J.; Plessis, E.D.; Claeys, M.; van Steen, E. Re-dispersion of cobalt on a model fischer–tropsch catalyst during reduction–oxidation–reduction cycles. *Chem. Catal. Chem.* **2012**, *4*, 1411–1419.

377. Weststrate, C.J.; Kizilkaya, A.C.; Rossen, E.T.R.; Verhoeven, M.W.G.M.; Ciobica, I.M.; Saib, A.M.; Niemanstsverdriet, J.W. Atomic and Polymeric Carbon on Co(0001): Surface Reconstruction, Graphene Formation, and Catalyst Poisoning. *J. Phys. Chem. C* **2012**, *116*, 11575–11583.

378. Weststrate, C.J.; Hauman, M.M.; Moodley, D.J.; Saib, A.M.; van Steen, E.; Niemanstsverdriet, J.W. Cobalt Fischer–Tropsch catalyst regeneration: The crucial role of the kirkendall effect for cobalt redispersion. *Top. Catal.* **2011**, *54*, 811–816.

379. Moodley, D.J.; van de Loosdrecht, J.; Saib, A.M.; Overett, M.J.; Datye, A.K.; Niemanstsverdriet, J.W. Carbon deposition as a deactivation mechanism of cobalt-based Fischer–Tropsch synthesis catalysts under realistic conditions. *Appl. Catal. A* **2009**, *354*, 102–110.

380. van de Loosdrecht, J.; Balzhinimaev, B.; Dalmon, J.-A.; Niemanstsverdriet, J.W.; Tsybulya, S.V.; Saib, A.M.; van Berge, P.J.; Visagie, J.L. Cobalt Fischer-Tropsch synthesis: Deactivation by oxidation?. *Catal. Today* **2007**, *123*, 293–302.

381. Saib, A.M.; Borgna, A.; van de Loosdrecht, J.; van Berge, P.J.; Niemanstsverdriet, J.W. In Situ surface oxidation study of a planar Co/SiO$_2$/Si(100) model catalyst with nanosized cobalt crystallites under model Fischer−Tropsch synthesis conditions. *J. Phys. Chem. B* **2006**, *110*, 8657–8664.

382. Moodley, D.J.; van de Loosdrecht, J.; Saib, A.M.; Niemanstsverdriet, J.W. The formation and influence of carbon on cobalt-based Fischer-Tropsch synthesis catalysts: An integrated review. In *Advances in Fischer-Tropsch Synthesis, Catalysts, and Catalysis*; Davis, B.H., Occelli, M.L., Eds.; CRC Press Taylor & Francis Group: Boca Raton, FL, USA, 2010; pp. 49–81.

383. Bezemer, G.L.; Herman, J.H.; Kuipers, P.C.E.; Oosterbeek, H.; Holewign, J.E.; Xu, X.; Kapteijn, F.; van Dillen, A.J.; de Jong, K.P. Cobalt particle size effects in the Fischer−Tropsch reaction studied with carbon nanofiber supported catalysts. *J. Am. Chem. Soc.* **2006**, *128*, 3956–3964.

384. Jacobs, G.; Ribeiro, M.C.; Ma, W.; Ji, Y.; Khalid, S.; Sumodjo, P.T.A.; Davis, B.H. Group 11 (Cu, Ag, Au) promotion of 15%Co/Al$_2$O$_3$ Fischer–Tropsch synthesis catalysts. *Appl. Catal. A* **2009**, *361*, 137–151.

Deactivation and Regeneration of Commercial Type Fischer-Tropsch Co-Catalysts—A Mini-Review

Erling Rytter and Anders Holmen

Abstract: Deactivation of commercially relevant cobalt catalysts for Low Temperature Fischer-Tropsch (LTFT) synthesis is discussed with a focus on the two main long-term deactivation mechanisms proposed: Carbon deposits covering the catalytic surface and re-oxidation of the cobalt metal. There is a great variety in commercial, demonstration or pilot LTFT operations in terms of reactor systems employed, catalyst formulations and process conditions. Lack of sufficient data makes it difficult to correlate the deactivation mechanism with the actual process and catalyst design. It is well known that long term catalyst deactivation is sensitive to the conditions the actual catalyst experiences in the reactor. Therefore, great care should be taken during start-up, shutdown and upsets to monitor and control process variables such as reactant concentrations, pressure and temperature which greatly affect deactivation mechanism and rate. Nevertheless, evidence so far shows that carbon deposition is the main long-term deactivation mechanism for most LTFT operations. It is intriguing that some reports indicate a low deactivation rate for multi-channel micro-reactors. *In situ* rejuvenation and regeneration of Co catalysts are economically necessary for extending their life to several years. The review covers information from open sources, but with a particular focus on patent literature.

Reprinted from *Catalysts*. Cite as: Rytter, E.; Holmen, A. Applying the Behavioural Family Therapy Model in Complex Family Situations. *Catalysts* **2015**, *5*, 478-499.

1. Introduction

In a gas-to-liquid (GTL) plant the high H_2/CO ratio obtained from reforming of natural gas to synthesis gas (syngas) obviates the need for shifting CO with steam to yield more hydrogen (and CO_2) for the FT unit. This is one main reason for using a cobalt catalyst instead of the much cheaper iron alternative as catalytic metal for the FT reaction. In addition, the cobalt catalyst is more active and has a simpler product slate of mainly paraffins and some α–olefins. However, both cobalt metal in itself, precious metal promoters as well as advanced overall formulations, make the catalyst inherently costly. Further, Co catalysts typically lose about half their activity within a few months. Assuming an economically acceptable catalyst lifetime of 2–3 years, this means that the catalyst cost will add several USD to the price per bbl of produced synthetic crude. Therefore, improving catalyst stability is a major focus among technology providers and plant operators. It follows logically that a basic understanding of the mechanisms involved in the deactivation process is vital to improving catalyst stability. Fortunately, it appears that, at least for most commercial Co catalysts, rejuvenation of catalyst activity is possible.

A comprehensive review of deactivation of Co FT catalysts appeared a few years ago [1]. This review discusses a wide variety of deactivation mechanisms comprising sintering; re-oxidation of cobalt, including surface oxidation; formation of stable compounds between cobalt and the support, e.g., cobalt aluminate; surface reconstruction; formation of carbon species on the cobalt surface; carbiding; and poisoning. However, less focus is given in the review to long-term deactivation under commercial conditions.

Historically, details of deactivation mechanisms and rates have been scarce particularly as only a few plants are operated commercially. Nevertheless, some data can be found in the patent literature, mainly based on operation of pilot or demonstration plants, and in conference presentations. A complicating factor is that an industrial process is typically operated at constant global production, *i.e.*, deactivation is counteracted by a steady increase in operating temperature. Only a few reports on deactivation under these most relevant conditions could be found. Fortunately, Sasol and its collaborator, Eindhoven University of Technology, have published extensive data on mechanisms and rates of deactivation at industrial conditions for their Co/Pt/alumina catalyst. Several of their papers have focused on long term deactivation due to polycarbon deposition [2].

We would also like to draw attention to an extensive report to US-DOE where long term experiments are reported focusing mainly on the effect of water [3]. Catalysts with the formulations Co(15 wt.%)/Re(0.2; 0.5; 1.0 wt.%)/γ-Al$_2$O$_3$, Co(10 wt.%)/Ru(0.2 wt.%)/TiO$_2$, Co(15 wt.%)/Pt(0;0.5 wt.%)/γ-Al$_2$O$_3$, Co(15 wt.%)/Ru(0.2;0.5;1.0 wt.%)/ γ-Al$_2$O$_3$ and Co(12.4 wt.%)/SiO$_2$ were tested for up to 3500 h in a 1L autoclave (CSTR-reactor). The authors claim that carbon deposition may be minimized by careful temperature control, and that deactivation caused by sintering and oxidation are the major concerns. These conclusions are controversial and have been disputed by several investigations; see later in this review and in our previous review [1].

Argyle *et al.* have fitted previously published activity *versus* time data to first or second order general power law rate expressions incorporating a limiting activity and have shown how parallel routes, e.g., sintering and carbon deposition deactivation, can be modeled separately. For example, their model predicts that during a 60 day run under typical FTS conditions a commercial Co catalyst loses about 30% activity within 10–15 days due to rapid sintering, while an additional 30% activity is lost gradually over the 60 day period due to carbon [4].

Causes of deactivation may depend on catalyst material and properties, e.g., support, promoters, dispersion, extent of reduction, *etc.*; reactor type; and especially operating conditions. It appears that after an initial break-in period during which cobalt is equilibrated with its reactor environment in terms of crystallite size, possibly crystal structure, and degree of reduction, a slow long term deactivation is observed. The origin of this latter deactivation period is discussed in terms of carbon formation and/or re-oxidation of the metal.

2. Catalyst Activity

To understand catalyst deactivation, it is first necessary to understand the factors that determine initial catalyst activity. Activity is largely dependent on the degree of reduction of the cobalt metal precursor and the average size of the cobalt crystallites, which together determine the surface density of catalytically active sites, *i.e.*, the dispersion of the metal. It has been verified that the turn-over-frequency (TOF) is rather constant for Co crystallites larger than 6–8 nm [5]. As activity falls off rapidly below this threshold, methods for making very high dispersion catalysts have limited relevance for FT-synthesis. Cobalt crystallite size and degree of reduction depend on several factors, including cobalt precursor, support material and its pretreatment; pore diameter, pore volume and available total surface area; method of impregnation or deposition; drying and calcination conditions; reduction conditions; *etc.* It is especially important to calcine and reduce the catalyst at optimum conditions, *i.e.*, optimal gas flow rates, temperature ramp and final temperature [6]. For example, overly high calcination and reduction temperatures result in large cobalt oxide and cobalt metal crystallites and, therefore, undesirably low dispersion due to over-sintering. For a given degree of reduction and crystallite size, the activity per kg catalyst is proportional to the cobalt loading. The loading of a commercial type catalyst may vary from 12 to 30 wt.% and is a compromise between several catalyst properties. For instance, a lower surface area and pore volume support will be able to accommodate less cobalt, but might be considerably more attrition resistant.

To optimize cobalt crystallite size is not particularly challenging as long as one is able to control preparation conditions. It is industrial practice to add a metal promoter that enhances the degree of reduction and maintains a targeted dispersion [7,8]. The literature provides no solid evidence that such metal promoters are able to enhance reaction rates or surface concentration of intermediates. Promoters used at a commercial or demo scale include platinum, rhenium and ruthenium. The promoter will add significantly to the cost of the catalyst; catalyst grade Re is today priced at *ca.* 3000 USD/kg and Pt at 45,000 USD/kg [9]. Typical loadings are up to 0.5 and 0.05 wt.% for Re and Pt, respectively. It can also be mentioned that a possible effect of cobalt being in the *fcc* or *hcp* crystallographic phase, with the latter being more active, has been reported [10]. However, studies on the actual configuration of an active cobalt crystal are needed to be able to correlate activity with atomic arrangement on the surface of a working catalyst. It is well documented that the catalyst support has a strong influence on the selectivity to C_{5+} of the process, but as long as known impurities like alkali, alkaline earth metals and sulfur are eliminated it is less clear how support chemistry and pore structure influence activity of a catalyst for a given cobalt crystallite size. However, different supports have varying interactions with cobalt oxide and therefore influence reducibility.

It is well known that reaction rate greatly depends on process conditions. Generally, rate increases with increasing temperature and overall pressure. Furthermore, indications are that the rate increases with H_2/CO ratio, possibly due to higher methane make, and decreases with increasing conversion as the partial pressure of syngas is reduced and a high level of product water may block active sites [11].

Of the CH_x monomers generated on the surface of a catalyst, CH_2 is probably the most abundant intermediate and is probably readily incorporated in the chain during polymerization. A smaller portion of the monomer will be hydrogenated all the way to methane. The growing chain can terminate by β–hydrogen abstraction and leave the surface as an olefin or be hydrogenated to an alkane. Olefins can also be hydrogenated in a secondary reaction. There is evidence from experiments at low conversion and small catalyst particle size that the primary product is dominated by olefins, but for practical purposes the olefin to paraffin ratio is well above two for C_3 and then diminishes rapidly with chain length [12].

3. Fischer-Tropsch Reactors

Apart from the type of FT-catalyst, selection of the FT-reactor, as well as how it is operated and incorporation in the XTL flowsheet, is the principal factor influencing catalyst deactivation. Comparison of properties of the main reactor types for low-temperature Fischer-Tropsch synthesis is given in Table 1. Conversion per path will vary, mainly with the propensity for heat removal and temperature control, whereas the temperature typically is between 200 and 250 °C and the pressure will be in the range 15–30 bar. H_2/CO ratio in the make-up gas from the syngas generator will be slightly below 2, whereas the outlet ratio of the FT-reactor will be considerable lower; down to 1.2, in some cases even below 1.0. Evidently, high temperature and low H_2/CO ratio are expected to promote deactivation, but reports on these effects are not available.

Table 1. Properties for different reactor types for Low-Temperature Fischer-Tropsch synthesis *.

Reactor	Conversion per path (%)	Capacity per reactor (bbl/day)	Characteristics
Tubular fixed-bed	30–35	≤6000	≤ 30,000 tubes with catalysts pellets or extrudates.
Slurry bubble column	55–65	≤25,000	Internal heat exchanger and optional product filter
Micro-channel	65–75	≤1000	Metal block with < 2mm diameter channels

* Based on open literature and patents for commercial Low Temperature Fischer-Tropsch (LTFT) synthesis with cobalt catalysts [13,14]. See also references in Table 2.

In a comparative study of reactor types for LTFT synthesis Guettel and Turek conclude that the productivity per reactor volume of a slurry bubble-column reactor or monolith reactor is up to one order of magnitude higher than for a fixed-bed or micro-channel reactor [15]. However, this conclusion is at variance with other work, reported for micro-channel reactors for which superior productivities have been reported [16].

Fixed-bed reactor. Due to the necessity of controlling the heat evolved during reaction, the design of a fixed-bed FT reactor is based on a multi-tube heat-exchange type of reactor where catalyst pellets are loaded into the tube bundles and the shell contains evaporating water. In this type of reactor axial temperature typically increases through a maximum of 5–20°°C, and it is imperative to avoid hot-spots which cause sintering. In order to minimize deactivation due to temperature effects several measures are taken. Once-through CO conversion is limited to 30–35%. Tube diameter is typically 2.5–5 cm and the size of the catalyst pellets or extrudates is relatively small, in the range of 1–3 mm. Small extrudates size is also important to secure good radial mixing and minimize diffusion limitations, thus maintaining high selectivity to liquids. It has been shown that above *ca.* 200 μm particle size the higher effective H_2/CO ratio in the inner part of the pellets significantly reduces C_{5+} yield [12]. Therefore, an egg-shell catalyst design is preferred where only the outer parts are impregnated with active metal. Another factor limiting the applicable superficial gas velocity and the global rate is pressure drop. In general, a challenge related to deactivation in a fixed-bed reactor is variations in the partial pressures of reactants and products along the tube length and within the catalyst particles.

H_2/CO ratio depends on several factors, but the make-up gas typically has a ratio slightly below 2 for maximizing C_{5+} selectivity. With recycle of the product gas to the FT-section the feed ratio to the reactor may be significantly lower, possibly 1.6–1.7, and there may be a gradual reduction from the inlet to around 1.4–1.5 at the outlet. Compared to a slurry reactor, the average gas composition may be richer in hydrogen and due to less efficient temperature control, the overall operating temperature usually lower. A consequence is lower reaction rates, but this is at least in part compensated for by a lower average partial pressure of product water and thereby a higher syngas pressure. The overall effect on deactivation, in particular carbon deposition, is complex and challenging to predict.

A distinct advantage of the tubular fixed-bed reactor is a well proven commercial design. Several tens of thousands of tubes can be incorporated within the reactor shell. Scale-up is comparatively easy, and optimization can be done in a single tube laboratory reactor. Operational experience with catalyst fouling or attrition and resulting difficulties with loading and unloading tubes are trade secrets, but it can be expected that an experienced operator is able to control these factors. Minimizing catalyst deactivation or being able to perform *in situ* regeneration is critical in order to reduce catalyst consumption and avoid an extensive unloading-reloading sequence. The liquid product is inherently separated from the catalyst and any need for removing residual particles and metal components will be low.

Slurry bubble-column reactor (SBCR). Catalyst particles are suspended in the liquid hydrocarbon product of the FT process and synthesis gas is bubbled through the slurry. Depending on the density of the catalysts particles, their diameter and the superficial gas velocity, there is a profile of solid concentration diminishing from bottom to top of the reactor. Gaseous components leave from the top of the reactor. Higher boiling products have to be removed from the reactor as liquid, and separated from the catalyst. Several methods for this purpose have been patented, both *in situ* and *ex situ* techniques. Broadly they can be classified as employment of filters, settling devices, magnetic separation and hydrocyclones. Sasol uses internal filters combined with secondary polishing filters of the product [14].

Settling of the catalyst should be avoided as overheating and consequently catalyst deactivation will occur. Particularly critical are the gas distribution system and, depending on design measures to prevent particle settling in stagnant regions below the nozzles. One way to improve overall liquid circulation, and thereby avoid settling, is to install so-called internal down-comers. A serious threat to the catalyst in a slurry operation is any upset in production, like a sudden reduction in syngas flow. Without adequate back-up systems such events will lead to settling and serious overheating in the catalyst mass due to continuous FT-synthesis with residual syngas. Similar conditions may occur in slurry separation devices like filters, but no public information is available on any effects on catalyst deactivation in these devices.

An SBCR operates preferentially in the churn turbulent flow regime for best distribution of catalyst particles as well as minimizing mass and heat transport restrictions. In the churn turbulent flow regime there is a mixture of smaller and larger bubbles that undergo frequent beak-up and coalescence. This mechanism prevents serious film transport restrictions on the catalyst slurry interphase, and with catalyst particles below 200 μm the H_2/CO ratio as well as water vapor pressure can be assumed constant over the entire reactor volume. Thus, deactivation should be more easily controlled compared to other reactor configurations. Further details on operation of SBCRs can be found in the book on Fischer-Tropsch technology by Steinberg and Dry [17].

Reasons for selection of a slurry bubble column reactor for Fischer-Tropsch synthesis include: (1) a comparatively simple construction; (2) high space-time-yield and catalyst efficiencies; (3) high heat transfer coefficients; and (4) isothermal conditions. Continuous catalyst regeneration of a slip stream is a viable option. Challenges include minimizing catalyst particle attrition and efficiently separating catalyst from the products. Efficient liquid and gas back-mixing and a high exit water concentration lead to high selectivity; the high exit water concentration is beneficial in reducing coke deposition. On the other hand reactant concentrations are lower than the average of a fixed-bed reactor resulting in comparatively lower global rates. Single pass conversion is typically in the range of 55–65%, significantly higher than for fixed-bed. Conversion is limited by the feasible height of the reactor, but there is also an upper conversion limit above about 75–80% for which the water-gas-shift activity will lead to possible catalyst oxidation and a steep increase in CO_2 yield [18]. Naturally, extensive recycle of syngas in the FT-section of the plant is necessary to obtain a very high overall CO conversion.

Micro-channel reactor. Micro-channel reactor technology holds great promise for process intensification due to outstanding heat and mass transfer rates [19,20]. Combined with highly active and stable catalysts, micro-channel reactors can achieve high volume based productivities. In some cases very high conversions (~90%) can be obtained while maintaining high C_5+ selectivity. Detailed studies of flow and temperature behavior have shown that a micro-channel reactor can operate isothermally and with very low pressure drop [19]. Except for the normal initial deactivation, the catalyst in the micro-channel reactor is remarkably stable even at very high conversions [21]. Observed rates of deactivation appear to be lower in the micro-channel reactor compared with the fixed bed laboratory reactor at similar conditions. Velocys and CompactGTL are operating microchannel demonstration plants [20].

3. Commercial Catalyst Formulations

Scale-up to commercial catalyst production is demanding, and little public information on the industrial manufacturing processes is available. Great care must be taken to obtain a homogeneous catalyst material, but the targeted distribution of cobalt on the support depends on the actual process. For slurry catalysts with diameters typically in the range 40–120 μm pore diffusion resistance of the syngas is negligible ensuring full utilization of the available surface area [12]. For micro-channel reactors the catalyst either will resemble a slurry catalyst or be impregnated onto special trays that are inserted into the channels. On the other hand, a fixed-bed catalyst is typically designed as an egg-shell or rim catalyst in which only the outer few hundred micrometers contain the active phase. As to the degree of reduction it has been shown by Sasol that the initial value is not critical, as the syngas will reduce the catalyst further under the first months of operation, thereby increasing CO conversion [2].

Cobalt FT catalysts can be classified according to supports and promoters used. Table 2 lists several commercial type catalysts. Precious metal promoters like Pt and Ru may facilitate hydrogenation activity and hence reduce propensity for carbon deposition as indicated, for example, by Exxon Mobil for ruthenium.

Some relative activities have been estimated and are included in the table. The values are based on fixed-bed data reported in the patent literature. As the process conditions vary considerably, an attempt to normalize the activities is made by using a simplified kinetic expression with activation energy of 110 kJ/mol, and partial pressures are average pressures in the reactor [22]. It is recognized that this comparison is only approximate, but still a guide. Activities are generally lower for a given catalyst operated in slurry compared to fixed-bed in spite of limited apparent diffusion limitations. The origin of this effect is not understood. Particularly low activities have been reported by Nippon and ENI/IFP, perhaps an indication of large cobalt crystals in the working catalyst.

It appears that Shell favors a titania based support over their previous zirconia modified silica system. Promoters are either Mn or V, the latter claimed to lower CO_2 make [23]. Titania has relatively large pores and moderately low surface areas, but is known to facilitate high selectivities to C_{5+} products. Fixed-bed catalysts contain modifiers like citric acid added prior to the forming step. In addition to Shell, it should be mentioned that BP is promoting its fixed-bed technology and claims that a CO_2 resistant support is vital [24].

Velocys/Oxford Catalyst and Compact GTL are offering micro-channel fixed-bed technologies with catalyst diameter or thickness of 0.1–0.5 mm. Velocys' carbon combustion preparation technique may very well take the edge off initial sintering during FT synthesis by optimizing cobalt crystallite size already at the catalyst manufacturing stage. Loading and off-loading the catalyst can be particular challenge for these systems, but efficient methods for this purpose are claimed by these companies.

The platinum promoted Sasol catalyst is prepared on γ-alumina (Puralox SCCa-2/150 or -5/150: pore volume 0.5 mL/g; surface area 150 m^2/g) and stabilized by impregnation with tetra-ethoxy-silane (TEOS) followed by calcination to give a surface concentration of *ca.* 2.5 Si atoms/nm^2. This procedure apparently modifies the surface so that the support becomes less prone to dissolution in the acidic product water. The GTL.F1/Statoil catalyst is based on a larger pore diameter γ-alumina modified with nickel and fired at high temperature to produce a nickel-aluminate (spinel)/α-alumina mixture. The pore properties resemble titania-based supports, but with very high attrition resistance. Also the ENI/IFP catalyst is supported on Si-modified γ-alumina, but probably strengthened by silanation and calcination giving a final SiO_2 content of 6–7 wt.%. Other support modification methods have been described by ENI/IFP in earlier patents, including formation of spinel compounds. It is unclear whether the catalyst formulation contains a reduction or other type of promoter. In their slurry catalyst development, Nippon has apparently adopted a silica based support formulation similar to that of Shell's previous fixed-bed catalyst, but using ruthenium as a promoter. No information on attrition resistance has been revealed, as is the case for most other slurry catalysts as well. Syntroleum used a catalyst similar to Sasol's, but also with a ruthenium promoter. Exxon Mobil, that pioneered titania as a support with an alumina binder, BP, Conoco-Phillips and Syntroleum have terminated their developments in FT-technology. However, BP is still licensing their FT technology and Exxon Mobil has announced that the technology is ready for commercialization should the right project be prioritized.

Table 2. Formulation of commercial type cobalt catalysts and their application *.

Technology provider	Support/ modifier	Promoter	Reactor type	Reactor scale (bbl/d)	Relative activity	Reference
Sasol	γ-Alumina/ Si **	Pt	Slurry	16,000		[25]
Shell	Titania	Mn; V	Fixed	6000	0.3	[26]
GTL.F1	Ni-aluminate/ α-Alumina	Re	Slurry	1000	0.9	[27]
ENI/IFP/ Axens	γ-alumina/ SiO₂;spinel	?	Slurry	20	0.19	[28]
Nippon Oil	Silica/ Zirconia	Ru	Slurry	500	0.16	[29]
Syntroleum	γ-Alumina/ Si **; La	Ru	Slurry	80		[30]
BP	ZnO	?	Fixed			[24]
Exxon Mobil	Titania/ γ-Alumina	Re	Slurry	200		[31]
Conoco-Phillips	γ-Alumina/ Boron	Ru/Pt/Re	Slurry	400	0.68	[32]
Compact GTL	Alumina?	Re?	Micro	500		[33]
Oxford cat. /Velocys	Titania-silica	Re	Micro	1000		[34]

* Deduced from open literature and patents. Actual commercial formulation may vary. ** Si from TEOS; tetraethoxy silane.

4. Causes of Deactivation

From our previous review on deactivation during LTFT synthesis the main causes of deactivation are sintering, re-oxidation of cobalt, formation of stable compounds between cobalt and the support, surface reconstruction, formation of carbon species on the cobalt surface, carbiding and poisoning [1]. In addition there can be a loss of catalyst material from the reaction zone due to attrition. The chemical environment is challenging with a number of reactive chemical species generated including significant amounts of water. In addition, the exothermicity of the reaction may lead to hot spots in the catalyst.

There appear to be two main "schools" for describing long-term deactivation mechanisms based on demo slurry operations, one favoring re-oxidation [35,36], and one poly-carbon formation on the surface [2]. It should be realized that both catalyst system and process conditions can affect the results.

In addition, an initial sintering stage may be observed if the fresh catalyst contains crystallites in the range of 6–12 nm. The consequences of severe deactivation can be a significant decline in the activity over a typical design period for catalyst life of two years to an estimated 25–30% of the initial value. In addition, slurry catalyst loss due to attrition can be significant. All previous experience considered catalyst replacement due to deactivation will contribute significantly to the operational costs of an FT plant.

In a study on the effect of impurities it was found [37], by impregnating 400 ppm of the impurity element from nitrate precursors, a poisoning effect which decreases in the following order:

$$Na > Ca > K > Mg > P,$$

Mn, Fe and Cl showing minimal effects. The latter is surprising as chlorine causes a 25% reduction in hydrogen chemisorption. Even alkali concentrations of less than 100 ppm have a large effect on the rate (site time yield) [38]. However, no effect of alkali on the hydrogen chemisorption was observed. The impact of alkali and alkaline earth elements is far stronger than any stoichiometric blocking of surface sites, and might be related to the strong electronegativity of the elements leading to blocking of steps on cobalt thus preventing CO dissociation [39]. Particular care must be taken to avoid alkali and alkaline earth elements in impregnation fluids and washing water as well as contaminants in the catalyst support. Sulfur as H_2S or $(CH_3)_2S$ added to the syngas gives a deactivation consistent with stoichiometric blocking of cobalt surface sites as *in situ* measurement of cobalt dispersion by H_2S is consistent with hydrogen derived dispersion of a fresh catalyst. The effect of ammonia appears to be strongly catalyst dependent, and reports vary from negligible influence to strong negative consequences.

Contributions to attrition of catalysts for three-phase slurry bubble column operation include mechanical abrasion and breakage of catalyst particles; chemical dissolution; and synergisms between these mechanisms. It appears that Sasol focuses more on avoiding chemical attack on their γ-alumina based support [40], whereas Statoil/GTL.F1 [41], IFP/ENI/Axens [28], and probably Exxon [42], have developed more mechanically robust slurry catalysts.

Sasol has reported deactivation profiles in several publications for their Co/Pt/modified γ-alumina catalyst [2]. We refer to the section below on carbon deposition and to the previous review for further details on the Sasol work and their extensive documentation of carbon deposits [1].

Statoil/GTL.F1 have disclosed deactivation profiles at several conferences for their attrition resistant catalyst of Co/Re/aluminate spinel catalyst. In a CSTR slurry reactor test over 3000 h the temperature was adjusted regularly, typically in 2–3 weeks intervals, to keep conversion reasonable constant [43]. Somewhat surprisingly, temperature was *decreased* from 222 °C to 215 °C during the operation meaning that the catalyst activity increased regularly. This is in line with the reported enhanced reduction during first months of slurry FT-synthesis. More surprisingly, the C_{5+} selectivity increased simultaneously by *ca.* 5%, significantly more than expected given the reduction in temperature. In another presentation on CSTR results, the rate of hydrocarbon formation increased up to 800 h time-on-stream (TOS) followed by decline towards end of test at 1600 h [44]. Characterization of a commercial catalyst after *ca.* one month operation in a commercial scale slurry-bubble column confirms an increase in degree of reduction [45]. This is concurrent with sintering probably facilitated by a high steam partial pressure [46].

5. Deactivation by Carbon Deposition

Higher hydrocarbons (waxes) are desired products from LTFT synthesis on Co catalysts. The hydrocarbons can accumulate on the surface and can slowly be converted to carbon or coke that blocks the active sites. By using TPR Lee *et al.* [47] could distinguish between several forms of carbon on the catalyst surface from CO disproportionation; they suggested that carbon was present in two forms: atomic and polymeric carbon. Support for stabilization of graphitic carbon on an *fcc*-Co(111) surface has been obtained by quantum-mechanical calculations [48]. DFT data indicate chemical bonding between graphene and cobalt, as also supported by other studies [49]. Direct STM evidence for formation of graphene on Co(0001) has been demonstrated through decomposition of ethylene [50]. It was found that carbon on the surface induces cobalt reconstruction and weakens CO and H_2 adsorptions.

Thus, there are ample investigations showing that carbon in different forms can interact with and block cobalt surfaces. Support for this deactivation mechanism comes from a few long term studies using commercial catalysts in pilot reactors. Sasol studied catalyst deactivation by periodically removing samples from a pilot slurry bubble column reactor operated for 6 months [2]. Wax was removed by inert solvent extraction before the catalyst samples were characterized by temperature programmed hydrogenation and oxidation, chemisorption, TEM and LEIS. Polymeric carbon was found both on the alumina support and on cobalt. This carbon is resistant to hydrogen treatment at temperatures above the FT synthesis temperature. The amount of polymeric carbon correlated well with observed long term deactivation. From XANES data they ruled out oxidation of cobalt during the run, but there was significant sintering taking place during the first 10–15 days on stream. Moodley *et al.* [2] concluded that accumulation of polymeric carbon was responsible for at least a part of long term catalyst deactivation.

Build-up of graphitic or polymeric carbon as deactivation mechanism was recognized by Syntroleum [51]. By TGA-MS they estimated that about 1% carbon was deposited on the catalyst. Carbon deposition on cobalt/ZnO has also been proposed by BP based on results from a demonstration plant and laboratory studies [52].

6. Deactivation by Re-Oxidation

Schanke *et al.* investigated the influence of water on deactivation of unpromoted or Re promoted alumina supported cobalt catalysts. Adding 20–28% steam to 50% syngas in the feed of a lab-scale fixed-bed reactor resulted in significant deactivation due to oxidation of highly dispersed cobalt crystals and surface cobalt atoms [53]. Although these experiments clearly show oxidation of cobalt, the conditions represent very high conversion levels (> 80%). Oxidation takes place within the stability range for bulk cobalt metal, and is presumed to be a consequence of surface reactivity of small crystallites. It is also evident that the effect of water depends critically on the support material used, e.g., samples of γ–alumina from different sources behave very differently. Even a positive effect of water on activity has been reported; see [1,3] for further discussion. From activity tests over a range of conversions in a slurry CSTR reactor Co/Re/γ-alumina catalyst activity is observed to slightly increase with conversion up to 85%; above 85% conversion it drops rapidly [18]. At high conversions high partial pressures of H_2O may oxidize small cobalt crystallites and promote aluminate formation thus enhancing WGS activity as shown by a significant CO_2 make. Similar results were found for a Pt promoted catalyst [54].

Very recently, the group of A.Y. Kodakov in cooperation with Total published evidence for surface oxidation of cobalt nano-crystallites in alumina supported, Pt promoted catalysts during FT synthesis [55]. However, surface oxidation was only clearly evident by STEM-EELS after an excursion to 340 °C and 100% CO conversion. B.H. Davis and coworkers exposed a freshly reduced catalyst directly to a water vapor pressure equivalent to 50% conversion [56]. They observed oxidation of a fraction of the smaller cobalt crystallites when supported on alumina or activated carbon, and recommended that careful crystallite size management is required for a commercial catalyst. Kliewer *et al.* studied redox transformations of cobalt catalysts by TEM in terms of agglomeration of the metal, mixed-oxide formation with the support and reversible oxidation followed by reduction under mild hydrogen treatment [57]. They claim that reactor and TEM studies show that nanoscale Co crystallites can oxidize to CoO during commercially relevant FT synthesis conditions in spite of bulk thermodynamic data that suggest otherwise [58]. The propensity for oxidation is enhanced by small Co crystallites and high CO conversion with attendant high water vapor pressure and a high H_2O to CO ratio in the reactor. The oxide crystallites thus formed can be fully reduced by hydrogen at standard FT temperature and pressure. Reported TEM images from Exxon show that the oxidized cobalt metal wets the support surface and thereby facilitates contact between nearby crystallites. This has also been illustrated in a presentation from Statoil in Figure 1 [59]. Both images show a thickness of a cobalt oxide crystallite of *ca.* 5 nm wetting the surface. Further, the TEM image indicates mixed orientation of several cobalt oxide crystallites and an amorphous layer at the support interphase. Upon re-reduction metal crystallites may agglomerate depending on the initial spatial distribution on the catalyst support.

That cobalt distribution can vary significantly with catalyst preparation procedure is illustrated in Figure 2 for a Co/Re/γ-alumina catalyst [43]. Cobalt dispersions are comparable, but we see that the distribution of clusters of the oxide varies significantly. Although it can be imagined that small well dispersed clusters are less prone to deactivation, this needs to be verified.

It is well known that mixed-oxides can be formed between cobalt and silica, alumina and titania. In part, a surface layer of mixed-oxide is formed during catalyst preparation from water solution followed by drying and calcination. It has also been claimed that enhanced mixed oxide formation takes place during FT-reaction at high conversion levels (> 70%) concurrent with oxidation of cobalt to Co^{2+}. In the case of silica supported catalysts, well defined crystalline needles of cobalt silicate are rapidly formed at higher conversions [60].

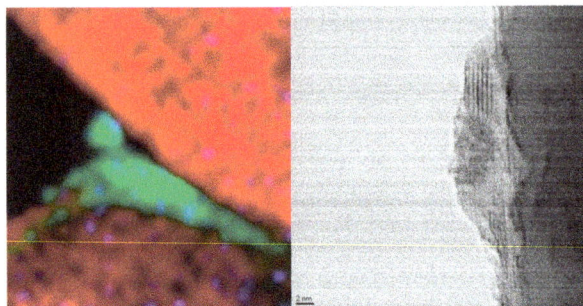

Figure 1. Wetting of cobalt oxide on a support. Left: EELS spectra; alumina: red; cobalt: green. Right: TEM.

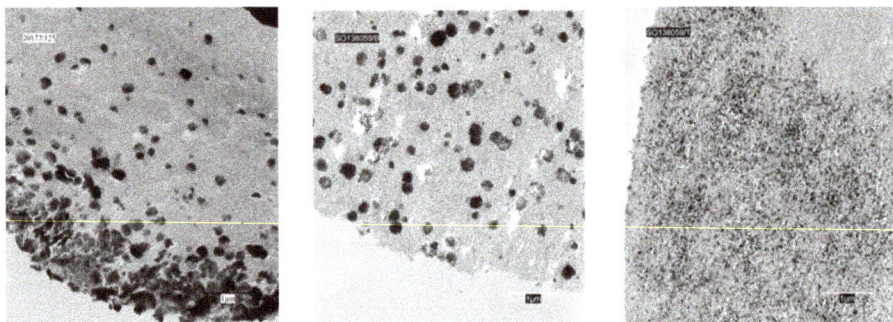

Figure 2. Cobalt oxide clusters on a γ-alumina. Bar shows 1 μm for all images.

Redox reactions of cobalt have been used deliberately to enhance the catalytic properties of FT catalysts by employing a reduction-oxidation-reduction (ROR) procedure [61,62]. An ROR treatment can increase Co dispersion presumably by forming hollow oxide domes during controlled oxidation that break up into smaller Co crystallites during re-reduction [63]. Whether the resulting Co crystals that will be in close proximity with each other experience agglomeration during the first months of commercial FT synthesis has apparently not been reported.

7. Catalyst Rejuvenation and Regeneration

Regeneration of cobalt LTFT synthesis catalysts is largely described in the patent literature. The options involve treatment of the catalyst with air (oxygen), hydrogen and/or CO and variations thereof in addition to procedures for removing produced wax. Therefore, regeneration addresses reversing the main deactivation processes of carbon deposition, metal oxidation and sintering by combustion, reduction and re-dispersion, respectively. A review of early reports on regeneration covering 1930–1952 has been presented [64]. It was concluded that there is no universal process for regenerating Co FT-catalysts; the art over this relatively short time period comprising conflicting results with patents covering a wide range of processes involving oxidation, reduction, combined oxidation-reduction, steam-reduction, operating at elevated temperatures and solvent extraction.

Commercial regeneration processes are either *in situ* in the FT-reactor itself or *ex situ* after removal of part of or the entire catalyst inventory. Indications are that Shell successfully regenerates their catalyst regularly, but it is undisclosed whether this takes place inside the tubes of the fixed-bed reactor or in a separate unit, whereas Sasol apparently removes part of the slurry from the reactor continuously for regeneration and re-deployment of the catalyst into the reactor. The latter approach allows continuous operation of the LTFT synthesis. Micro-channel reactors pose special challenges depending on the catalyst configuration. *In situ* regeneration is an option, or the catalyst can be removed for external treatment either by unloading the catalyst particles or removing multi-channel trays with catalyst attached.

A summary of regeneration procedures from some of the main industrial companies that are or recently have been involved in Fischer-Tropsch technology development is given in Table 3. The table is representative of available information but should not be assumed to be complete or up-to-date; nevertheless it illustrates what is probably the preferred approach of each individual company. In the first column the type of commercial regeneration process and primary FT-reactor type are listed and whether the regeneration is intermittent or continuous, while data in the other columns refer to the actual test protocol and results described in the patent literature. Note that reactor type can be different in columns one and three.

It is clearly possible to regenerate a deactivated catalyst to a level close to the original activity by different combinations of wax removal, hydrogenation and combustion of carbonaceous deposits; followed by re-reduction if needed. There is unfortunately little information available on the long-term performance of regenerated catalysts.

Sasol has published a brief summary of their procedure for removing most of the wax, followed by combustion of the remaining carbonaceous species and complete reactivation [65]. Specifically, Sasol describes dewaxing by hydrogenolysis with pure hydrogen for 2 h at 220 °C and reduction for 2 h at 350 °C [66]. After passivation with CO_2, the catalyst is subjected to oxidation with air in a fluidized-bed calcination unit at 250 °C for 6 h under a pressure preferably of *ca.* 10 bar. Re-reduction is performed at 425 °C and 98% of the original activity is regained. In cooperation with Eindhoven University, Sasol has investigated mechanisms of deactivation and regeneration for both model and commercial catalysts [67]. After a heptane wash and reoxidation, several FT cycles were demonstrated with no apparent permanent loss in activity. The oxidation step is described in terms of the Kirkendall effect involving spreading of a Co oxide film during oxidation followed by breaking up of the film to form small re-dispersed Co crystallites.

It appears that Shell is aiming at *in situ* regeneration in the tubes of their fixed-bed FT reactor, although an external post FT-reaction step is part of their described procedure [69]. Regeneration is based on a procedure that most likely comes from their Bintulu plant in Malaysia. After wax removal a mild hydrogenation and oxidation is conducted. The catalyst is then taken out of the FT-reactor, treated with a concentrated aqueous ammonia solution and subsequently with CO_2, giving Co ammonium carbonate. Most likely the latter procedure gives cobalt ammine carbonate complexes suitable for re-dispersing cobalt [76].

Table 3. Summary of regeneration concepts and procedures based on patents and presentations from LTFT technology companies.

Technology owner Regeneration configuration*	Catalyst TOS	FT test reactor	Wax removal step	Primary hydrogenation	Calcination/ Oxidation/ Re-dispersion	Regeneration effect (activity)
Sasol [66,68] Slurry; continuous Ex situ	Co/Pt/ alumina -	Slurry**	H_2 strip at 220 °C or xylene wash	H_2 at 350 °C	Air in fluid-bed	98%
Shell [69] Fixed-bed; intermittent in situ	Co/Mn/ Titania - Years	Fixed-bed Full scale	Gas oil wash	Diluted H_2	Diluted O_2 at 270 °C; NH_3/CO_2	> 100% < 100% (800 h TOS)
GTL.F1 [70] Slurry; continuous Ex situ	Co/Re/ aluminate -	Slurry**	Draining or cyclohexane wash	No	Diluted air in fluid-bed	98–115%
ExxonMobil [71] Slurry; continuous Ex situ	Co/Re/titania - A few days	Fixed-bed (lab)	Filtration + H_2 strip	No	Diluted O_2 in fixed-bed	~100%
Nippon Oil [72] Slurry; continuous Ex situ	Co/ zirconia-silica -	Slurry (lab)	"de-oiling"	No	Steaming at 200 °C; 25 bar in fixed-bed	95%
ConocoPhillips [73] Slurry; continuous Ex situ	Co/Ru or Re/ (F-)alumina - 1014 h	Fixed-bed (lab) or slurry.**	No	7% H_2/steam at 300 °C/ 3 bar	No	80–95%
Syntroleum [74] Slurry; continuous Ex situ	Co/alumina - 4000 h	Slurry**	N_2 at 316–343 °C	No	Diluted O_2 in fluid-bed	"Good performance"
Oxford Catalyst/ Velocys [75] Micro-channel; in situ intermittent	Co/Re/ titania-silica?	Micro-channel	N_2 flow?	H_2 in situ	O_2 in situ	~100%

* Estimated. ** Size of slurry reactor is undisclosed, but the technology providers have had both lab and pilot/demo units in operation.

Regeneration may re-disperse cobalt, possibly even to a level higher than for a freshly reduced catalyst [70]. A promoted cobalt catalyst on a spinel support was subjected to an extended run in a slurry-bubble column reactor as described by Schanke et al. [14]. Actual time on stream was not reported, but samples from four different TOS's were analyzed and regenerated. Data from the patent have been plotted in Figure 3. All activity data are from standard runs in a laboratory fixed-bed reactor at 210 °C, 20 bar pressure and H_2/CO ratio of 2:1, and after a conventional reduction protocol with hydrogen at 350 °C. Fresh and used catalysts, all embedded in wax, were drained to remove excessive wax at 85 °C before activity testing. A successive reduction in activity to *ca.* 50% was experienced. If the spent catalyst was calcined at 300 °C to burn off excessive wax and carbon deposits before testing, the activity is only slightly reduced compared to the fresh catalyst. Interestingly, even an enhanced activity was experienced if the wax was removed by cyclohexane/n-heptane solvent extraction before calcination. Pore volumes and surface areas of freshly calcined catalysts and used catalysts subject to wax draining and oxidation are unchanged. From these data it appears that hydrogenation by itself has moderate or no effect on regeneration, but that burning off coke deposits completely rejuvenates the catalyst.

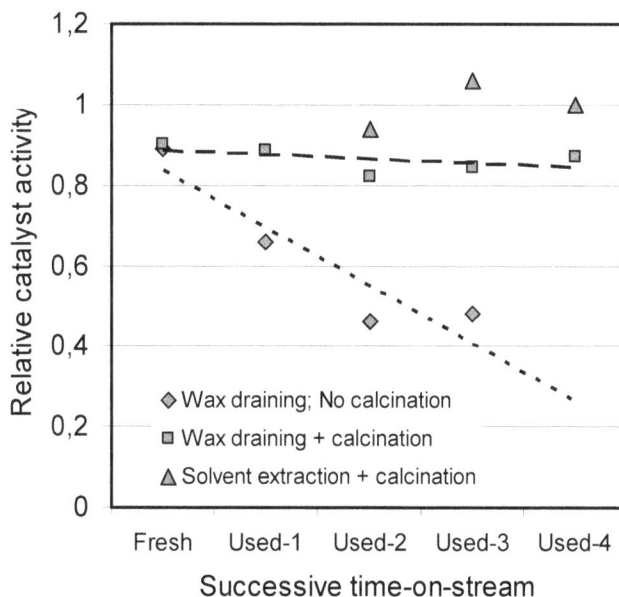

Figure 3. Catalyst activity before and after oxidation of carbon deposits of a catalyst used in a slurry bubble column reactor (data from ref. [70]).

In an early patent, Exxon researchers describe successful rejuvenation by atmospheric hydrogen treatment at typical FT synthesis temperatures of 200–230 °C of spent Co/Ru/titania catalysts [77]. The promoter is needed to facilitate the rejuvenation and is claimed also to inhibit carbon deposits [78]. However, examples are only for FT runs of 10–30 days at 50–60% CO conversion in lab-scale, fixed-bed reactors, and therefore may only addresses hydrogenolysis of very heavy waxes that partially block pores and possibly hydrogenation of oligomeric carbons on the surface of the catalyst. In a later patent, as shown in the table above, Exxon Mobil has demonstrated that reduction with hydrogen after oxidative regeneration can be performed in the slurry FT-reactor itself at mild process conditions of 200 °C and 20 bar, thus resembling the FT synthesis conditions. The company has a number of patents describing regeneration procedures, including adding more active metal after combustion of carbon deposits [79].

As pointed out above, - deposits of heavy hydrocarbon waxes reside in the pores of a used catalyst that should be largely removed before regeneration. Hydrogen treatment may reduce part of the wax through hydrogenation, but may also leave residual components at the surface. ConocoPhillips claims to have designed a suitable reactivation procedure that both removes heavy hydrocarbons and reduces the active metal [73]. For a 19 wt.% Co/0.1 wt.% Ru on alumina catalyst run for 40 days at standard FT-conditions, best regeneration results were obtained in a fixed-bed using a mixture of 93% steam/7% H_2 at 300 °C and 3 bar where 95% of initial activity was regained. Similar experiments for a catalyst composition of 19 wt.% Co/0.1 wt.% Re on fluorinated alumina in a slurry reactor were less successful as 71% of initial activity was obtained, up from 33% after the FT period.

In their continuous regeneration of a spent slurry catalyst, Syntroleum focuses on removing as much wax as possible by nitrogen stripping at 300–350 °C and 3 bar, followed by calcination. Improved cobalt dispersion and reducibility are claimed, but no actual performance data are reported for the regenerated catalyst [74]. Nippon Oil has followed an alternative approach where they hydrothermally treat the spent catalyst with steam at elevated pressure [72]. Much of the activity is regained, although it appears that the TOS between regenerations is relatively short in view of the high activity level of *ca.* 75% before regeneration, thus moderating build-up of polymeric carbonaceous deposits.

Oxford Catalyst/Velocys have presented interesting long-term operation and regeneration performances for their micro-channel reactor [75]. Velocys' catalyst and reactor operate at much higher space velocities and productivities than conventional catalyst/large-scale FT reactors. To compensate for deactivation, temperature is increased from 205 °C to 232 °C after *ca.* 650 days TOS, accompanied by a slight increase in CO conversion from 71.2 to 73.6%. Naturally, the temperature increase is accompanied by a reduction in C_{5+} selectivity, in this case by as much as from 87.9% to 75.8%. Both activity and selectivity are fully restored after *in situ* hydrogen treatment followed by calcination with oxygen and re-reduction. The first step of hydrogenation probably re-reduces smaller cobalt crystallites and partly removes some wax and deposits. The hydrogenation step can be carried out under conditions resembling FT-synthesis and can thus be carried out at the plant location and with the catalyst loaded in the reactor. However, burning off the carbonaceous deposits is far more demanding and probably requires removing the catalyst from the reactor. These results are qualitatively in agreement with what can be expected from large scale fixed-bed and slurry bubble column operations although the rate of deactivation for the Velocys' catalyst is quantitatively smaller than those reported in available literature for large scale reactors. The alternative micro-channel LTFT provider, CompactGTL, appears to have only a regeneration patent directed at removing accumulated ammonia deposits [80].

8. Conclusions

An attempt to visualize the main catalyst deactivation mechanisms that impact FT catalyst activity during commercial operation is shown in Figure 4. The regions shown are approximate with respect to TOS and conversion, and there certainly will be some overlap. The following takes place:

- In the initial phase, (a few weeks) there will be driving forces towards both an increase in activity by reduction as well as a decrease due to sintering. The net effect can be positive or negative and will to a large degree depend on the catalyst formulation, pretreatment and FT reactor environment. These mechanisms are well documented for slurry operations, but are less evident for fixed-bed and micro-reactors.

- Sintering is favored by small Co crystallites and high conversion levels due to enhanced water activity, whereas reduction is facilitated by low conversion and water vapor pressure.
- To a large extent oxidation of cobalt depends on operating conditions and may proceed at high conversion conditions, particularly for smaller cobalt crystals and may involve interaction with the support material.
- Polycarbon deposition is the principal long-term cause of deactivation for all LTFT catalysts and reactor types since commercial catalysts are stabilized against oxidation by operation at realistic conversions, while reduction and sintering are short term phenomena which determine steady-state activity after just 1–2 months of TOS.

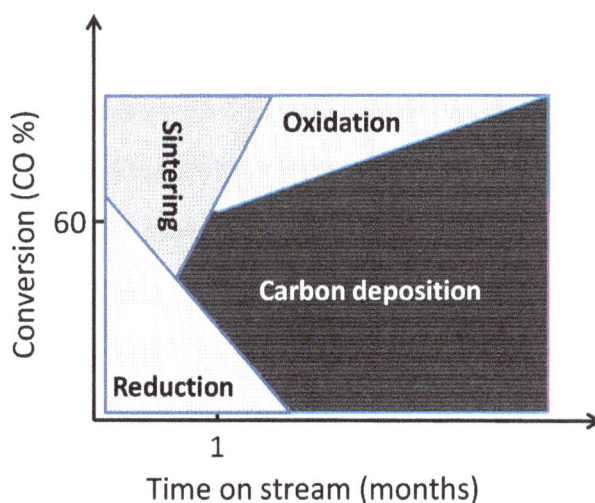

Figure 4. Main reactions taking place during equilibration and deactivation of Co Fischer-Tropsch catalysts.

Although deactivation during the first 6–18 months of operation can cause 30–60% loss in activity in low-temperature Fischer-Tropsch synthesis with cobalt based catalysts, fortunately it appears possible to regenerate the catalyst to approach the activity of a freshly equilibrated catalyst. In a slurry reactor system a slip stream can be taken from the reactor for continuous rejuvenation. For a fixed-bed reactor, including micro-channel reactors, an intermittent *in situ* rejuvenation and *in situ* regeneration procedures can presumably be implemented as long as removal of wax and combustion of carbonaceous deposits can be controlled to prevent large temperature excursions. Nevertheless, based on patent literature, *ex situ* regeneration procedures appear to be the norm for most FT processes.

Acknowledgement

The financial support from inGAP centre of research-based innovation, supported by the Norwegian research council under contract no. 174893, is acknowledged with gratitude.

146

References and Notes

1. Tsakoumis, N.E.; Rønning, M.; Borg, Ø.; Rytter, E.; Holmen, A. Deactivation of cobalt based Fischer-Tropsch catalysts. A review. *Catal. Today* **2010**, *154*, 162–182e.
2. Moodley, D.J.; van de Loosdrecht, J.; Saib, A.M.; Overett, M.J.; Datye, A.K.; Niemantsverdriet, J.W. Carbon deposition as a deactivation mechanism of cobalt-based Fischer–Tropsch synthesis catalysts under realistic conditions. *Appl. Catal.* **2009**, *354*, 102–110.
3. Davis, B.H.; Iglesia, E. Technology Development for Iron and Cobalt Fischer-Tropsch Catalysts, Final Technical Report. Available online: http://www.fischer-tropsch.org/DOE/DOE_reports/40308/FC26-98FT40308-f/FC26-98FT40308-f_toc.htm (accessed on 13 March 2015).
4. Argyle, M.D.; Frost, T.S.; Bartholomew, C.H. Cobalt Fischer–Tropsch Catalyst Deactivation Modeled Using Generalized Power Law Expressions. *Top. Catal.* **2014**, *57*, 415–429.
5. Bezemer, G.L.; Bitter, J.H.; Kuipers, H.P.C.E.; Oosterbeek, H.; Holewijn, J.E.; Xu, X.; Kapteijn, F.; van Dillen, A.J.; de Jong, K.P. Cobalt particle size effects in the Fischer-Tropsch reaction studied with carbon nanofiber supported catalysts. *J. Am. Chem. Soc.* **2006**, *128*, 3956–3964.
6. Borg, Ø. Role of Alumina Support in Cobalt Fischer-Tropsch Synthesis. Ph.D. Theses, NTNU, Norway, April 2007.
7. Diehl, F.; Khodakov, A.Y. Promotion of Cobalt Fischer-Tropsch Catalysts with Noble Metals: A Review. *Oil Gas Sci. Technol. Rev. IFP.* **2009**, *64*, 11–24.
8. Ma, W.; Jacobs, G.; Keogh, R.A.; Bukur, D.B.; Davis, B.H. Fischer-Tropsch synthesis: Effect of Pd, Pt, Re, and Ru noble metal promoters on the activity and selectivity of a 25%Co/Al$_2$O$_3$ catalyst. *Appl. Catal. A* **2012**, *437–438*, 1–9.
9. Information could be found at http://www.metal-pages.com/metalprices/rhenium/ and http://www.metal-pages.com/metals/platinum/metal-prices-news-information/.
10. Ducreux, O.; Rebours, B.; Lynch, J.; Roy-Auberger, M.; Bazin, D. Microstructure of Supported Cobalt Fischer-Tropsch Catalysts. *Oil Gas Sci. Technol. Rev. IFP.* **2009**, *64*, 49–62.
11. Blekkan, E.A.; Borg, Ø.; Frøseth, V.; Holmen, A. Fischer-Tropsch synthesis on cobalt catalysts: The effect of water. In *Catalysis*; Royal Society of Chemistry: Cambridge, UK, 2007; Volume 20, pp. 1–21e.
12. Rytter, E.; Eri, S.; Skagseth, T.H.; Schanke, D.; Bergene, E.; Myrstad, R.; Lindvåg, A. Catalyst Particle Size of Cobalt/Rhenium on Porous Alumina and the Effect on Fischer–Tropsch Catalytic Performance. *Ind. Eng. Chem. Res.* **2007**, *46*, 9032–9036
13. Zennaro, R. The ENI-IFP/Axens GTL technology for the conversion of natural gas. Proceedings of 6th World GTL Summit, London, UK, 17–19 May 2006.
14. Schanke, D.; Wagner, M.; Taylor, P. Proceedings of 1st annual gas processing symposium, Doha, Qatar, 10–12 January 2009; Alfadala, H., Reklaitis, G.V.R., El-Halwagi, M.M., Eds.; Elsevier: Amsterdam, The Netherlands, 2009.
15. Guettel, R.; Turek, T. Comparison of different reactor types for low temperature Fischer–Tropsch synthesis: A simulation. *Chem. Eng. Sci.* **2009**, *64*, 955–964.

16. LeViness, S. presentation, Energy Frontiers International, 22nd October 2012. Available online: http://www.velocys.com/arcv/press/ppt/EFI%202012%20Presentation%20121019_1_rev2.pdf (accessed on 3 December 2014).

17. Steynberg, A.P.; Dry, M.E.; Davis, B.H.; Breman, B.B. Fischer-Tropsch Reactors. *Stud. Surf. Sci. Catal.* **2004**, *152*, 64–195.

18. Schanke, D.; Lian, P.; Eri, S.; Rytter, E.; Sannæs, B.H.; Kinnari, K.J. Optimisation of Fischer-Tropsch Reactor Design and Operation in GTL Plants. *Stud. Surf. Sci. Catal.* **2001**, *136*, 239–244.

19. Myrstad, R.; Eri, S.; Pfeifer, P.; Rytter, E.; Holmen, A. Fischer-Tropsch synthesis in a microstructured reactor. *Catal. Today* **2009**, *147S*, S301–S304.

20. Holmen, A.; Myrstad, R.; Venvik, H.J.; Zhu, J.; Chen, D. Monolithic, microchannel and carbon nanofibers/carbon felt reactors for syngas conversion by Fischer-Tropsch synthesis. *Catal. Today* **2013**, *216*, 150–157.

21. Yang, J.; Eiras, S.B.; Myrstad, R.; Pfeifer, P.; Venvik, H.J.; Holmen, A. Fischer-Tropsch synthesis over high-loading Co-based catalysts in a microreactor. *Prep. ACS Div. Energy Fuels* **2014**, *59*, 828–830.

22. Rytter, E.; Ochoa-Fernández, E.; Fahmi, A. In *Catalytic Process Development for Renewable Materials*; Imhof, P., van der Waal, J.-K., Eds.; Wiley: Weinheim, Germany, 2013; pp. 265–308.

23. Creyghton, E.J.; Mesters, C.M.A.M.; Reynhout, M.J.; Verbist, G.L.M.M. Process for preparing a catalyst. Patent WO2008071640 A2, June 2008.

24. Freide, J.J.H.M.F.; Collins, J.P.; Nay, B.; Sharp, C. A history of the BP Fischer-Tropsch catalyst from laboratory to full scale demonstration in Alaska and beyond. *Stud. Surf. Sci. Catal.* **2007**, *163*, 37–44.

25. Van Berge, P.J.; Barradas, S. Catalysts. U.S. Patent 7365040 B2, March 2008.

26. Dogterom, R.J.; Mesters, C.M.A.M.; Reynhout, M.J. Process for preparing a hydrocarbon synthesis catalyst. Patent WO2007068731 A1, June 2007.

27. Rytter, E.; Skagseth, T.H.; Wigum, H.; Sincadu, N. Enhanced strength of alumina based Co Fischer-Tropsch catalyst. Patent WO 2005072866, August 2005.

28. Bellussi, G.; Carluccio, L.C.; Zennaro, R.; del Piero, G. Process for the preparation of Fischer-Tropsch catalysts with a high mechanical, thermal and chemical stability. Patent WO2007009680 A1, January 2007.

29. Ikeda, M.; Waku, T.; Aoki, N. Catalyst for Fisher-Tropsch synthesis and method for producing hydrocarbon. Patent WO2005099897 A1, October 2005.

30. Inga, J.; Kennedy, P.; LeViness, S. Fischer-Tropsch process in the presence of nitrogen contaminants. Patent WO2005071044, August 2005.

31. Soled, S.L.; Fiato, R.A.; Iglesia, E. Cobalt-ruthenium catalyst for Fischer-Tropsch synthesis. Eur. Pat. Appl. 87310896, June1989.

32. Srinivasan, N.; Espinoza, R.L.; Coy, K.L.; Jothimurugesan, K. Fischer-Tropsch catalysts using multiple precursors. U.S. Patent 6822008 B2, November 2004.

33. Compact GTL. The modular gas solution. Available online: http://www.compactgtl.com (accessed on 24 February 2015).

148

34. Velocys. Think Smaller. Available online: http://www.velocys.com/resources/ Velocys_Booklet.pdf (accessed on 24 February 2015).

35. Kliewer, C.E.; Soled, S.L.; Kiss, G. Characterizing Intrinsic Deactivation in Cobalt-Catalyzed Fischer-Tropsch Synthesis. Available online: http:///www.nycsweb.org/files/37637807.pdf (accessed on 24 February 2015).

36. Kliewer, C.E.; Kiss, G.; Soled, S.L. Characterizing Intrinsic Deactivation in Cobalt-Catalyzed Fischer-Tropsch Synthesis. *Microsc. Microanal.* **2010**, *16*, 1258–1259.

37. Borg, Ø.; Hammer, N.; Enger, B.C.; Myrstad, R.; Lindvåg, O.A.; Eri, S.; Skagseth, T.H.; Rytter, E. Effect of biomass-derived synthesis gas impurity elements on cobalt Fischer–Tropsch catalyst performance including *in situ* sulphur and nitrogen addition. *J. Catal.* **2011**, *279*, 163–173.

38. Balonek, C.M.; Rane, A.H.L.S.; Schmidt, E.R.L.; Holmen, A. Effect of alkali metal impurities on Co-Re catalysts for Fischer-Tropsch synthesis from biomass-derived syngas. *Catal. Lett.* **2010**, *138*, 8–13.

39. Lillebø, A.H. Conversion of biomass derived synthesis gas into liquid fuels via the Fischer-Tropsch synthesis process: Effect of alkali and alkaline earth metal impurities and CO conversion levels on cobalt based catalysts. Ph.D. Thesis, NTNU, Norway, September 2014.

40. Van Berge, P.J.; van de Loosdrecht, J.; Barradas, S. Method of treating an untreated catalyst support and forming a catalyst precursor and catalyst from the teated support. U.S. Patent 6875720 B2, October 2005.

41. Rytter, E.; Eri, S.; Schanke, D.; Wigum, H.; Skagseth, T.H.; Borg, Ø.; Bergene, E. Development of an Attrition Resistant Fischer–Tropsch Catalyst for Slurry Operation. *Topics Catal.* **2011**, *54*, 801–810.

42. Behrmann, W.C.; Davis, S.M.; Mauldin, C.H. Catalyst preparation. Eur. Patent 535790 A1, August 1992.

43. Rytter, E.; Eri, S.; Schanke, D.; Wigum, H.; Skagseth, T.H.; Sincadu, N.; Bergene, E. Proceedings of 9th Novel Gas Conversion Symposium, Lyon, France, 30 May–3 June 2010.

44. Rytter, E.; Schanke, D.; Eri, S.; Wigum, H.; Skagseth, T.H.; Lian, P.; Sincadu, N. Optimization of Statoil's Fischer-Tropsch Co/Re/alumina catalyst. *Prep. ACS Div. Petro. Chem.* **2004**, *49*, 182–183.

45. Tsakoumis, N.E.; Dehghan-Niri, R.; Rønning, M.; Walmsley, J.C.; Borg, Ø.; Rytter, E.; Holmen, A. X-ray absorption, X-ray diffraction and electron microscopy study of spent cobalt based catalyst in semi-commercial scale Fischer–Tropsch synthesis. *Appl. Catal. A* **2014**, *479*, 59–89.

46. Bartholomew, C.H. Sintering kinetics of supported metals: New perspectives from a unifying GPLE treatment. *Appl. Catal. A* **1993**, *107*, 1–57.

47. Lee, D.; Lee, J.-H.; Ihm, A.-K. Effect of Carbon Deposits on Carbon Monoxide Hydrogenation over Alumina Supported Cobalt Catalyst. *Appl. Catal.* **1988**, *36*, 199–207.

48. Swart, J.C.W.; van Steen, E.; Ciobícã, I.M.; van Santen, R.A. Interaction of graphene with FCC-Co(111). *Phys. Chem. Chem. Phys.* **2009**, *11*, 803–807.

49. Tan, K.F.; Xu, J.; Chang, J.; Borgna, A.; Saeys, M.A. Carbon deposition on Co catalysts during Fischer–Tropsch synthesis: A computational and experimental study. *J. Catal.* **2010**, *274*, 121–129

50. Weststrate, C.J.; Kızılkaya, A.C.; Rossen, E.T.R.; Verhoeven, M.W.G.M.; Ciobîcă, I.M.; Saib, A.M.; Niemantsverdriet, J.W. Atomic and Polymeric Carbon on Co(0001): Surface Reconstruction, Graphene Formation, and Catalyst Poisoning. *J. Phys. Chem. C* **2012**, *116*, 11575–11583.

51. Gruver, V.; Young, R.; Engman, J.; Robota, H.J. The Role of Accumulated Carbon in Deactivating Cobalt Catalysts during FT Synthesis in a Slurry-Bubble-Column Reactor. *Prep. Am. Chem. Soc., Div. Petrol. Chem.* **2005**, *50*, 164–169.

52. Freide, J.J.H.M.F.; Gamlin, T.D.; Hensman, J.R.; Nay, B.; Sharp, C. Deactivation of Cobalt-Based Catalyst in the F-T Process. *J. Nat. Gas. Chem.* **2004**, *13*, 1–9.

53. Schanke, D.; Hilmen, A.M.; Bergene, E.; Kinnari, K.; Rytter, E.; Ådnanes, E.; Holmen, A. Study of the deactivation mechanism of Al_2O_3-supported cobalt Fischer-Tropsch catalyst. *Catal. Lett.* **1995**, *34*, 269–284.

54. Li, J.; Zhan, X.; Zhang, Y.; Jacobs, G.; Das, T.; Davis, B.H. Fischer–Tropsch synthesis: effect of water on the deactivation of Pt promoted Co/Al_2O_3 catalysts. *Appl. Catal. A* **2002**, *228*, 203–212.

55. Lancelot, C.; Ordomsky, V.V.; Stéphan, O.; Sadeqzadeh, M.; Karaca, H.; Lecroix, M.; Curulla-Ferré, D.; Luck, F.; Fongarland, P.; Griboval-Constant, A.; Kodakov, A.Y. Direct Evidence of Surface Oxidation of Cobalt Nanoparticles in Alumina-Supported Catalysts for Fischer-Tropsch Synthesis. *ACS Catal.* **2014**, *4*, 4510–4515.

56. Jermwongratanachai, T.; Jacobs, G.; Shafer, W.D.; Ma, W.; Pendyala, V.R.R.; Davis, B.H.; Kitiyanan, B.; Khalid, S.; Cronauer, D.C.; Kropf, A.J.; Marshall, C.L. Fischer–Tropsch Synthesis: Oxidation of a Fraction of Cobalt Crystallites in Research Catalysts at the Onset of FT at Partial Pressures Mimicking 50% CO Conversion. *Top. Catal.* **2014**, *57*, 479–490.

57. Kliewer, C.E.; Kiss, G.; DeMartin, G.J. Ex situ transmission electron microscopy: A fixed-bed approach. *Microsc. Microanal.* **2006**, *12*, 135–144.

58. Kubaschewski, O.; Alcock, C.B. *Metallurgical Thermochemistry*, 5th ed.; Pergaman: New York, NY, USA, 1979.

59. Rytter, E.; Schanke, D.; Eri, S.; Wigum, H.; Skagseth, T.H.; Bergene, E. Some QA and optimization issues during development of the Statoil FT-catalyst. Available online: http://www.globalspec.com/reference/43550/203279/qa-and-optimization-issues-during-development-of-the-statoil-ft-catalyst (accessed on 24 February 2015).

60. Kiss, G.; Kliever, C.E.; DeMartin, G.J.; Culross, C.C.; Baumgartner, J.E. Hydrothermal deactivation of silica-supported cobalt catalysts in Fischer–Tropsch synthesis. *J. Catal.* **2003**, *217*, 127–140.

61. Eri, S.; Goodwin, J.G., Jr.; Marcelin, G.; Riis, T. Catalyst for Production of Hydrocarbons. U.S. Patent 4801573, January 1987.

62. Kobylinski, T.P.; Kibby, C.L.; Pannell, R.B.; Leigh, E.L. Synthesis Gas Conversion using ROR Activated Catalyst. Eur. Patent 253924, July 1988.

63. Ernst, B.; Bensaddik, A.; Hilaire, L.; Chaumette, P.; Kiennemann, A. Study on a cobalt silica catalyst during reduction and Fischer-Tropsch reaction: In situ EXAFS compared to XPS and XRD. *Catal. Today* **1998**, *39*, 329–341.

64. Arcuri, K.B.; LeViness, S.C. The Regeneration of Hydrocarbon Synthesis Catalyst—A Partial Review of the Related Art Published during 1930 to 1952. Available online: http://www.fischer-tropsch.org/primary_documents/presentations/AIChE%202003%20Spring %20National%20Meeting/Paper%2086a%20Arcuri%20Regeneration.pdf (accessed on 24 February 2015).

65. Weststrate, C.; Hauman, M.; Moodley, D.; Saib, A.; Steen, E.; Niemantsverdriet, J. Cobalt Fischer–Tropsch Catalyst Regeneration: The Crucial Role of the Kirkendall Effect for Cobalt Redispersion. *Topics Catal.* **2011**, *54*, 811–815.

66. Van de Loosdrecht, J.; Saib, A.M. Catalysts. Patent WO 2008/139407, November 2008.

67. Saib, A.M.; Gauché, J.L.; Weststrate, C.J.; Gibson, P.; Boshoff, J.H.; Moodley, D.J. Fundamental Science of Cobalt Catalyst Oxidation and Reduction Applied to the Development of a Commercial Fischer-Tropsch Regeneration Process. *Ind. Eng. Chem. Res.* **2014**, *53*, 1816–1824.

68. Van de Loosdrecht, J.; Booysen, W.A. Catalysts. U.S. Patent 2011/0245355, June 2011.

69. Bezemer, G.L.; Nkrumah, S.; Smits, J.T.M. Process for Regenerating a Catalyst. U.S. Patent 2012/0165417, December 2011.

70. Rytter, E.; Eri, S.; Skagseth, T.H.; Borg, Ø. Fischer-Tropsch Catalyst Regeneration. U.S. Patent 0210939, August 2013.

71. Soled, S.L.; Baumgartner, J.E.; Kiss, G. *In situ* catalyst regeneration/activation process. U.S. Patent 6900151, November 2003

72. Ono, H.; Nagayasu, Y.; Hayasaka, K. Method for Manufacturing a Regenerated Fischer-Tropsch Synthesis Catalyst and Hydrocarbon Manufacturing Method. U.S. Patent 2012/0322899, December 2012.

73. Wright, H.A.; Raje, A.P.; Espinoza, R.L. Pressure Swing Catalyst Reneration Procedure for Fischer-Tropsch Catalyst. U.S. Patent 6962947, November 2005.

74. Huang, J.-R.; Arcuri, K.; Agee, K.; Schubert, P.F. Process for Regenerating a Slurry Fischer-Tropsch Catalyst. U.S. Patent 6989403, April 2004.

75. LeViness, S. 245th ACS National Meeting, April 2013, New Orleans. Available online: http://www.velocys.com/arcv/press/ppt/AIChE%202013%20Presentation%2004–30–13%20- %20LeViness%20v1_rev1.pdf (accessed on 10 March 2105).

76. Bailey, S.; Gray, G.; Lok, M.C. Methods for the production of cobalt catalysts supported on silicon dioxide and their use. Patent WO 2001/062381, August 2001.

77. Iglesia, E.; Soled, S.L.; Fiato, R.A. Cobalt-ruthenium catalysts for Fischer-Tropsch synthesis and process for their preparation. U.S. Patent 4738948, April 1988.

78. Iglesia, E.; Soled, S.L.; Fiato, R.A.; Via, G.H. Bimetallic Synergy in Cobalt Ruthenium Fischer-Tropsch Synthesis Catalysts. *J. Catal.* **1993**, *143*, 345–368.

79. Daage, M.; Koveal, R.J.; Chang, M. Catalyst Regeneration. U.S. Patent 6800579, January 2002.

80. Minnie, O.R.; Ndifor, E.N.; Kavian, S.; Smith, B.; Knowles, R. Process for the regeneration of a Fischer-Tropsch Catalyst. Patent WO 2013/093423, June 2013.

Influence of Reduction Promoters on Stability of Cobalt/γ-Alumina Fischer-Tropsch Synthesis Catalysts

Gary Jacobs, Wenping Ma and Burtron H. Davis

Abstract: This focused review article underscores how metal reduction promoters can impact deactivation phenomena associated with cobalt Fischer-Tropsch synthesis catalysts. Promoters can exacerbate sintering if the additional cobalt metal clusters, formed as a result of the promoting effect, are in close proximity at the nanoscale to other cobalt particles on the surface. Recent efforts have shown that when promoters are used to facilitate the reduction of small crystallites with the aim of increasing surface Co^0 site densities (e.g., in research catalysts), ultra-small crystallites (e.g., <2–4.4 nm) formed are more susceptible to oxidation at high conversion relative to larger ones. The choice of promoter is important, as certain metals (e.g., Au) that promote cobalt oxide reduction can separate from cobalt during oxidation-reduction (regeneration) cycles. Finally, some elements have been identified to promote reduction but either poison the surface of Co^0 (e.g., Cu), or produce excessive light gas selectivity (e.g., Cu and Pd, or Au at high loading). Computational studies indicate that certain promoters may inhibit polymeric C formation by hindering C-C coupling.

Reprinted from *Catalysts*. Cite as: Jacobs, G.; Ma, W.; Davis, B.H. Influence of Reduction Promoters on Stability of Cobalt/γ-Alumina Fischer-Tropsch Synthesis Catalysts. *Catalysts* **2014**, *4*, 49-76.

1. Introduction

Fischer-Tropsch synthesis (FTS) making use of cobalt catalysts is the core of the gas-to-liquids (GTL) process [1,2]. Due to the high H_2/CO syngas ratio derived from reforming of natural gas, additional water gas shift is not required to adjust the ratio upward, and internal water-gas shift (WGS) activity is undesirable. This is one benefit of cobalt catalysts relative to iron catalysts for GTL, as the former typically possess low intrinsic WGS activity. Because cobalt is much more expensive than its iron counterpart, and because the reaction occurs on the surface, it is important to disperse the cobalt metal particles in order to improve usage efficiency.

A typical cost effective way to do so is to impregnate the pre-calcined support with a cobalt nitrate solution by wet or dry (incipient wetness) impregnation followed by drying, air calcination to decompose the cobalt nitrate precursor to cobalt oxide, and reduction (e.g., 10 h in hydrogen gas at 350 °C) to Co^0 crystallites (typically in the range of 5 to 20 nm). The surface of Co particles provides the catalytically active sites for Fischer-Tropsch synthesis.

However, with typical reduction of supported cobalt at low temperatures (e.g., 350–400 °C) appropriate for obtaining active small crystallites of 6–15 nm, a sizeable fraction (typically in the range of 15–70%) of the cobalt remains in the oxide form, mainly as CoO. The fraction of unreduced cobalt is larger for supports such as alumina which interact strongly with cobalt oxide and for low cobalt loadings, e.g., less than 10–15% on such supports. The extent of the interaction increases with decreasing loading of cobalt. At low loadings (e.g., < 5%), 60 to 80% of the cobalt is present as CoO strongly bound to the support surface, *i.e.*, a surface cobalt aluminate, $CoO*Al_2O_3$, which requires very high temperatures to reduce [3]. At high loadings (e.g., 15%–30%Co), cobalt will be present primarily as Co_3O_4, which reduces in two steps: $Co_3O_4 + H_2 = 3CoO + H_2O$ and $3CoO + 3CoO = Co + 3H_2O$, for which maximum rates of reduction occur at about 300–350 and 500–650 °C, respectively [4]. Thus, following reduction of an unpromoted 15–30% Co/alumina at 350–400 °C for 5–15 h, a significant fraction (30–60%) of CoO typically remains [5–8]. Since higher extents of reduction (80–90%) of cobalt are highly desirable, *i.e.*, correlated with higher activity on a per g catalyst basis, as well as improved C_5+ selectivity, there is considerable interest and widespread application of noble metal promoters, which facilitate the reduction of cobalt oxides and increase the surface density of cobalt active sites.

This article reviews a number of stability issues associated with the application of promoters for cobalt FTS catalysts. Examples are provided to demonstrate a number of considerations for selecting a noble metal for Co catalysts. The main point of the article is that each promoter has its own advantages and set of issues that must be addressed and, in some cases, still defined.

2. Results and Discussion

Figure 1 compares temperature programmed reduction (TPR) TPR profiles of a number of noble metal and Group IB-promoted 15%Co/γ-Al_2O_3 catalysts pertinent to this review. The commonly used promoters are Pt, Re, and Ru and the solid line profiles are at close to atomically equivalent loadings. Pt and Ru facilitate the reduction of both steps of cobalt oxide reduction, while Re catalyzes the reduction of primarily the second step. This has been explained by Re oxide reducing at a higher temperature than Pt and Ru and that a reduced form of the promoter is required to obtain the promoting effect [7]; however, further confirmation of this is needed as the oxidation state remains in question [9]. Both Pt and Re appear to be more effective at facilitating reduction relative to Ru, but higher loadings of Ru can be used to further facilitate reduction, as indicated by the dashed profile at 0.5%Ru loading. Similar trends were reported in the TPR peak locations in a recent investigation of promoter characteristics by Cook *et al.* [9], as shown in Figure 2.

Cu, Ag, and Au (Group IB) promoters are also effective at promoting the reduction of cobalt oxides, as described in Figure 1 (top) [10]. However, note that the loadings indicated by the solid lines are approximately three times higher on an atomic basis than those of the commonly used promoters (Pt, Re, Ru) shown in the lower part of the figure. The costs of Ag and Cu are, whether on a weight or atomic basis, much lower than any of the other promoters shown. Therefore, it was of interest to explore their ability to facilitate reduction at even higher loadings. Increasing the loading by a factor of 3.3 resulted in further and important shifts of both TPR peaks of cobalt oxide reduction to lower temperatures (Figure 1, top).

Figure 1. TPR profiles demonstrate the effectiveness of Cu for facilitating reduction of cobalt oxides. Curve labels: unpromoted 15%Co/Al$_2$O$_3$ (thick solid) and Cu-promoted with 0.033%Cu (thick dashed), 0.49%Cu (thin solid), and 1.63%Cu (thin dash-dotted) by weight. (Reproduced with permission from [7] and [10] Copyright 2002, 2009, Elsevier).

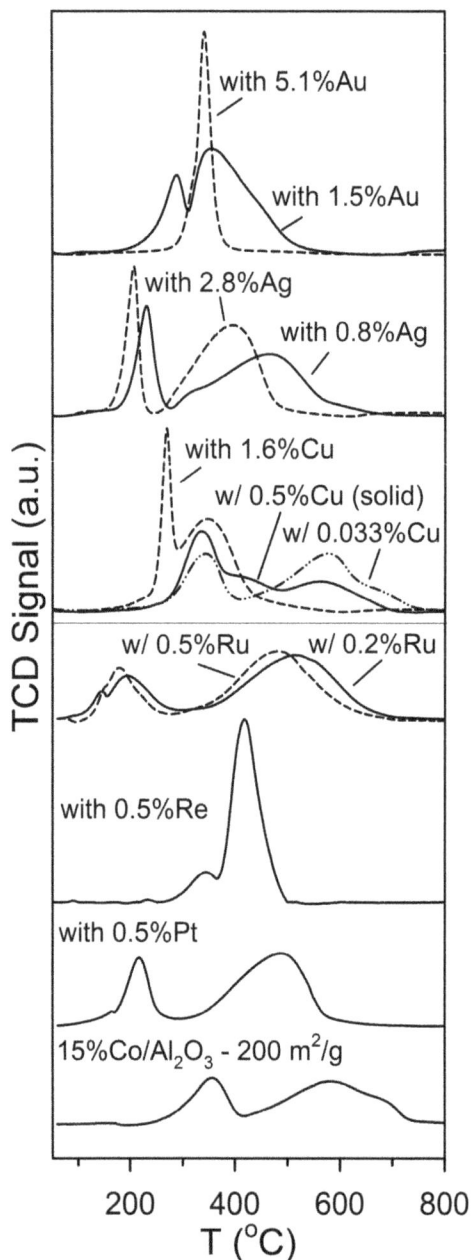

Figure 2. TPR profiles by Cook *et al.* [9] show that Pt and Re are more efficient than Ru in facilitating reduction of cobalt oxides over 25%Co/γ-Al₂O₃ catalysts with equivalent atomic loading (*i.e.*, noble metal / Co ratio was 0.007).

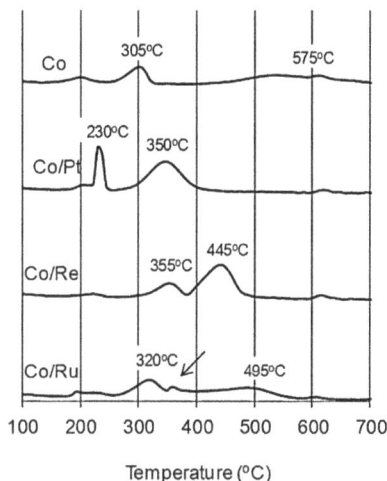

Temperature (°C)

The choice of promoter metal and its loading may influence the stability of cobalt catalysts in a number of ways. The first section examines how reduction promoter type and loading influence the activity and selectivity of cobalt catalysts, while the second section discusses how promoters may exacerbate deactivation rates through oxidation and a possible complex sintering mechanism, the two of which are not mutually exclusive. A brief summary of the application of computational methods is also provided, which discusses the location of promoter relative to cobalt, the resistance or sensitivity of cobalt to oxidation depending on size, and how promoters may hinder carbon formation pathways. Finally, in adding a second catalytic metal to the catalyst, the ability to regenerate the catalyst in a simple and effective manner becomes an important concern.

2.1. Influence of Promoter Choice and Loading on Catalyst Activity and Selectivity

2.1.1. Example #1—Copper

The first example demonstrates a relatively inexpensive metal that is highly effective for promoting the reduction of cobalt oxides: Cu, which is a common promoter in iron carbide FTS catalysts [11,12]. As of this writing, Cu is approximately 0.015% of the cost of Pt on a mass basis and would seem to be an ideal candidate as a promoter.

With increases in extent of reduction of cobalt (from 49.8% for 15%Co/Al₂O₃ to 53.2, 69.4, and 93.3% for 0.033%, 0.49%, and 1.63% Cu-promoted 15%Co/Al₂O₃ catalysts, respectively) the active metal site densities with Cu addition increased also, and hydrogen chemisorption uptakes measured by TPD increased from 72 μmol H_2/g_{cat} for the unpromoted 15%Co/Al₂O₃ catalyst to 82, 140, and 172 μm H_2/g_{cat} for the 0.033%, 0.49%, and 1.63% Cu-promoted 15%Co/Al₂O₃ catalysts [10]. However, the metal dispersions do not account for the partitioning of metal type on

156

the surface of Co particles, or the influence of the presence of Cu on the ensembles of Co required for conducting the synthesis. Surface enrichment by Cu has been detected in bimetallic Cu-Co catalysts before [13].

Table 1 shows a comparison of X_{CO} at the same weight hourly space velocity of two Cu-promoted 15%Co/Al$_2$O$_3$ catalysts relative to the unpromoted 15%Co/Al$_2$O$_3$ catalyst. Despite increases in metal site densities as measured by hydrogen TPD, a decrease in X_{CO} is observed, which is exacerbated at higher Cu loading [10]. These results suggest a poisoning of surface Co sites, likely due to enrichment of Cu at the surface. This finding is further supported by the changes in selectivity that occur when comparing the catalysts to the unpromoted catalyst at a similar conversion level. Table 2 shows that at the lower Cu promoter loading, the methane is slightly increased, C$_5$+ is slightly decreased [10]. However, increasing the Cu promoter loading further leads to a prohibitive increase in methane selectivity (21.6% versus 9.2%) and a precipitous drop in C$_5$+ selectivity (47.7% versus 81.6%) [10].

Table 1. X_{CO} for two Cu promoted 15%Co/Al$_2$O$_3$ catalysts at a SV of 4.2 NL/g$_{cat}$/h relative to the unpromoted catalyst. Conditions: 220 °C, 1.6 MPa, H$_2$/CO = 2.0 (adapted with permission from [10] Copyright 2009, Elsevier).

Catalyst	TOS (h)	X_{CO} (%)	SV (NL/g$_{cat}$/h)
15%Co/Al$_2$O$_3$	26–98	28.7	4.2
0.49%Cu-15%Co/Al$_2$O$_3$	30–99	27.9	4.2
1.63%Cu-15%Co/Al$_2$O$_3$	25–104	14.2	4.2

Table 2. Two comparisons of product selectivity at similar X_{CO} levels for two Cu promoted 15%Co/Al$_2$O$_3$ catalysts relative to the unpromoted catalyst. Conditions: 220 °C, 1.6 MPa, H$_2$/CO = 2.0 (adapted with permission from [10] Copyright 2009, Elsevier).

Catalyst	X_{CO} (%)	SV (NL/g$_{cat}$/h)	S(CH$_4$)	S(C$_5$+)	S(CO$_2$)
15%Co/Al$_2$O$_3$	47.8	2.0	8.9	80.6	0.82
0.49%Cu-15%Co/Al$_2$O$_3$	50.6	1.7	9.9	76.6	0.83
15%Co/Al$_2$O$_3$	28.7	4.2	9.2	81.6	0.67
1.63%Cu-15%Co/Al$_2$O$_3$ *	29.9	1.0	21.6	47.7	1.51

* Due to the low activity of the 1.63%Cu promoted catalyst, a separate comparison was made at lower X_{CO}, as it was not possible to decrease SV further.

2.1.2. Example #2—Silver and Gold

A comparison between Ag and Au shows that, in the case of Ag promoted 15%Co/Al$_2$O$_3$, the catalyst achieves higher activity (Table 3) and C$_5$+ selectivity (Table 4) than the unpromoted catalyst at both high and low loadings of promoter. A 15% Co/Al$_2$O$_3$ catalyst promoted with 1.51% Au also performs better than the unpromoted catalyst with both an increase in productivity and a slight improvement in selectivities [10]. However, at a higher Au loading (5.05%) the catalyst performed poorly with a steep drop in productivity (from X_{CO} of 51.7 at 1.51%Au to an X_{CO} of 14.1 at 5.05%Au at SV = 4.2, Table 3) and adverse impacts on selectivity (C_5+ is 60.1% compared to 81.6% for the unpromoted catalyst at X_{CO} of ~28%, Table 4) [10]. Thus, noble metal loading of promoter is important, not only from the standpoint of cost.

In a recent detailed kinetic investigation [14], which was a collaboration between CAER and Texas A&M University in Qatar, modeling results point to the presence of two kinds of sites on the Co FTS catalyst for the production of methane—FTS sites from standard Anderson-Schulz-Flory kinetics and additional sites for methanation. The results of the Au promoted catalyst at the lower loading, and the Ag promoted catalysts at both low and high loadings, suggest that these Group IB promoters assist in either blocking methanation sites or controlling the relative surface fugacity of hydrogen relative to adsorbed CO and intermediates on the surface of the cobalt catalyst.

Table 3. X_{CO} for two Ag and Au-promoted 15%Co/Al$_2$O$_3$ catalysts at a SV of 4.2 NL/g$_{cat}$/h relative to to the unpromoted catalyst. Conditions: 220 °C, 1.6 MPa, H$_2$/CO = 2.0 (adapted with permission from [10] Copyright 2009, Elsevier).

Catalyst	TOS (h)	X_{CO} (%)	SV (NL/g$_{cat}$/h)
15%Co/Al$_2$O$_3$	26–98	28.7	4.2
1.51%Au-15%Co/Al$_2$O$_3$	26–57	51.7	4.2
5.05%Au-15%Co/Al$_2$O$_3$	30–84	14.1	4.2
0.83%Ag-15%Co/Al$_2$O$_3$	20–47	50.4	4.2
2.76%Ag-15%Co/Al$_2$O$_3$	22–92	46.9	4.2

Table 4. Product selectivities for Ag and Au-promoted 15%Co/Al$_2$O$_3$ catalysts at X_{CO} values comparable to the unpromoted catalyst. Conditions: 220 °C, 1.6 MPa, H$_2$/CO = 2.0 (adapted with permission from [10] Copyright 2009, Elsevier).

Catalyst	X_{CO} (%)	SV (NL/g$_{cat}$/h)	S(CH$_4$)	S(C$_5+$)	S(CO$_2$)
15%Co/Al$_2$O$_3$	47.8	2.0	8.9	80.6	0.82
1.51%Au-15%Co/Al$_2$O$_3$	50.0	4.2	8.0	83.7	0.83
0.83%Ag-15%Co/Al$_2$O$_3$	50.4	4.2	7.7	83.6	0.94
2.76%Ag-15%Co/Al$_2$O$_3$	46.9	4.2	7.6	85.0	0.87
15%Co/Al$_2$O$_3$	28.7	4.2	9.2	81.6	0.67
5.05%Au-15%Co/Al$_2$O$_3$*	27.1	1.0	18.0	60.1	1.68

* Due to the low activity of the 5.05%Au promoted catalyst, a separate comparison was made at lower X_{CO}, as it was not possible to decrease SV further.

2.1.3. Example #3—Common Promoters (Pt, Re, Ru)

Slight differences in selectivity can also be achieved with the commonly used reduction promoters, which are Pt, Re, and Ru [15], as compiled in Table 5. Ruthenium itself is catalytically active for the FTS reaction, and higher alpha values have been measured in the hydrocarbon distribution [16]. Therefore, it is not surprising that lower methane and higher C_5+ were achieved with a 0.26%Ru-25%Co/Al$_2$O$_3$ catalyst relative to an unpromoted one. Re promoter was also found to give slightly better selectivities, in agreement with the work of Borg *et al.* [17] (Table 6). However, an atomically equivalent amount of Pt slightly (though not prohibitively) worsened the selectivities, and an attempt to use Pd to replace Pt resulted in a significantly poorer product distribution.

Table 5. Product selectivities[*] for $25\%Co/Al_2O_3$ catalysts containing commonly used promoters (Pt, Re, Ru) or Pd at X_{CO} values comparable an unpromoted reference catalyst. Conditions: 220 °C, 2.2 MPa, $H_2/CO = 2.1$ (adapted with permission from [15] Copyright 2012, Elsevier).

Catalyst	X_{CO} (%)	SV (NL/g_{cat}/h)	$S(CH_4)$	$S(C_5+)$
$25\%Co/Al_2O_3$	49.4	4.3	7.9	83.4
$0.26\%Ru$-$25\%Co/Al_2O_3$	51.3	7.6	7.0	86.8
$0.48\%Re$-$25\%Co/Al_2O_3$	49.6	8.0	7.2	86.0
$0.50\%Pt$-$25\%Co/Al_2O_3$	48.0	5.6	8.3	83.0
$0.27\%Pd$-$25\%Co/Al_2O_3$	50.3	4.9	11.5	75.9

[*] $S(CO_2)$ ranged from 0.35–0.75% in all catalysts. All data taken within first 81 h on-stream.

Table 6. Product selectivities from data taken at 210 °C, 2.0 MPa, $H_2/CO = 2$, and X_{CO} of 43–44% (adapted with permission from [17] Copyright 2009, Elsevier) over $20\%Co/\gamma$-Al_2O_3 catalysts without or with 0.5%Re using narrow pore (7.4 nm), medium pore (12.3 nm), and wide pore (16.7 nm) supports.

Catalyst	Hydrocarbon selectivity (%)		
	C_1	C_2-C_4	C_5+
Co/NPA	9.0	9.9	81.1
CoRe/NPA	8.8	9.5	81.7
Co/MPA	8.6	8.7	82.8
CoRe/MPA	8.4	8.3	83.4
Co/WPA	8.0	7.5	84.5
CoRe/WPA	8.0	7.2	84.9

In terms of differences in catalyst structure, the three commonly used promoters have, in a number of cases, been observed to be in atomic contact with Co (e.g., as an alloy), with no presence of promoter-promoter coordination at relatively low loadings (Re [18–20], Ru [21], Pt [22–24]). This is not always the case (e.g., Ru [9,16]), indicating that loading and preparation method are also factors that influence coordination environment. Pretreatment is also a factor. For example, Iglesia *et al.* [25] noted with Ru-Co/TiO$_2$ catalysts that, with increasing calcination temperature, total coordination of Ru with neighbors increases to suggest sintering, but that most of the increase in coordination is due to Ru taking on coordination with Co; thus, calcination promoted mixing of the two metals.

Moreover, Chonco *et al.* [26] have recently demonstrated with physical mixtures of Pt/Al$_2$O$_3$ and Co/Al$_2$O$_3$ that atomic coordination of the promoter to cobalt is not always required to obtain a reduction promoting effect. In our work, unlike the Re, Ru, and Pt promoted Co/alumina catalysts at low promoter loadings, Pd promoter exhibited some promoter-promoter (*i.e.*, Pd-Pd) coordination, suggesting the presence of well dispersed islands of Pd that likely gave rise to excessive hydrogenation activity [15] and rapid deactivation relative to the other three promoted catalysts [15].

Considering commercial research, a patent by Sasol researchers [1] examined Ru and Re promoters (of a catalyst containing 30 g Co and 100 g Al_2O_3) at 0.41 g and 3.0 g levels, respectively, *versus* a catalyst containing just 0.05 g of Pt and reported slightly higher productivity with the Pt promoted catalyst (0.349 kg HC/kg cat/h at X_{CO} = 87 vol.% with Pt *versus* 0.307 and 0.281 kg HC/kg cat/h for Ru and Re at X_{CO} = 77 and 70%, respectively). The conditions were 220 °C, 2.0 MPa, H_2/CO = 2/1; space velocity of 2.0 m_n^3/h/kg catalyst. Under the same conditions, similar productivity (~0.29 kg HC/kg cat/h at X_{CO} = 73 vol.%) was observed with a 0.28 g Ru as with 0.05 g Pt with catalysts of lower loading (20 g Co and 100 g Al_2O_3). The results appear to indicate that Pt is a very efficient promoter.

In a patent by Conoco researchers [27], the benefit of adding Re on selectivity was highlighted. Examples 27 through 31 in that patent compare 1%Re promoted 20%Co/Al_2O_3 catalysts with an unpromoted catalyst. Higher conversions (77–100%) and C_5+ productivities (240–270 g/h/kg_{cat}) for the Re promoted catalysts relative to the unpromoted catalyst (X_{CO} of 65% and C_5+ productivity of 170 g/h/kg_{cat}) were reported. In addition, improvements in selectivities were observed as well, including improvements in alpha (0.89–0.90 with Re promotion *versus* 0.88 for the unpromoted catalyst) and decreases in methane (9–15 wt.% for Re promoted *versus* 18 wt.% for the unpromoted catalyst). Many other examples of Re promotion only or in combination with other elements are also highlighted in the patent report.

2.1.4. Example #4—Impact of Loading for Pt and Ag Promoted Catalysts

The final example is provided to show that, just because the promoter forms promoter-promoter bonds (*i.e.*, as in the case of Pd described in the previous subsection), it should not immediately be ruled out. Ag, by itself, is a catalyst that is only weakly active for hydrogenation, and its addition as a promoter does result in significant Ag-Ag coordination, but the resulting activity and selectivity of the Co catalyst is improved. Figure 3 (left) compares the Ag K-edge EXAFS Fourier transform magnitude spectra of Ag-promoted 25%Co/Al_2O_3 catalysts as a function of Ag loading. While a lower distance peak for Ag-Co coordination is suggested (and confirmed by EXAFS fittings), with increasing loading of Ag, the general trend in EXAFS fittings for Ag-promoted 25%Co/Al_2O_3 catalysts was increasing Ag-Ag coordination (higher distance peak), such that the N_{Ag-Co}/N_{Ag-Ag} ratio decreased from 0.59 at 0.276%Ag loading to 0.16 at 2.76%Ag loading [28].

Figure 3 (right) compares the Pt L_{III}-edge EXAFS Fourier transform magnitude spectra of Pt-promoted 25%Co/Al_2O_3 catalysts. A single low distance peak indicates primarily Pt-Co coordination, with no visible Pt-Pt coordination being evident. As shown in Table 7, as loading was increased for the Pt promoted catalyst, a slight negative impact on selectivity occurred, with a slight decrease in C_5+ and a slight increase in the WGS rate [28]. For all Pt-promoted catalysts, slightly lower C_5+ selectivities were observed compared to the unpromoted catalyst. With the Ag-promoted catalysts, C_1 and C_5+ selectivities were slightly improved at all loadings [28]. Thus, the presence of a weakly hydrogenating metal [29], Ag, did not adversely affect selectivity to a significant degree, even when excessive amounts of promoter were added [28].

In summary, the above examples show that (1) metals that facilitate reduction of cobalt oxides do not automatically increase X_{CO} on a per gram of catalyst basis; (2) type and loading of promoter influence activity and selectivity such that a metal that may promote X_{CO} (on a per g basis) at lower loading may or may not poison or adversely impact surface fugacities (and selectivity) at higher loadings; and (3) the intrinsic hydrogenation rate of the promoter is an important factor to consider, as it may adversely or beneficially influence selectivity.

2.2. Influence of Promoter Addition on Oxidation and Complex Sintering of Cobalt

2.2.1. Reoxidation of Small Cobalt Crystallites at the Onset of FTS at Realistic Conversions

The primary aims of adding a metal promoters are to (1) lower reduction temperature thereby increasing extent of reduction to Co metal and (2) boost active site densities by facilitating the reduction of cobalt oxide crystallites in strong interaction with the alumina support, such that clusters of cobalt metal crystallites can be formed to provide the active surface for carrying out the FTS reaction. Thus, when a promoter is added, the additional gain in active site density will be due in large part to the reduction of smaller cobalt oxide species having stronger interactions with the support. Depending on the loading of cobalt and method of preparation, if smaller cobalt metal crystallites (e.g., <2–4.4 nm [30,31]) within cobalt clusters are formed, they may be susceptible to reoxidation [30,32]. Some researchers have recently indicated that Co clusters less than 6–8 nm have lower intrinsic activity [33,34]. Additional investigations are needed in this area. At commercially relevant FTS conditions, a problem was identified by us in defining intrinsic activity; as chemisorption is conducted on freshly activated catalysts, any oxidation of such small Co clusters that occurs at commercially relevant conditions can mask a measurement of intrinsic activity at the level of the active site [35]. Therefore, it is important to take into account the oxidation state of Co in the working FTS catalyst.

To probe the role of oxidation, a recent XANES study was made whereby freshly reduced unpromoted and Pt-promoted cobalt/alumina catalysts were subjected to $H_2:CO:H_2O$ mixtures typical of the 50% conversion condition of a slurry phase reactor [36] for one hour. A lower-than-commercial loading of 10% cobalt was utilized in order to favor the formation of small cobalt crystallites after activation that fall in the range of being susceptible to reoxidation. The average cobalt cluster size (*i.e.*, cluster of crystallites) was ~5 nm [36]. Even though the 10%Co/Al$_2$O$_3$ catalyst was reduced at 550 °C as opposed to 400 °C for the 0.5%Pt-10%Co/Al$_2$O$_3$ catalyst, the extent of reduction from XANES indicated that the Pt-promoted catalyst had a higher extent of reduction, as defined by the intensity of the white line. When switching to conditions to mimic 50% conversion, the white line intensity in the XANES spectra of both catalysts increased significantly (Figure 4), but the change was more severe for the Pt-promoted catalyst [36]. This reoxidation occurred rapidly, is confined to a fraction of cobalt, and is not associated with the initial decay period commonly observed in FTS reaction tests, which may take on the order of days to establish. Reoxidation of the small cobalt crystallites (<2–4.4 nm) has been recently verified by in-situ XRD and magnetometer investigations [37].

Figure 3. (left) Ag promoted 25%Co/Al₂O₃ catalysts displayed EXAFS peaks that could only be fitted well by including both Ag-Co and Ag-Ag coordination; (right) Pt promoted 25%Co/Al₂O₃ catalysts displayed a single peak in the first coordination shell that could be fitted well by only including Pt-Co coordination (adapted with permission from [28] Copyright 2013, Elsevier).

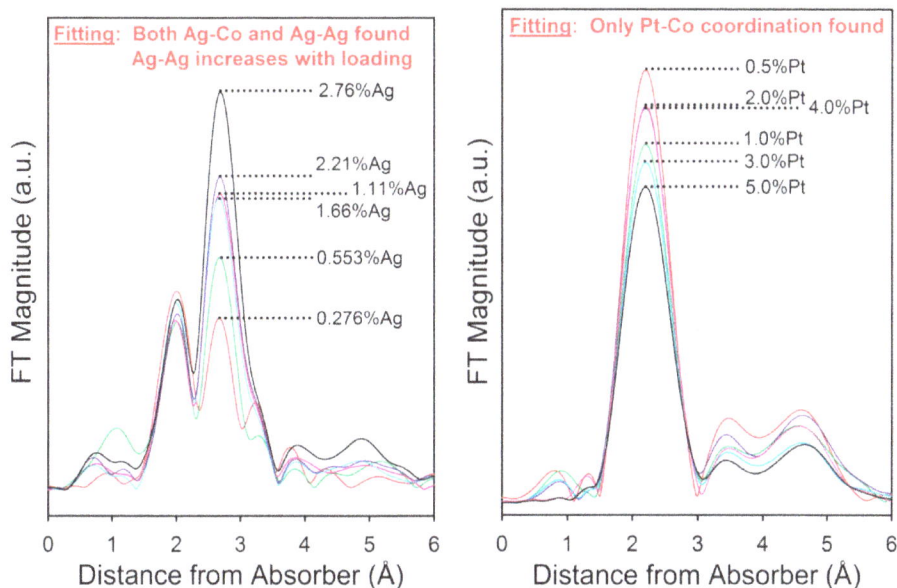

Table 7. Product selectivity * of 25%Co/Al₂O₃ catalysts having different loadings of Ag and Pt (adapted with permission from [28] Copyright 2013, Elsevier).

Catalyst	X_{CO} (%)	SV (NL/g$_{cat}$/h)	S(CH$_4$)	S(C$_5$+)	S(CO$_2$)
25%Co/Al₂O₃	51.0	3.4–4.2	8.3	82.5	0.8
0.5%Pt-25%Co/Al₂O₃	52.0	1.7–12	9.1	81.2	1.1
2.0%Pt-25%Co/Al₂O₃	45.0	9.0–12	9.1	81.9	1.1
5.0%Pt-25%Co/Al₂O₃	52.5	10–16	9.5	80.7	3.2
0.276%Ag-25%Co/Al₂O₃	46.4	8.8–12	7.4	84.1	0.4
1.11 %Ag-25%Co/Al₂O₃	48.1	9.3–12	7.3	83.7	0.4
2.76%Ag-25%Co/Al₂O₃	44.5	7.0–12	7.6	84.1	0.6

* All data taken within first 58 h on-stream. T = 220 °C; P = 2.2 MPa; H₂/CO = 2.1.

An extreme case in terms of small Co cluster size was also recently examined where Co particles were infused in the pores of KL-zeolite support by a CVD method [35] to produce a 0.5%Pt-5%Co/KL catalyst with 1 nm cobalt particles. The catalyst, following activation in hydrogen, exhibited an extent of reduction of 75%. However, after exposure to FTS conditions (220 °C, 1.8 MPa, H₂/CO 1.95, SV of 3.0 NL/g$_{cath}$), the extent of reduction fell to 33%, as measured by XANES spectroscopy. A loss in Co-Co coordination and growth of Co-O coordination was quantified by EXAFS. X_{CO} fell to 3.0%, and the resulting product selectivities were very poor (CH_4 selectivity of 29.6%, C_5+ selectivity of 49.8%, and CO_2 selectivity of 3.4%).

Figure 4. (**left**) Pt promoted and (**right**) unpromoted 10%Co/Al$_2$O$_3$ catalyst after (solid) hydrogen activation (550 °C for unpromoted and 400 °C for Pt-promoted) and (dashed) after exposure to H$_2$, CO, and H$_2$O partial pressure ratios mimicking X_{CO} = 50% at 20.7 bar (adapted with permission from [36] Copyright 2014, Springer).

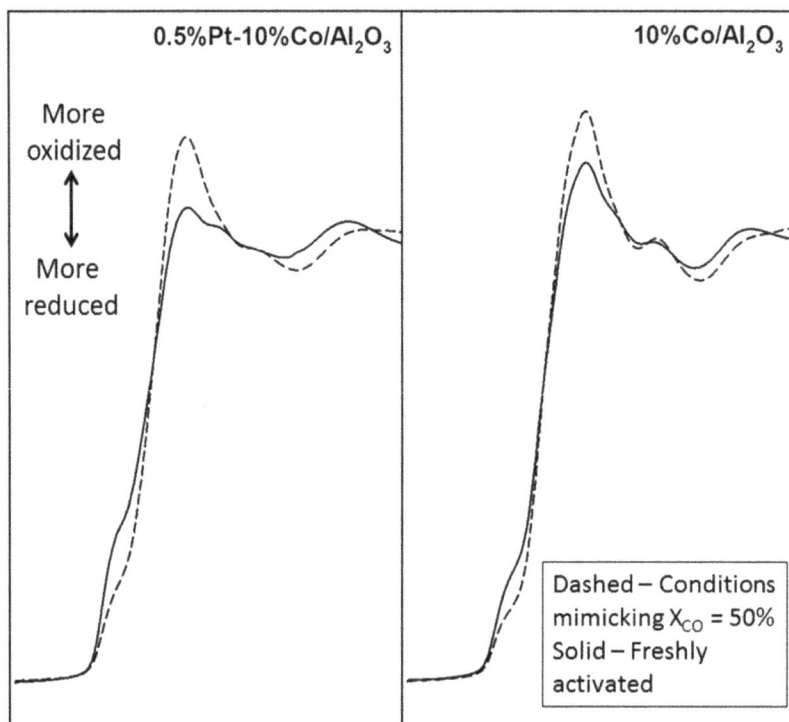

2.2.2. Sintering and Co Support Compound Formation during Initial Deactivation Period Prior to Leveling-off Period

The mechanisms of sintering during FTS on Co catalysts during typical commercial operation are not well understood and hence will not be discussed in detail in this review. There is evidence to indicate that H$_2$O accelerates sintering of metallic particles [38–41]; on this aspect, a mechanism involving surface oxidation, coalescence, and formation of larger clusters has been postulated by Sadeqzadeh *et al.* [39]. Further research is needed in this area.

Unreduced cobalt oxide in the working FT catalyst could be problematic since it may coalesce, reduce, and provide a driving force for the sintering of cobalt metal particles; this leads to net reduction with time onstream in the initial catalyst decay period (*i.e.*, which follows the onset period of ~1 h as described in the previous section) [32,42]. Moreover, small cobalt oxide species can react with the support and contribute to the formation of cobalt aluminates [42,43]. Since a promoter can facilitate the reduction of smaller species that may, depending on size, be susceptible to reoxidation at high conversion (as previously described), any cobalt oxide formed at the onset can contribute to either net reduction/sintering with time on-stream or cobalt aluminate formation. Net reduction and changes in Co-Co coordination consistent with sintering were observed for a 0.2%Re-15%Co/Al2O3 research catalyst at CAER during 2000 h of operation (Figure 5) [44,45]. Sintering, *i.e.*, growth of Co metal crystallites, and carbon formation were observed by Saib *et al.* [31] during a 56-day commercial test of a 0.05%Pt/20%Co catalyst. Sintering was determined to be rapid, reaching completion within the initial 15–16 days based on analysis by synchrotron-XRD of samples withdrawn from the reactor during this period; an increase in average crystallite diameter from 9 to 14 nm was observed. Formation in the catalyst of unreactive surface carbons, which restructured or poisoned the catalysts, occurred relatively more slowly over the 56-day period. Carbon deposits were analyzed by TPR during the latter part of this run [46]. During a second 140-day run [32], extent of reduction (EOR) was determined by periodically removing catalyst samples and analyzing them by XANES (Figure 6). EOR was determined to increase from 53 to 89% over the 140-day run. A small amount of cobalt aluminate formation was also observed, as it was detected in a used commercial 20%Co/0.05%Pt/Al2O3 catalyst [31]. Cobalt aluminate was also observed in used 0.2%Re-15%Co/Al2O3 research catalyst samples [44,47] (Figure 7).

Figure 5. (left) XANES and **(right)** EXAFS spectra as a function of time for a 0.2%Re-15%Co/Al2O3 catalyst. T = 220 °C, 2.0 MPa, SV = 5 SL/h/gcat, H2/CO = 2:1. Adapted with permission from [44] (Copyright 2003, Elsevier) and [45] (2006).

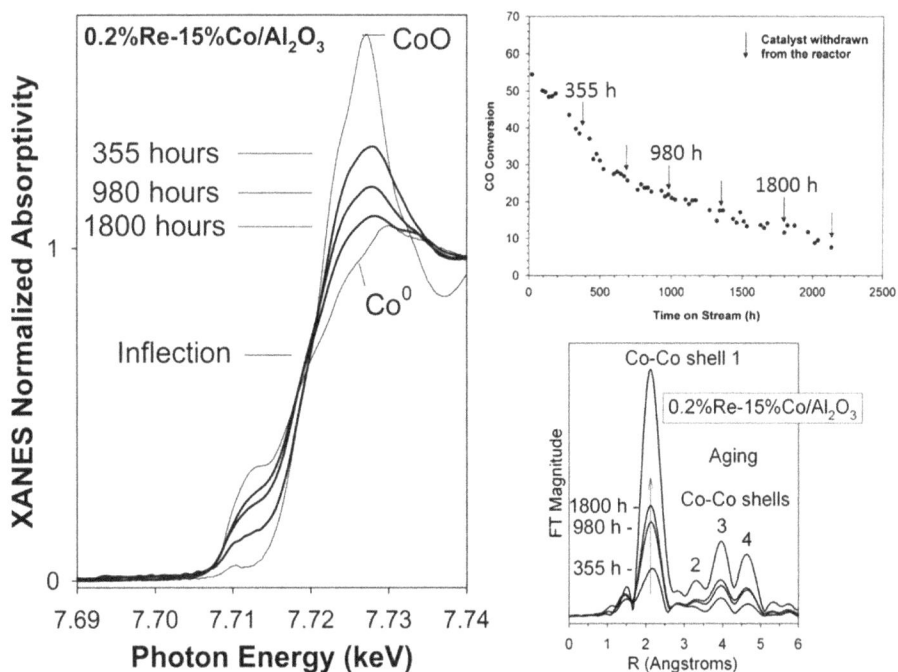

The impact of the promoter on cobalt aluminate formation from initially reduced small Co crystallites is difficult to assess, because unpromoted cobalt/alumina catalysts contain more residual unreduced CoO after activation, and this residual oxide, which is inactive for FTS, can also react with the support to form cobalt aluminate. Thus, there is a question as to how much cobalt aluminate is formed from the oxidation of very small (e.g., <2 – 4.4 nm) crystallites of cobalt metal at the onset of CoO formation (e.g., within the first hour, as shown in Figure 7 [44,45]) and its subsequent reaction with alumina, and how much is formed from the reaction of residual CoO (*i.e.*, leftover due to incomplete reduction) with the support, as described by Sasol researchers [43]. The cobalt aluminate was detected by them in XANES derivative spectra, as shown in Figure 8. The results from the previous section suggest that the former is dependent on Co crystallite size and $P(H_2O)/P(syngas)$ ratio, which is in turn influenced by conversion level. In a recent kinetic investigation, excursions of a Ru-promoted 25%Co/Al_2O_3 catalyst (average cluster size of 5.0 nm) to high CO conversion levels (e.g., $X_{CO} > 75\%$) resulted in the oxidation of a fraction of cobalt clusters [21] (as demonstrated by changes in the lineshape of the XANES derivative spectra) and increases in CO_2 and CH_4 selectivities (Figure 9). The oxidized cobalt is active for WGS, and the additional H_2 produced therefrom tends to increase the C_1 product. Thus, there appears to be an unfavorable synergism in the selectivities of CO_2 and C_1 when this threshold is surpassed (Figure 9) [21].

Crystallite size sensitivity for cobalt aluminate formation was also suggested in water co-feeding studies. Although H_2O co-feeding can lead to improvements in activity and selectivity for certain cobalt catalysts [48], when a Pt promoter was utilized to facilitate the reduction of Co oxides in a 15%Co/Al_2O_3 catalyst (average cluster diameter = 5.6 nm), at 28 vol.% added H_2O the catalyst underwent significant cobalt aluminate formation, as demonstrated in Figure 10 (left) and (right) [49,50] along with catastrophic deactivation (75% drop in X_{CO}). An unpromoted 25%Co/Al_2O_3 catalyst with larger cluster size (11.8 nm average diameter) [40] was more resistant to this phenomenon. Thus, on the one hand, a promoter is very useful for boosting Co° site densities during activation when the support interaction with Co oxides is high. On the other hand, if the strongly interacting Co oxides are reduced and form tiny Co^0 crystallites on the surface, they are more sensitive to H_2O. Higher loadings can help to make the catalyst more robust by increasing cobalt size, a technique that has been implemented commercially, and thereby reoxidation and subsequent Co aluminate formation may be largely avoided. With a commercial catalyst stabilized against these processes, only up to ~3% cobalt aluminate was formed during realistic FTS conditions [31]. However, it should be noted that when exposed to 1.0 MPa H_2O an increase was observed to 10% cobalt aluminate [43]; thus, water co-feeding or operating at high conversions may have drawbacks, depending on catalyst type and conditions utilized. A schematic of the structural changes discussed for research catalysts, including reoxidation of tiny (<2–4.4 nm) cobalt crystallites at the startup of Fischer-Tropsch synthesis at industrially relevant conversions, is shown in Figure 11 [42].

Figure 6. XANES analysis of a series of used $Co/Pt/Al_2O_3$ catalyst samples retrieved from a 100 barrel/day slurry bubble column reactor operated at 220 °C, 2.0 MPa, $(H_2 + CO)$ and conversions between 50% and 70%, feed gas composition = 50 vol.% H_2 and 25 vol.% CO, $P(H_2O)/P(H_2) = 1-1.5$, $P(H_2O) = 0.4-0.6$ MPa. Reproduced with permission from [32]. Copyright (2006) Elsevier.

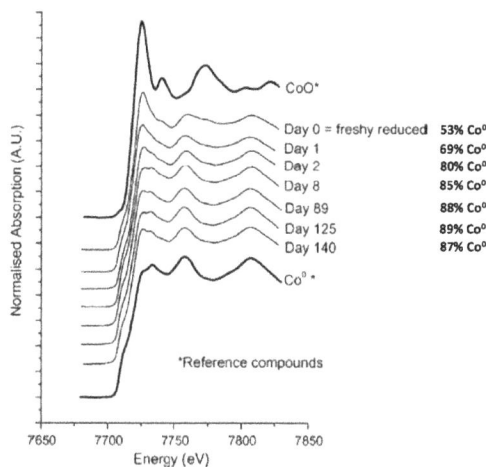

Figure 7. From the run shown in Figure 5, XANES derivative spectra of **(left)** freshly reduced/passivated catalysts, which could be fit with Co^0 and CoO, and **(right)** used catalyst samples, which could only be fitted with Co^0, CoO, and $CoAl_2O_4$. Adapted with permission from [44] (Copyright 2003, Elsevier) and [45] (2006).

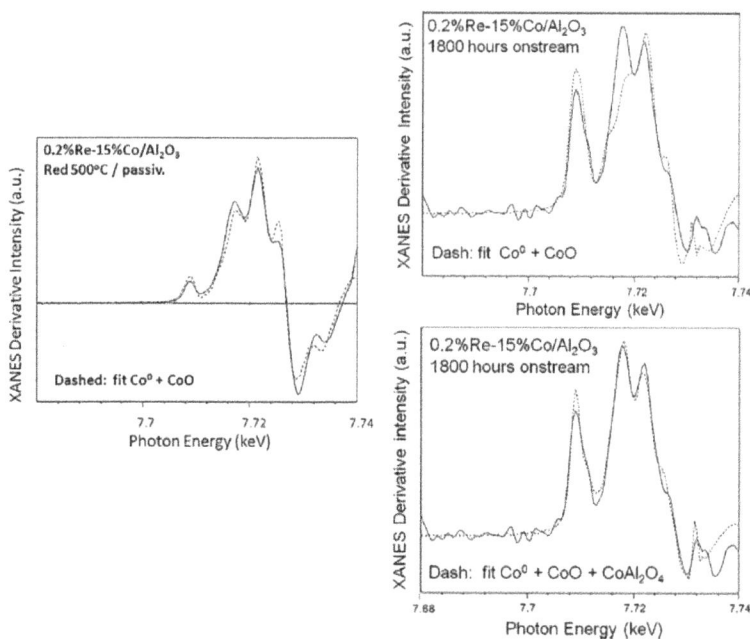

Figure 8. Formation of a minor cobalt aluminate component at 1.0 MPa H₂O by increasing conversion in a continuously stirred tank reactor (CSTR) reactor run at 230 °C, 2.0 MPa, 50 vol.% H₂, 25 vol.% CO and 25 vol.% inerts. Reproduced with permission from [43]. Copyright (2011) Elsevier.

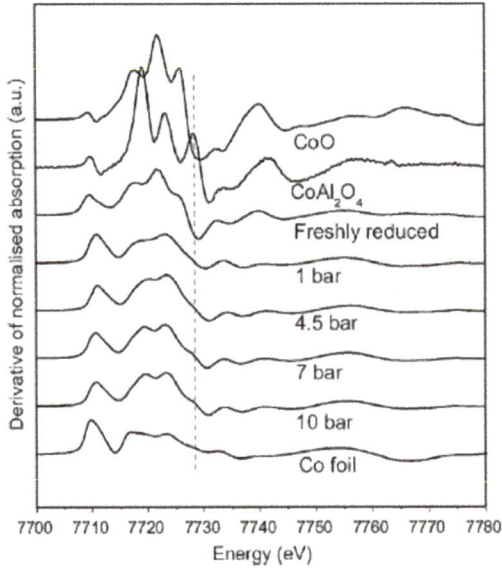

Figure 9. Changes in CO_2 and CH_4 selectivities as a function of conversion over 0.27%Ru-25%Co/Al₂O₃ catalyst (Co cluster size of ~5 nm by hydrogen chemisorption/pulse reoxidation) at 220 °C, 1.5 MPa, H₂/CO = 2.1 and SV = 0.3–15 NL/g$_{cat}$h (reproduced with permission from [21] Copyright 2011, Springer).

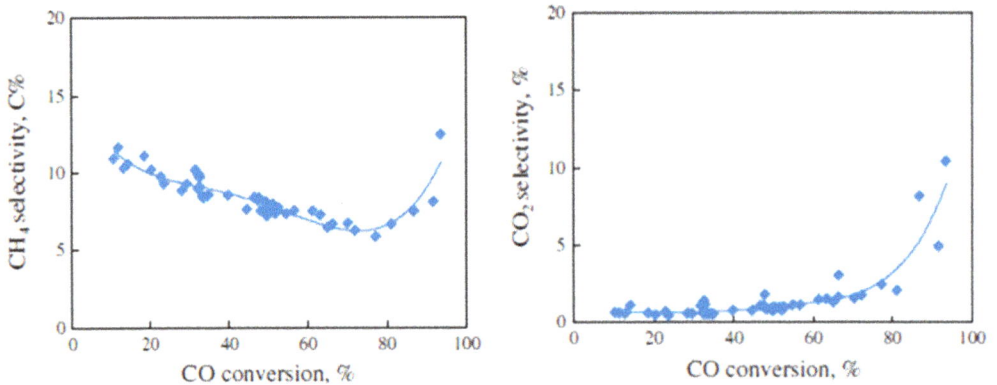

Figure 10. (left) Co-feeding of H_2O over 0.5%Pt-15%Co/Al_2O_3 at $T = 210$ °C, $P = 2.0$ MPa, H_2:CO = 2:1, SV = 8 SL/g_{cat}h reveals that irreversible deactivation (and a minor reversible effect) is observed at 28% H_2O addition; **(right)** XANES analysis of the used catalyst reveals formation of cobalt aluminate through reaction of the CoO formed with the support. Reprinted with permission from [40,49,50]. Copyright (2002, 2003, 2004) Elsevier.

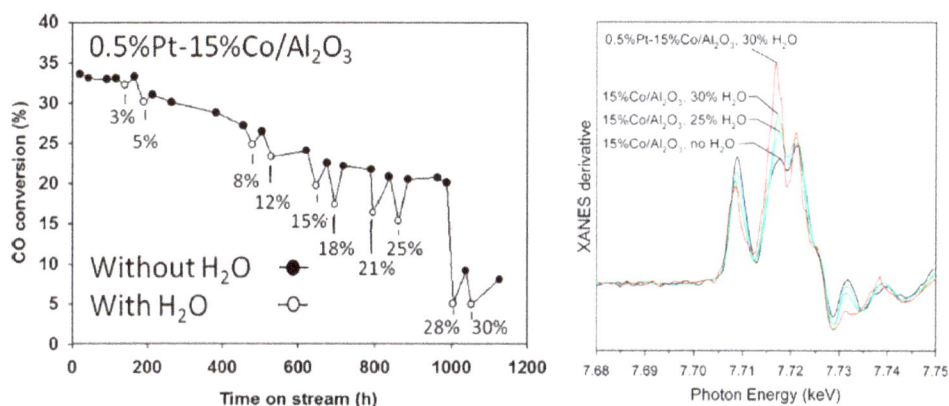

Figure 11. Proposed explanation for the deactivation of alumina-supported cobalt nanoparticles in research catalysts as a function of time on-stream. Adapted with permission from [42]. Copyright (2013) Elsevier [42]. The figure emphasizes why commercial catalysts adopt larger crystallite diameters for the purpose of stability.

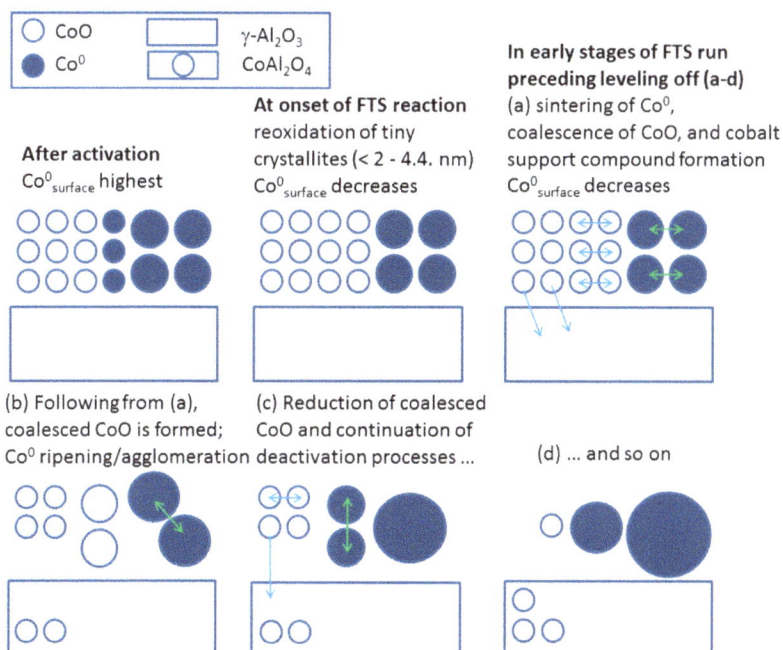

2.3. Regeneration

In recent years, there has been interest in carrying out oxidation-reduction cycles in order to explore the potential for regenerating metal promoted cobalt catalysts. In an earlier work [16], we simulated regeneration with 2%Ru-15%Co/alumina catalysts having relatively high Ru loadings. Part of the aim was to see if any promoter mixing or separation occurred at the atomic level, and the resulting influence on selectivity. To do so, TPR, EXAFS/XANES, and transmission electron microscopy with elemental mapping were applied. It was difficult to detect atomic mixing at the promoter loadings utilized, though elemental mapping showed that the metals were in close proximity to one another at the scale of nanometers. With reduction-oxidation cycles, TPR profiles revealed that while the first step of the TPR shifted to slightly higher temperatures, the second peak (*i.e.*, CoO to Co^0) shifted to slightly lower temperatures [16] for these heavily Ru-loaded catalysts.

Westrate *et al.* [51] compared Pt and Ru promoted Co/alumina catalysts subjected to oxidation-reduction treatments. Following oxidation treatment both the cobalt and promoter phases are well mixed and in an oxidized state. This finding is in agreement with our EXAFS results for a Ag-promoted Co/alumina catalyst [10]. Upon reduction, noble metals form bonds with cobalt metal, again in agreement with our earlier findings for Pt, Re, and Ru promoted catalysts at low promoter loadings [42]. During simulated regeneration by oxidation, the promoter separates from the cobalt phase and is found inside a ring of Co_3O_4, in "Kirkendall voids" as shown in Figure 12. Re-reduction of this state leads to a decrease in the promoter concentration at the surface of Co particles, as observed by XPS.

Figure 12. Pt promoter (oxide form) is separated from Co_3O_4 and found within Kirkendall voids within the Co_3O_4 particle. Reproduced with permission from [51]. Copyright (2013) Elsevier.

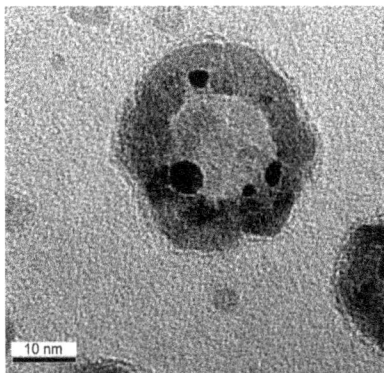

A key point regarding regeneration is whether or not the promoter continues to facilitate cobalt oxide reduction once an oxidation (e.g., carbon burn-off) step has been conducted. The proximity of the promoter to cobalt is important; although this will likely vary with promoter chemistry (see earlier comment regarding van Steen's group's use of physical mixtures to demonstrate that atomic contact may not be necessary in all cases [26]). To probe this attribute further, a TPR and XANES investigation was carried out to screen a number of promoter metals [52]. An example comparing two promoters—Pt and Au—is provided in Figure 13 to demonstrate the methodology. The oxidation-reduction cycle involved a 4 h calcination in flowing air at 350 °C followed by reduction for 10 h in hydrogen at 350 °C. It is evident (Figure 13) that Pt improves its effectiveness after three oxidation-reduction cycles, while Au is no longer effective after the first cycle. In the case of Au, the TPR profiles move to higher temperatures with increasing reduction-oxidation cycle number, suggesting that Au is separating from the Co (e.g., by sintering). At the same time, XANES shows that the oxidation state of Co following oxidation reduction moves toward higher oxidation state in the activated Au-promoted catalyst. The TPR profiles for Pt-promoted Co/alumina shift slightly to lower temperature. Corresponding XANES spectra demonstrate that the cobalt is largely reduced after activation following several simulated regeneration cycles. Thus, Pt is a more effective promoter for long-term use, although a different regeneration method at different conditions might be more effective in the case of Au. Both XANES and TPR data revealed that Re retains its ability to facilitate reduction even after 3 oxidation-reduction cycles. With Ru, the XANES results indicated that Ru was also effective after 3 ORcycles, although a slight shift to higher temperature was observed for the CoO to Co^0 TPR peak position in the preliminary study [52].

Figure 13. (Top) TPR profiles after RO cycles and **(Bottom)** XANES profiles after RO cycling demonstrate that Pt is more effective at continuing to facilitate reduction after simulate regeneration. Reproduced with permission from [52]. Copyright (2014) Elsevier.

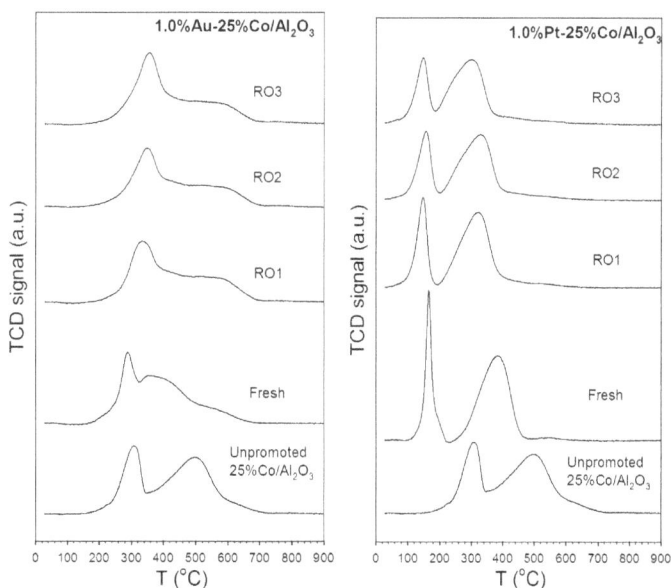

Figure 13. *Cont.*

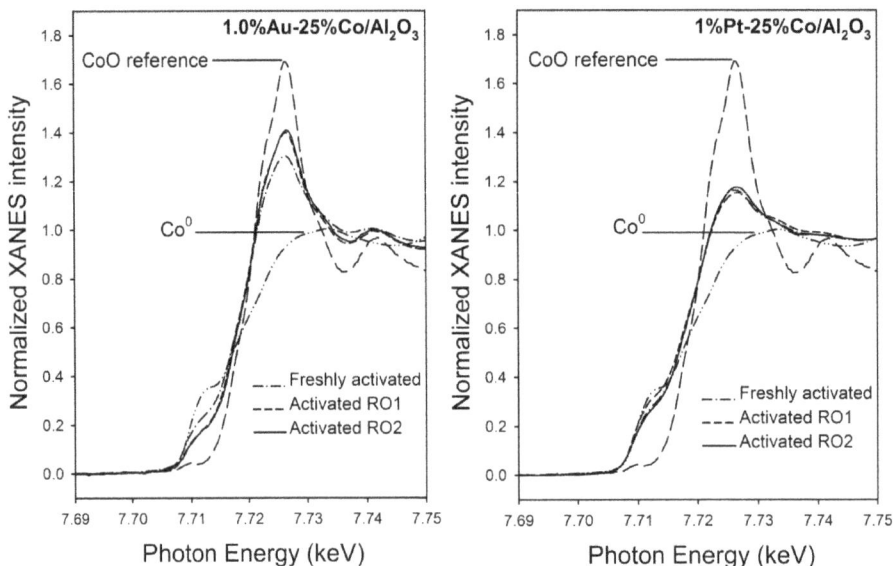

2.4. Modeling

2.4.1. Modeling of Site Suppression and Deactivation

Although this manuscript is focused on catalyst structure and its influence on FTS stability, a brief word should also be made regarding modeling. One puzzling aspect about the water effect during co-feeding and kinetic investigations is that it can be either positive or negative depending on the nature of the cobalt catalyst. Figure 10 displays results for a 0.5%Pt promoted 15%Co/Al$_2$O$_3$ catalyst having an average cobalt cluster size of 5.6 nm (*i.e.*, crystallites must be equal to below this value), and the water effect is negative but exhibits a significant degree of reversibility at levels below 28 vol.% added H$_2$O. The reversibility was, based on EXAFS/XANES investigations, suggested to be due to oxidation and re-reduction of small cobalt crystallites. Because oxygen was bound to the cobalt sites, the behavior could be modeled in terms of a kinetic parameter based on adsorption inhibition of reactants. On the other hand, Co/silica catalysts having larger cobalt clusters exhibited a positive effect [53], and positive effects on cobalt catalysts have been suggested to be due to water increasing the concentration of surface active carbon species (e.g., unsupported Co and Co/titania [38]) or removing heavy wax from catalyst pores leading to a higher available site density (e.g., Co/silica with varying pore size [54]). Returning to cobalt/alumina catalysts, interestingly, by aging the catalyst sufficiently (*i.e.*, the catalyst is significantly deactivated from its initial condition) [55] or utilizing catalysts with 10+ nm size [56], the deactivation rate becomes low, and the positive kinetic effect occurring on metallic cobalt particles can be observed. The main point is that the water effect can be modeled using a simple power law expression with a water effect parameter, *m*, and the magnitude and sign (*i.e.*, positive [57] or negative [55,57]) of *m* provides valuable information about the structure of the catalyst.

$$r(FT) = kP(CO)^{a}P(H_2)^{b}/[1 + m\,P(H_2O)/P(H_2)] \qquad (1)$$

A number of the phenomena associated with the stability and deactivation of cobalt catalysts have been discussed herein, and these and other aspects of stability (e.g., carbon deposition and carbide formation) are discussed in the Editor's book on Catalyst Deactivation [58] and a review article [59]. At this point in time, a number of important aspects and trends regarding the structure sensitivity of cobalt catalysts, and especially experimental research catalysts, are known. However, moving forward, there is a great deal of focus on developing catalysts that make the most efficient use of cobalt. Co crystallites should be small enough to maximize active site density, but also large enough that crystallites will be stabilized against reoxidation and sintering. Researchers are also exploring preparation methods to disperse and adequately position metal clusters in spatially favorable ways (*i.e.*, as far apart as possible). Some methods include freeze-drying [60], vapor phase impregnation [61], coating the support with carbon [62], using solvents such as ethylene glycol [63], optimizing drying temperature or calcination chemistry with dilute NO/N_2 [64], bypassing calcination altogether [53], or locking metal particles onto the support so that they cannot find one another and undergo agglomeration [65]. Thus, using advanced preparation methods to obtain spatially uniform distribution of cobalt crystallites is critical, since in conventional cobalt/alumina catalysts prepared by impregnation, cobalt clusters can be within close vicinity to one another, or in grapelike clusters, which have been described as "graveyards" for cobalt active sites during reaction testing [66]. Thus, thinking toward the future, it will be of increasing importance to model deactivation mechanisms and quantify how much each mechanism contributes to overall deactivation of the catalyst. Robust models which address the chemistry of promoters could enable the performance of new research catalysts to be compared to current commercial catalyst formulations. For example, a recently published forward thinking article by Argyle *et al.* [67] addresses modeling of the contributions of several deactivation mechanisms to overall deactivation rate in Co catalyzed FTS.

2.4.2. Computational Methods Based on First Principles

Related to promoters of FTS catalysts, computational methods based on first principles have been useful in describing the location of promoter with respect to cobalt, for providing insight into the requirements for cobalt oxidation by H_2O, and for determining how the promoter may influence carbon deposition.

Computational methods are being utilized to elucidate the preferential location for the occupancy of the promoter with respect to the cobalt atoms that make up the cluster. For example, a combined study making using of Low Energy Ion Scattering (*i.e.*, on a 1%Re-12%Co/Al$_2$O$_3$ catalyst) and computational DFT modeling (*i.e.*, on a Co$_{13}$Re cluster) determined that there is a preference of Re promoter to occupy subsurface sites, where it coordinates with a maximum number of cobalt atoms [68]. This is in agreement with the results of some EXAFS investigations, where direct Re-Co atomic contact has been observed [19,20].

Molecular modeling has also been conducted to examine surface oxidation of larger cobalt particles (e.g., as utilized in commercial catalysts) by H_2O, and the pathway was ruled out as a significant chemical transformation mechanism for deactivating sites under commercial FTS conditions [31].

Computational methods have also been used extensively to gain insight into the role of carbon deposits in catalyst deactivation. For example [69], HR-TEM and computational DFT modeling were applied to elucidate the preferred occupancies of carbon over different cobalt surfaces. The stability of various carbon species under reaction conditions was evaluated. Extended graphene islands and a surface carbide were found to be 99 and 79 kJ/mol more stable than surface CH_2 groups. Both carbon phases were suggested to initiate and grow from step sites. Saib *et al.* [31] have recently reviewed carbon formation in detail on cobalt FTS catalysts, including the application of computational methods. They also analyzed used catalysts from a commercial slurry bubble column reactor and, following wax extraction by THF, carried out temperature programmed hydrogenation and oxidation measurements to characterize the carbons. The least reactive species toward hydrogenation, which reacted at 430 °C, was assigned to polymeric carbons. A model [70] showing the location of small carbon oligomers of the fcc Co(111) surface, the precursors of polymeric carbon, was described. Note that polymeric carbon was found on both cobalt and the support. The authors [31] also reviewed the role of subsurface carbon [69,71], where theoretical modeling has indicated that subsurface carbon hinders CO adsorption and dissociation processes on associated Co atoms, and the requirements under which carbon induces clock surface reconstruction [72]. This, in turn, may cause deactivation of sites via shape changes or, on the other hand, induce the formation of active sites (e.g., B5 sites [31]: 3-fold sites that more easily dissociate CO; or triangular nanoscale islands having step edges similar to C7 sites [73], as observed by scanning tunneling microscopy). The restructuring of cobalt by strong CO chemisorption (*i.e.*, roughening—leading to more active sites) was described by Schulz *et al.* [74] as to be in competition with sintering.

Some computational studies have focused on defining how promoters of cobalt catalysts may impact carbon formation. Recently, the mechanisms for carbon compound formation on unpromoted and Pt or Ru promoted Co surfaces were investigated [75]. The activation energies for carbon-carbon and carbon-carbon-carbon coupling reactions were found to be larger on Pt or Ru promoted Co surfaces relative to unpromoted Co surface. The results suggest that carbon formation and thus, carbon compound (e.g., polymeric carbon formation, may be inhibited by the presence of the promoters. The authors also found that the promoters did not change the activation energy of C diffusion to the subsurface.

3. Experimental Section

Typical catalyst preparation method: the support used was Sasol Catalox-150 γ-Al_2O_3. It was first calcined at 400 °C in a muffle furnace for 4 h. A slurry impregnation method was performed, whereby the ratio of the volume of loading solution used to the weight of alumina was 1:1, such that approximately 2.5 times the pore volume of solution was used to prepare the catalyst in two steps [7]. Due to the solubility limit of cobalt nitrate, multiple impregnation steps were used. After each impregnation step, the catalyst was dried under vacuum in a rotary evaporator from 80 to 100 °C. Promoter precursors used were tetraammine palladium (II) nitrate, tetraammine platinum

(II) nitrate, rhenium oxide (Re_2O_7), ruthenium nitrosyl nitrate, silver nitrate, $HAuCl_4$, and copper nitrate. The promoters were added dropwise to achieve incipient wetness impregnation. After final drying at 80–100 °C, the final catalysts were calcined at 350 °C under flowing air for 4 h.

Typical CSTR reaction test: the catalyst (15 g) was ground and sieved to 170–325 mesh before loading into a fixed-bed reactor for 10–15 h of ex situ reduction at 350 °C and atmospheric pressure using a gas mixture of H_2/He with a molar ratio of 1:3. The reduced catalyst was then transferred to a 1-L continuously stirred tank reactor (CSTR) which was previously charged with 315 g of melted Polywax 3000, under the protection of a N_2 inert gas. The transferred catalyst was further reduced *in situ* at 230 °C at atmospheric pressure using pure hydrogen for another 10 h before starting the FT reaction. In this study, the FT conditions were 220 °C, 1.5–2.2 MPa, H_2/CO = 2.0–2.1. The space velocity varied between 1.0 and 16 NL/g-cat/h. in order to give about 50% CO conversion in different tests. This allowed us to fairly compare the differences in hydrocarbon selectivity data resulting from the promoter effect.

4. Conclusions

There are a number of stability issues that must be considered when selecting metal reduction promoters for use in Fischer-Tropsch synthesis catalysts. If tiny cobalt crystallites (<2–4.4 nm) are formed by facilitating the reduction of cobalt oxides that are strongly interacting with the support, they may undergo reoxidation at the onset of FTS at high conversion. Any cobalt oxide either left unreduced or formed from reoxidation of tiny cobalt crystallites can participate in a complex sintering mechanism that involves agglomeration of cobalt oxides, re-reduction, and sintering of the metal. Promoters can also exacerbate sintering if the cobalt metal clusters formed as a result of the promoting effect are in close proximity to other cobalt particles on the surface. Not all metals that facilitate cobalt reduction promote activity on a per gram catalyst basis; some will poison the surface (e.g., Cu). A poor choice of promoter (or poor choice in loading) can also lead to excessive hydrogenation activity and raise the light gas selectivity (e.g., Pd or Cu; Au at high loading). Furthermore, certain metals (e.g., Au) that promote cobalt oxide reduction can separate from cobalt during oxidation-reduction (regeneration) cycles. Therefore, they may not be effective for long-term use, or they may require non-standard regeneration treatments. Computational studies suggest that certain promoters (e.g., Pt or Ru) may hinder deactivation by carbon by increasing the energy barrier for carbon-carbon coupling reactions, while subsurface C formation was not found to be affected.

Acknowledgments

The authors would like to acknowledge the support of the Commonwealth of Kentucky. We would also like to thank the Editor for helpful comments.

Conflicts of Interest

The authors declare no conflict of interest.

References

1. Espinoza, R.L.; Visagie, J.L.; van Berge, P.J.; Bolder, F.H. Fischer-Tropsch catalysts containing iron and cobalt. US Patent No. 5,733,839, March 1998.

2. Van Berge, P.J.; Barradas, S.; van de Loosdrecht, J.; Visagie, J.L. Advances in the cobalt catalyzed Fischer-Tropsch synthesis. *Erdöl Erdgas Kohle* **2001**, *117*, 138–142.

3. Wang, W.-J.; Chen, Y.-W. Influence of metal loading on the reducibility and hydrogenation activity of cobalt/alumina catalysts. *Appl. Catal.* **1991**, *77*, 223–233.

4. Jacobs, G.; Ji, Y.; Davis, B.H.; Cronauer, D.C.; Kropf, A.J.; Marshall, C.L. Fischer-Tropsch synthesis: Temperature programmed EXAFS/XANES investigation of the influence of support type, cobalt loading, and noble metal promoter addition to the reduction behavior of cobalt oxide particles. *Appl. Catal. A* **2007**, *333*, 177–191.

5. Reuel, R.C.; Bartholomew, C.H. The stoichiometries of H_2 and CO adsorptions on cobalt: Effects of support and preparation. *J. Catal.* **1984**, *85*, 63–77.

6. Vada, S.; Hoff, A.; Adnanes, E.; Schanke, D.; Holmen, A. Fischer-Tropsch synthesis on supported cobalt catalysts promoted by platinum and rhenium. *Top. Catal.* **1995**, *2*, 155–162.

7. Jacobs, G.; Das, T.K.; Zhang, Y.-Q.; Li, J.; Racoillet, G.; Davis, B.H. Fischer-Tropsch synthesis: support, loading, and promoter effects on the reducibility of cobalt catalysts. *Appl. Catal. A* **2002**, *233*, 263–281.

8. Farrauto, R.J.; Bartholomew, C.H. *Fundamentals of Industrial Catalytic Processes*; John Wiley & Sons: Chichester, UK, 2003; p. 700.

9. Cook, K.M.; Poudyal, S.; Miller, J.T.; Bartholomew, C.H.; Hecker, W.C. Reducibility of alumina-supported cobalt Fischer-Tropsch catalysts: Effects of noble metal type, distribution, retention, chemical state, bonding, and influence on cobalt crystallite size. *Appl. Catal. A* **2012**, *449*, 69–80.

10. Jacobs, G.; Ribeiro, M.C.; Ma, W.; Ji, Y.; Khalid, S.; Sumodjo, P.T.A.; Davis, B.H. Group 11 (Cu, Ag, Au) promotion of 15%Co/Al_2O_3 Fischer-Tropsch synthesis catalysts. *Appl. Catal. A* **2009**, *361*, 137–151.

11. Aldossary, M.A.; Sharma, P.; Ojeda, M.; Gupta, M.; Fierro, J.L.; Spivey, J.J. Effect of different Cu loading on Fe-Mg catalyst for Fischer-Tropsch synthesis. In Proceedings of ACS National Meeting & Exposition, Philadelphia, PA, USA, August 19–23, 2012. Curran Associates, Inc.: Red Hook, NY, USA, 2012; Volume 57, p. 944.

12. Li, S.; Li, A.; Krishnamoorthy, S.; Iglesia, E. Effects of Zn, Cu, and K promoters on the structure and on the reduction, carburization, and catalytic behavior of iron-based Fischer-Tropsch synthesis catalysts. *Catal. Lett.* **2001**, *77*, 197–205.

13. Cesar, D.V.; Perez, C.A. Quantitative XPS analysis of bimetallic Cu-Co catalysts. *Phys. Status Solidi A* **2001**, *187*, 321–326.

14. Todic, B.; Bhatelia, T.; Froment, G.F.; Ma, W.; Jacobs, G.; Davis, B.H.; Bukur, D.B. Kinetic model of Fischer-Tropsch synthesis in a slurry reactor on Co/Re/Al_2O_3 catalyst. *Ind. Eng. Chem. Res.* **2013**, *52*, 669–679.

15. Jacobs, G.; Ma, W.; Gao, P.; Todic, B.; Bhatelia, T.; Bukur, D.B.; Khalid, S.; Davis, B.H. Fischer-Tropsch synthesis: differences observed in local atomic structure and selectivity with Pd compared to typical promoters (Pt, Re, Ru) of Co/Al$_2$O$_3$ catalysts. (Special Issue in honor of the late Prof. Laszlo Guczi 1932–2012). *Top. Catal.* **2012**, *55*, 811–817.

16. Jacobs, G.; Sarkar, A.; Ji, Y.; Luo, M.-S.; Dozier, A.; Davis, B.H. Fischer-Tropsch synthesis: assessment of the ripening of cobalt clusters and mixing between Co and Ru promoter via oxidation-reduction cycles over lower Co-loaded Ru-Co/Al$_2$O$_3$ catalysts. *Ind. Eng. Chem. Res.* **2008**, *47*, 672–680.

17. Borg, O.; Hammer, N.; Eri, S.; Lindvag, O.A.; Myrstad, R.; Blekkan, E.A.; Ronning, M.; Rytter, E.; Holmen, A. Fischer-Tropsch synthesis over un-promoted and Re-promoted gamma-Al$_2$O$_3$ supported cobalt catalysts with different pore sizes. *Catal. Today* **2009**, *142*, 70–77.

18. Bazin, D.; Borko, L.; Koppany, Zs.; Kovacs, I.; Stefler, G.; Sajo, L.I.; Schay, Z.; Guczi, L. Re-Co/NaY and Re-Co/Al$_2$O$_3$ bimetallic catalysts: *In situ* EXAFS and catalytic activity. *Catal. Lett.* **2002**, *84*, 169–182.

19. Jacobs, G.; Chaney, J.A.; Patterson, P.M.; Das, T.K.; Davis, B.H. Fischer-Tropsch synthesis: Study of the promotion of Re on the reduction property of Co/Al$_2$O$_3$ catalysts by *in situ* EXAFS/XANES of Co K and Re L$_{III}$ edges and XPS. *Appl. Catal. A* **2004**, *264*, 203–212.

20. Ronning, M.; Nicholson, D.G.; Holmen, A. *In situ* EXAFS study of the bimetallic interaction in a rhenium-promoted alumina-supported cobalt Fischer-Tropsch catalyst. *Catal. Lett.* **2001**, *72*, 141–146.

21. Ma, W.; Jacobs, G.; Ji, Y.; Bhatelia, T.; Bukur, D.B.; Khalid, S.; Davis, B.H. Fischer-Tropsch synthesis: Influence of CO conversion on selectivities, H$_2$/CO usage ratios, and catalyst stability for a Ru promoted Co/Al$_2$O$_3$ catalyst using a slurry phase reactor. *Top. Catal.* **2011**, *54*, 757–767.

22. Guczi, L.; Bazin, D.; Kovacs, I.; Borko, L.; Schay, Z.; Lynch, J.; Parent, P.; Lafon, C.; Stefler, G.; Koppany, Z.; Sajo, I. Structure of Pt-Co/Al$_2$O$_3$ and Pt-Co/NaY bimetallic catalysts: Characterization by *in situ* EXAFS, TPR, XPS and by activity in CO (Carbon Monoxide) Hydrogenation. *Top. Catal.* **2002**, *20*, 129–139.

23. Jacobs, G.; Chaney, J.A.; Patterson, P.M.; Das, T.K.; Maillot, J.C.; Davis, B.H. Fischer-Tropsch synthesis: Study of the promotion of Pt on the reduction property of Co/Al$_2$O$_3$ catalysts by *in situ* EXAFS of Co K and Pt L$_{III}$ edges and XPS. *J. Synch. Rad.* **2004**, *11*, 414–422.

24. Sadeqzadeh, M.; Karaca, H.; Safonova, O.V.; Fongarland, P.; Chambrey, S.; Roussel, P.; Griboval-Constant, A.; Lacroix, M.; Curulla-Ferré, D.; Luck, F.; *et al.* Identification of the active species in the working alumina-supported cobalt catalyst under various conditions of Fischer–Tropsch synthesis. *Catal. Today* **2011**, *164*, 62–67.

25. Iglesia, E.; Soled, S.L.; Fiato, R.A.; Via, G.H. Bimetallic synergy in cobalt-ruthenium Fischer-Tropsch synthesis catalysts. *J. Catal.* **1993**, *143*, 345–368.

26. Chonco, Z.H.; Nabaho, D.; Claeys, M.; van Steen, E. The role of reduction promoters in Fischer-Tropsch catalysts for the production of liquid fuels. In Proceedings of 23rd Meeting of the North American Catalysis Society, 2–7 June 2013, Louisville, KY, USA.

27. Ionkina, O.; Subramanian, M.A.; Chao, W.; Makar, K.M.; Manzer, L.E. Fischer-Tropsch processes and catalysts with promoters. *US Patent* 20020010221A1, January 2002.

28. Jermwongratanachai, T.; Jacobs, G.; Ma, W.; Shafer, W.D.; Gnanamani, M.K.; Gao, P.; Kitiyanan, B.; Davis, B.H.; Klettlinger, J.L.S.; Yen, C.H.; *et al.* Fischer-Tropsch synthesis: Comparisons between Pt and Ag promoted Co/Al_2O_3 catalysts for reducibility, local atomic structure, catalytic activity, and oxidation-reduction (OR) cycles. *Appl. Catal.* **2013**, *464–465*, 165–180.

29. Redjala, T.; Remita, H.; Apostolescu, G.; Mostafavi, M.; Thomazeau, C.; Uzio, D. Bimetallic Au-Pd and Ag-Pd clusters synthesized by gamma or electron beam radiolysis and study of the reactivity/structure relationships in the selective hydrogenation of but-1,3-diene. *Oil Gas Sci. Technol.* **2006**, *61*, 789–797.

30. Van Steen, E.; Claeys, M.; Dry, M.E.; van de Loosdrecht, J.; Viljoen, E.L.; Visagie, J.L. Stability of nanocrystals: thermodynamic analysis of oxidation and re-reduction of cobalt in water/hydrogen mixtures. *J. Phys. Chem. B* **2005**, *109*, 3575–3577.

31. Saib, A.M.; Moodley, D.J.; Ciobica, I.M.; Hauman, M.M.; Sigwebela, B.H.; Weststrate, C.J.; Niemantsverdriet, J.W.; van de Loosdrecht, J. Fundamental understanding of deactivation and regeneration of cobalt Fischer-Tropsch synthesis catalysts. *Catal. Today* **2010**, *154*, 271–282.

32. Saib, A.M.; Borgna, A.; van de Loosdrecht, J.; van Berge, P.J.; Niemantsverdriet, J.W. XANES study of the susceptibility of nano-sized cobalt crystallites to oxidation during realistic Fischer-Tropsch synthesis. *Appl. Catal.* **2006**, *312*, 12–19.

33. Bezemer G.L, Bitter J.H.; Kuipers H.P.C.E.; Oosterbeek H.; Holewijn J.E.; Xu X.D.; Kapteijn, F.; van Dillen, A.J.; de Jong, K.P. Cobalt particle size effects in the Fischer-Tropsch reaction studied with carbon nanofiber supported catalysts. *J. Am. Chem. Soc.* **2006**, *128*, 3956–3964.

34. Borg, Ø.; Dietzel, P.D.C.; Spjelkavik, A.I.; Tveten, E.Z.; Walmsley, J.C.; Diplas, S.; Eri, S.; Holmen, A.; Rytter, E. Fischer-Tropsch synthesis: cobalt particle size and support effects on intrinsic activity and product distribution. *J. Catal.* **2008**, *259*, 161–164.

35. Azzam, K.; Jacobs, G.; Ma, W.; Davis, B.H. Effect of cobalt particle size on the catalyst intrinsic activity for Fischer-Tropsch synthesis. *Catal. Lett.* **2014**, *144*, 389–394.

36. Jermwongratanachai, T.; Jacobs, G.; Shafer, W.D.; Ma, W.; Pendyala, V.R.R.; Davis, B.H.; Kitiyanan, B.; Khalid, S.; Cronauer, D.C.; Kropf, A.J.; Marshall, C.L. Fischer-Tropsch synthesis: Oxidation of a fraction of cobalt crystallites in research catalysts at the onset of FT at partial pressures mimicking 50% CO conversion. *Top. Catal.* in press.

37. Fischer, N.; Clapham, B.; Feltes, T.E.; van Steen, E.; Claeys, M. The reoxidation of cobalt Fischer-Tropsch catalysts. In Proceedings of Syngas 2012 Convention, Cape Town, South Africa, 1–4 April 2012.

38. Bertole, C.J.; Mims, C.A.; Kiss, G. The effect of water on the cobalt-catalyzed Fischer-Tropsch synthesis. *J. Catal.* **2002**, *210*, 84–96.

39. Sadeqzadeh, M.; Hong, J.; Fongarland, P.; Curulla-Ferre, D.; Luck, F.; Bousquet, J.; Schweich, D.; Khodakov, A.Y. Mechanistic modeling of cobalt based catalyst sintering in a fixed bed reactor under different conditions of Fischer-Tropsch synthesis. *Ind. Eng. Chem. Res.* **2012**, *51*, 11955–11964.

40. Jacobs, G.; Das, T.K.; Patterson, P.M.; Luo, M.; Conner, W.A.; Davis, B.H. Fischer-Tropsch synthesis: Effect of water on Co/Al$_2$O$_3$ catalysts and XAFS characterization of reoxidation phenomena. *Appl. Catal. A* **2004**, *270*, 65–76.

41. Soled, S.; Kliewer, C.; Kiss, G.; Baumgartner, J. Reversible and irreversible changes in Co Fischer-Tropsch catalysts during synthesis. In Proceedings of 21st Meeting of the North American Catalysis Society, San Francisco, CA, USA, 7–12 June 2009.

42. Jacobs, G.; Ma, W.; Gao, P.; Todic, B.; Bhatelia, T.; Bukur, D.B.; Davis, B.H. The application of synchrotron methods in characterizing iron and cobalt Fischer-Tropsch synthesis catalysts. *Catal. Today* **2013**, *214*, 100–139.

43. Moodley, D.J.; Saib, A.M.; van de Loosdrecht, J.; Welker-Nieuwoudt, C.A.; Sigwebela, B.H.; Niemantsverdriet, J.W. The impact of cobalt aluminate formation on the deactivation of cobalt-based Fischer-Tropsch synthesis catalysts. *Catal. Today* **2011**, *171*, 192–200.

44. Das, T.K.; Jacobs, G.; Patterson, P.M.; Conner, W.A.; Davis, B.H. Fischer-Tropsch synthesis: Characterization and catalytic properties of rhenium promoted cobalt alumina catalysts. *Fuel* **2003**, *82*, 805–815.

45. Jacobs, G.; Sarkar, A.; Ji, Y.; Patterson, P.M.; Das, T.K.; Luo, M.; Davis, B.H. Fischer-Tropsch synthesis: characterization of interactions between reduction promoters and Co for Co/Al$_2$O$_3$–based GTL catalysts. In Proceedings of *AIChE Annual Meeting*, San Francisco, CA, USA, 12–17 November 2006.

46. Karaca, H.; Hong, J.; Fongarland, P.; Roussel, P.; Griboval-Constant, A.; Lacroix, M.; Hortmann, K.; Safonova, O.V.; Khodakov, A.Y. *In situ* XRD investigation of the evolution of alumina-supported cobalt catalysts under realistic conditions of Fischer-Tropsch synthesis. *Chem. Commun.* **2010**, *46*, 788–790.

47. Jacobs, G.; Patterson, P.M.; Zhang, Y.-Q.; Das, T.K.; Li, J.; Davis, B.H. Fischer-Tropsch synthesis: Deactivation of noble metal-promoted Co/Al$_2$O$_3$ catalysts. *Appl. Catal. A* **2002**, *233*, 215–226.

48. Kim, C.J. Water addition for increased CO/H$_2$ hydrocarbon activity over catalysts comprising * e There is virtually no difference. carried out for thousands of hours, the time ranges specified are very close to each othecobalt, ruthenium, and mixtures thereof which may include a promoter metal. U.S. Patent 5,227,407, July 1993.

49. Li, J.; Zhan, X.; Zhang, Y.-Q.; Jacobs, G.; Das, T.K.; Davis, B.H. Fischer-Tropsch synthesis: Effect of water on the deactivation of Pt promoted Co/Al$_2$O$_3$ catalysts. *Appl. Catal. A* **2002**, *228*, 203–212.

50. Jacobs, G.; Das, T.K.; Patterson, P.M.; Li, J.; Sanchez, L.; Davis, B.H. Fischer-Tropsch synthesis: XAFS studies of the effect of water on a Pt-promoted Co/Al$_2$O$_3$ catalyst. *Appl. Catal. A* **2003**, *247*, 335–343.

51. Weststrate, C.J.; Saib, A.M.; Niemantsverdriet, J.W. Promoter segregation in Pt and Ru promoted cobalt model catalysts during oxidation-reduction treatments. *Catal. Today* **2013**, *215*, 2–7.

52. Jermwongratanachai, T.; Jacobs, G.; Shafer, W.D.; Pendyala, V.R.R.; Ma, W.; Gnanamani, M.K.; Hopps, S.; Thomas, G.A.; Kitiyanan, B.; Khalid, S.; *et al.* Fischer-Tropsch synthesis: TPR and XANES analysis of the impact of oxidation-reduction (OR) cycles on the reducibility of Co/alumina catalysts with different promoters (Pt, Ru, Re, Ag, Au, Rh, Ir). *Catal. Today* 2014, in press.

53. Li, J.; Jacobs, G.; Das, T.K.; Zhang, Y.-Q.; Davis, B.H. Fischer-Tropsch synthesis: Effect of water on the catalytic properties of a Co/SiO$_2$ catalyst. *Appl. Catal. A* **2002**, *236*, 67–76.

54. Dalai, A.K.; Das, T.K.; Chaudhari, K.V.; Jacobs, G.; and Davis, B.H. Fischer-Tropsch synthesis: Water effects on Co supported on wide and narrow-pore silica. *Appl. Catal. A* **2005**, *289*, 135–142.

55. Ma, W.; Jacobs, G.; Sparks, D.E.; Spicer, R.L.; Davis, B.H.; Klettlinger, J.L.S.; Yen, C.H. Fischer-Tropsch synthesis: Kinetics and water effect study over 25%Co/Al$_2$O$_3$ catalysts. *Catal. Today* 2014, in press.

56. Logdberg, S.; Boutonnet, M.; Walmsley, J.C.; Jaras, S.; Holmen, A.; Blekkan, E.A. Effect of water on the space-time yield of different supported cobalt catalysts during Fischer-Tropsch synthesis. *Appl. Catal. A* **2011**, *393*, 109–121.

57. Ma, W.; Jacobs, G.; Sparks, D.E.; Gnanamani, M.K.; Pendyala, V.R.R.; Yen, C.H.; Klettlinger, J.L.S.; Tomsik, T.M.; Davis, B.H. Fischer-Tropsch synthesis: support and cobalt cluster size effects on kinetics over Co/Al$_2$O$_3$ and Co/SiO$_2$ catalysts. *Fuel* **2011**, *90*, 756–765.

58. Studies in Surface Science and Catalysis. *Catalyst Deactivation 1991: Proceedings of the 5th International Symposium*; Bartholomew, C.H., Butt, J.B., Eds.; Elsevier: Amsterdam, The Netherlands, 1991; Volume 68.

59. Bartholomew, C.H. Mechanisms of catalyst deactivation. *Appl. Catal. A* **2001**, *212*, 17–60.

60. Eggenhuisen, T.M.; Munnik, P.; Talsma, H.; de Jongh, P.E.; de Jong, K.P. Freeze-drying for controlled nanoparticle distribution in Co/SiO$_2$ Fischer-Tropsch catalyst. *J. Catal.* **2013**, *297*, 306–313.

61. Graham, U.M.; Jacobs, G.; Gnanamani, M.; Lipka, S.; Shafer, W.D.; Swartz, C.; Jermwongratanachai, T.; Chen, R.; Rogers, F.; Davis, B.H. Fischer Tropsch Synthesis: High Oxygenate-Selectivity of Cobalt Catalysts supported on Hydrothermal Carbons. *ACS Catal.* submitted for publication, 2014.

62. van de Loosdrecht, J.; Datt, M.; Visagie, J.L. Carbon coated supports for cobalt based Fischer-Tropsch catalysts. *Top. Catal.* 2014, in press.

63. Rane, S.; Borg, O.; Yang, J.; Rytter, E.; Holmen, A. Effect of alumina phases on hydrocarbon selectivity in Fischer-Tropsch synthesis. *Appl. Catal. A* **2010**, *388*, 160–167.

64. Sietsma, J.R.A.; den Breejen, J.P.; de Jongh, P.E.; van Dillen, J.; Bitter, J.H.; de Jong, K.P. Highly active cobalt-on-silica catalysts for the Fischer-Tropsch synthesis obtained via a novel calcination. *Stud. Surf. Sci. Catal.* **2007**, *167*, 85–90.

65. Lu, J.; Elam, J.W.; Stair, P.C. Synthesis and stabilization of supported metal catalysts by atomic layer deposition. *Accts. Chem. Res.* **2013**, *46*, 1806–1815.

66. Soled, S.L.; Kiss, G.; Kliewer, C.; Baumgartner, J.; El-Malki, E.-M. Learnings from Co Fischer-Tropsch catalyst studies. Abstracts of Papers, ENFL-412, 245th ACS National Meeting & Exposition, New Orleans, LA, USA, 7–11 April 2013.

67. Argyle, M.D.; Frost, T.S.; Bartholomew, C.H. Cobalt Fischer Tropsch catalyst deactivation modeled using generalized power law expressions. *Top. Catal.* 2014, in press.

68. Bakken, V.; Bergene, E.; Rytter, E.; Swang, O. Bimetallic cobalt/rhenium systems: Preferred position of rhenium through an interdisciplinary approach. *Catal. Lett.* **2010**, *135*, 21–25.

69. Tan, K.F.; Xu, J.; Chang, J.; Borgna, A.; Saeys, M. Carbon deposition on Co catalysts during Fischer-Tropsch synthesis: A computational and experimental study. *J. Catal.* **2010**, *274*, 121–129.

70. Swart, J.C.W.; Ciobica, I.M.; van Santen, R.A.; van Steen, E. Intermediates in the formation of graphitic carbon on a flat FCC-Co(111) surface. *J. Phys. Chem. C* **2008**, *112*, 12899–12904.

71. Burghgraef, H. A quantum chemical study of CH and CC bond activation on transition metals. PhD Thesis, Eindhoven University of Technology: Eindhoven, The Netherlands, June 1995.

72. Ciobica, I.M.; van Santen, R.A.; van Berge, P.J.; van de Loosdrecht, J. Adsorbate induced reconstruction of cobalt surfaces. *Surf. Sci.* **2008**, *602*, 17–27.

73. Wilson, J.; de Groot, C. Atomic-scale restructuring in high-pressure catalysis. *J. Phys. Chem.* **1995**, *99*, 7860–7866.

74. Schulz, H.; Nie, Z.; Ousmanov, F. Construction of the Fischer–Tropsch regime with cobalt catalysts. *Catal. Today* **2001**, *71*, 351–360.

75. Balakrishnan, N.; Joseph, B.; Bhethanabotla, V.R. Effect of Pt and Ru promoters on deactivation of Co catalysts by C deposition during Fischer-Tropsch synthesis: A DFT study. *Appl. Catal. A* **2013**, *462–463*, 107–115.

Deactivation of Pd Catalysts by Water during Low Temperature Methane Oxidation Relevant to Natural Gas Vehicle Converters

Rahman Gholami, Mina Alyani and Kevin J. Smith

Abstract: Effects of H_2O on the activity and deactivation of Pd catalysts used for the oxidation of unburned CH_4 present in the exhaust gas of natural-gas vehicles (NGVs) are reviewed. CH_4 oxidation in a catalytic converter is limited by low exhaust gas temperatures (500–550 °C) and low concentrations of CH_4 (400–1500 ppmv) that must be reacted in the presence of large quantities of H_2O (10–15%) and CO_2 (15%), under transient exhaust gas flows, temperatures, and compositions. Although Pd catalysts have the highest known activity for CH_4 oxidation, water-induced sintering and reaction inhibition by H_2O deactivate these catalysts. Recent studies have shown the reversible inhibition by H_2O adsorption causes a significant drop in catalyst activity at lower reaction temperatures (below 450 °C), but its effect decreases (water adsorption becomes more reversible) with increasing reaction temperature. Thus above 500 °C H_2O inhibition is negligible, while Pd sintering and occlusion by support species become more important. H_2O inhibition is postulated to occur by either formation of relatively stable $Pd(OH)_2$ and/or partial blocking by OH groups of the O exchange between the support and Pd active sites thereby suppressing catalytic activity. Evidence from FTIR and isotopic labeling favors the latter route. Pd catalyst design, including incorporation of a second noble metal (Rh or Pt) and supports high O mobility (e.g., CeO_2) are known to improve catalyst activity and stability. Kinetic studies of CH_4 oxidation at conditions relevant to natural gas vehicles have quantified the thermodynamics and kinetics of competitive H_2O adsorption and $Pd(OH)_2$ formation, but none have addressed effects of H_2O on O mobility.

Reprinted from *Catalysts*. Cite as: Gholami, R.; Alyani, M.; Smith, K.J. Deactivation of Pd Catalysts by Water during Low Temperature Methane Oxidation Relevant to Natural Gas Vehicle Converters. *Catalysts* **2015**, *5*, 561-594.

1. Introduction

Natural gas, an abundant energy resource with worldwide proven reserves of over 204.7 trillion m^3 [1], is used primarily for electricity generation and heating. The composition of natural gas (NG) is highly variable, but CH_4 typically accounts for 80–90% of the components of NG. CH_4 has the highest H/C ratio among all hydrocarbon fuels and during combustion, generates the lowest amount of CO_2 per unit of energy. The amount of SO_2 generated during NG combustion is also relatively low because the S content of NG is significantly lower than that of gasoline or diesel fuels. These environmental benefits, together with a relatively low cost of NG, have resulted in an increased interest in its use as a transportation fuel. Currently there are >16 million natural gas vehicles (NGVs) in operation around the world, and their numbers are growing at about 20% annually [2]. However, a significant concern for the wide-spread implementation of NG as a fuel for combustion engines is that unburned CH_4, expelled in the engine exhaust, is a significant greenhouse gas with potency more than 25xs that of CO_2.

The transportation sector is a major contributor to air pollution through the combustion of gasoline and diesel fuels, accounting for ~77% of CO emissions, ~47% of hydrocarbon emissions and ~60% of NO_x emissions in the USA [3]. The exhaust gas of a conventional gasoline powered spark-ignition internal combustion engine (SI-ICE) consists mostly of N_2 (70–75%), CO_2 (11–13%) and water (10-12%) with about 1–2% of pollutants, specifically unburned hydrocarbons, CO and NO_x [4,5]. The pollutants must be removed before the exhaust gas is emitted to the atmosphere so as to meet increasingly stringent worldwide emission standards. The pollutants present in the engine exhaust are dependent on the engine air/fuel (A/F) ratio. For example, if the A/F ratio is above the stoichiometric value for complete combustion (A/F = 14.6), the concentration of reducing agents (hydrocarbons and CO) in the exhaust gas decreases whereas the concentration of oxidizing agents (O_2 and NO_x) increases. Consequently, several different strategies have been developed to control engine emissions, depending on the operating conditions and the target emission levels [5]. Typically, a gasoline engine management system controls the A/F ratio or the exhaust gas composition (using an oxygen sensor connected to a secondary air supply) near the stoichiometric value. A single three-way catalyst (TWC) bed, placed in the exhaust gas flow, ensures simultaneous oxidation of the CO and hydrocarbons and the reduction of the NO_x. Alternatively, dual-bed systems combine a NO_x reduction catalyst bed with a separate oxidation catalyst and secondary air to remove the CO and hydrocarbons. Under lean-burn conditions a gasoline engine may operate with sufficiently high A/F ratios so as to obtain a significant reduction in CO and NO_x emissions and improved fuel efficiency. The function of the catalyst in this case is limited to the oxidation of mainly hydrocarbons, while the NO_x emissions are captured using a NO_x trap followed by desorption and reduction in a TWC during an occasional near stoichiometric excursion of the engine. Although lean-burn engines improve fuel efficiency, the exhaust gas temperature is significantly lower than from conventional gasoline powered engines, and consequently, catalysts with high oxidation activity at relatively low temperatures are needed for this application [5].

Modern TWC converters used in gasoline ICEs contain Pt, Rh and Pd, dispersed on a washcoat applied to a cordierite ceramic monolith or metal monolith [3,5]. The monolith usually has a honeycomb structure with 1 mm square channels to accommodate the high gas throughputs from the exhaust with minimal pressure drop. The washcoat, a mix of several metal oxides (Al_2O_3, CeO_2, ZrO_2), is applied to increase the metal support surface area (Al_2O_3), to improve thermal stability (ZrO_2) and to provide enhanced oxygen storage capacity (CeO_2) that widens the operating range for optimal oxidation and reduction by the catalyst. The metal composition of the converter varies with application but typically contains 5–20:1 of Pt:Rh with a total metal loading of 0.9–2.2 g L^{-1}. Pd may be used to replace all or part of the Pt for cost savings [5].

Exhaust gas emissions from NGVs are difficult to control because low concentrations of CH_4 (400–1500 ppmv) must be oxidized in the presence of high concentrations of H_2O (10-15 vol.%) and CO_2 (15 vol.%) at relatively low exhaust gas temperatures (450–550 °C). The greater strength of the C-H bond in CH_4 (450 kJ/mol) relative to other hydrocarbons [6] implies that catalysts with high CH$_4$ oxidation activity must be used. NGVs operate near the stoichiometric point or under lean-burn conditions [7,8]. Stoichiometric NGV engines are primarily used in light-duty passenger cars, whereas lean-burn engines are more common in heavy-duty vehicles such as buses. Over the past ~20 years, conventional converter technologies have been adapted for NGVs using Pd catalysts (which have the highest activity for CH$_4$ oxidation [7,9,10]) to adequately reduce (by 50–60%) the CH$_4$ content in NGV exhausts at <500 °C in the presence of high H$_2$O concentrations. Commercial catalysts for SI-NG engines also typically incorporate a CeO$_2$/ZrO$_2$ solid solution for high O$_2$ adsorption capacity, which serves to buffer O$_2$ concentration during the rapid switching between slightly oxidizing and reducing conditions close to a stoichiometric mixture (e.g., [7,11]).

Several papers and reviews have assessed the activity and deactivation of Pd catalysts for CH$_4$ oxidation, supported on Al$_2$O$_3$, SiO$_2$, ZrO$_2$, CeO$_2$, and zeolites; promoted with noble metals, e.g., Pt and Rh, and with transition metal oxides, e.g., oxides of Co, Ni, and Sn [6,7,10–20]. Studies have largely focused on CH$_4$ oxidation on supported Pd catalysts containing 0.5 to 5% Pd (typical Pd loadings in commercial SI-NG monolithic coated catalysts are about 3–7 g.L^{-1}, equivalent to about 1.5–4 wt.% loading in a monolith washcoat) at temperatures ranging from 450 to 600 °C and at CH$_4$ concentrations of 0.04 to 1 vol.% (0.04 to 0.15 vol.% for commercially representative tests). High activity for CH$_4$ oxidation appears to be favored by Pd loadings of 3–5 g L^{-1} and dispersions lower than about 0.12–0.15 [7]. Pd-O sites associated with Pd/CeO$_2$ surfaces appear to have the highest activity for CH$_4$ oxidation [21,22].

Mechanisms and kinetics of CH$_4$ oxidation over Pd/PdO catalysts have elicited continued debate in the literature [6,13,14,23], for which data interpretation is complicated by the transitions that Pd catalysts undergo during thermal pre-treatment and reaction [24]. Furthermore, the high concentration of H$_2$O in the NGV exhaust and the typically transient reaction conditions that result from cycling between oxidizing and reducing conditions in the NG engine [6,11] are known to significantly impact catalyst activity and stability.

The present review is focused on the inhibition and deactivation effects of H$_2$O, especially at the relatively low temperatures representative of CH$_4$ oxidation over Pd catalysts in a NG engine. Although previous reviews have addressed the issue of Pd catalyst stability in the presence or absence of H$_2$O [4,12,20,25], and several catalyst deactivation mechanisms are possible at the exhaust gas conditions [26], several unresolved issues remain. More recent studies of the past decade have provided new insights into the effects of H$_2$O, especially at lower temperatures, and these are the focus of the present review. Note, however, that in many cases, fresh catalysts in powder form have been evaluated using ideal fixed-bed micro-reactors and simulated exhaust gas under steady state operating conditions. Tests of monolith catalysts with promoters suitably aged and operated with A/F frequency and amplitude modulation that occur in a vehicle are few [7,11]. Nonetheless, interpretation of data from ideal catalyst studies allows direct links to be drawn between fundamental catalyst properties and catalyst performance for CH$_4$ oxidation, whereas in real systems this may be more difficult to achieve.

2. Effects of H₂O on CH₄ Oxidation over Pd Catalysts

Water is a major component of the engine exhaust and is also a product of the combustion that occurs in the catalytic converter. In TWCs, H_2O acts as an oxidizing agent for CO conversion by the water-gas-shift reaction and for steam reforming of hydrocarbons [4]. H_2O also significantly affects the thermal stability of the metals (Pt, Rh and Pd) present in the TWC as well as the support, mostly through sintering mechanisms [4,27,28] and by changes in the Pd oxidation state during hydrothermal aging [29]. Water may also act as a reaction inhibitor by adsorption onto the catalyst.

Bounechada *et al.* [11] reported on the activity of a Pd-Rh (Pd/Rh = 39/1) TWC converter supported on stabilized Al_2O_3, promoted with Ce-Zr (Zr/Ce = 3.5) and wash coated on a ceramic honeycomb monolith, tested under fuel-lean ($\lambda > 1$), stoichiometric ($\lambda = 1.00$), and fuel-rich ($\lambda < 1$) conditions (gas composition: 0.15 vol.% CH_4, 0.6% CO, 0.1% H_2, 10% H_2O, 10.7% CO_2, 0.13% NO, 0–1.14% O_2; λ was varied by changing feed O_2 concentration; GHSV = 50,000 h^{-1}). At stationary conditions (constant λ; steady-state experiment), the CH_4 conversion was observed to continuously decrease under both stoichiometric (52 to 43% after 0.5 h reaction) and fuel-lean (from 62 to 59% after 0.5 h reaction) conditions, even though injecting a fuel-rich pulse during fuel-lean stationary operation increased the CH_4 conversion to its initial value at the onset of reaction. The authors attributed the deactivation under fuel-lean conditions to the inhibition effect of H_2O on the CH_4 oxidation reaction, whereas under stoichiometric conditions, partial reduction of PdO due to the lack of oxygen, may lead to a loss in PdO active sites for CH_4 oxidation. The authors also claimed that the presence of high oxygen capacity metals (Ce and Zr) in the catalyst made the reduction of PdO improbable under stoichiometric conditions. Under fuel-rich conditions, H_2O acts as an oxidant through water-gas shift and steam reforming reactions.

2.1. Water Concentration and Reaction Temperature Effects on CH₄ Oxidation Activity of Pd Catalysts

With the growing interest in NGVs, recent studies have focused on effects of H_2O on Pd catalysts during CH_4 combustion [16,18,30–38]. Deactivation or inhibition effects of H_2O are dependent upon several factors including catalyst formulation, reaction temperature, catalyst time-on-stream history, and H_2O concentration. Table 1 summarizes selected data that show effects of H_2O added to the feed gas during CH_4 light-off experiments over Pd catalysts. The light-off temperature (here reported as the temperature corresponding to 30% CH_4 conversion during temperature programmed reaction, T_{30}) increases as the H_2O concentration increases, showing a clear inhibition effect that increases in magnitude with increasing H_2O concentration.

In several cases the effects of H_2O have been examined by measuring the CH_4 conversion at steady-state, with and without H_2O added to the feed gas. A typical set of data, reported by Persson *et al.* [35], is shown in Figure 1 using several Pd/Al_2O_3 catalysts reacted at 500 °C. These data also show that added H_2O significantly suppresses CH_4 conversion, but the effect is at least partially reversible. Similar effects of H_2O addition have been reported in the literature, as summarized in Table 2. These reports confirm that H_2O acts as an inhibitor of CH_4 oxidation over

Pd catalysts and that upon removal of the H_2O from the CH_4/O_2 reactant, the inhibition is partially reversible [31,33].

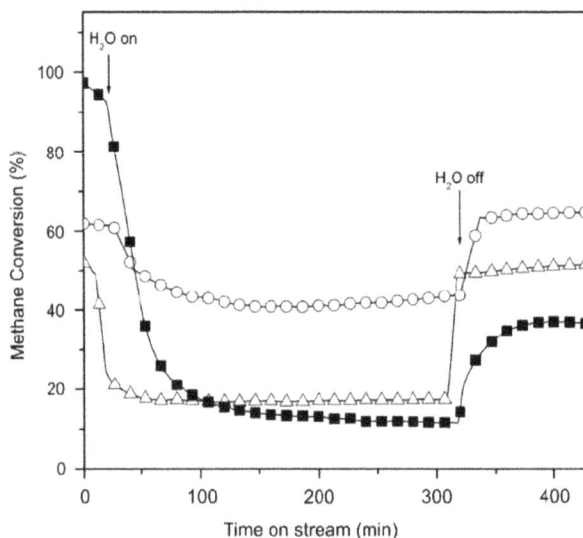

Figure 1. Effect of water vapor on the activity for CH_4 combustion over Pd/Al_2O_3 (■); 2:1 $PdPt/Al_2O_3$ (Δ); and 1:1 $PdPt/Al_2O_3$ (o) at 500 °C; 5 vol.% of steam was added to the 1.5% CH_4/air feed gas, GHSV = 100,000 h^{-1}, for 5 h. [35] Copyright© 2007 Elsevier.

Reaction temperature is another key variable affecting the role of H_2O addition. Although the data of Table 2 cannot be compared directly because of the different operating conditions, they do show that at 600 °C, the decrease in CH_4 conversion with H_2O addition is much less significant than at lower temperatures (400 °C). Several authors have proposed that the deactivation is related to the reaction of H_2O with active PdO sites [16,18,31,40,41], $PdO + H_2O \rightarrow Pd(OH)_2$, resulting in the formation of inactive $Pd(OH)_2$, as first proposed by Cullis *et al.* [40]. Burch *et al.* [31] also reported a strong inhibitory effect of water on Pd catalysts up to 450 °C. However, at higher temperatures the negative influence of water on the activity was very small, suggesting that above 450 °C the reverse reaction ($Pd(OH)_2 \rightarrow PdO + H_2O$) occurs. Eriksson *et al.* [41] observed a significant decrease in CH_4 conversion over a much wider range of temperatures (200–800 °C) after adding 18% H_2O to a CH_4/O_2 feed over a Pd/ZrO_2 catalyst, which was likely due to the relatively high H_2O concentration used in this study. Different results were reported by Kikuchi *et al.* [16] when adding 1 vol.% H_2O during CH_4 oxidation over a Pd/Al_2O_3 catalyst, *i.e.*, a decrease in activity was observed up to about 450 °C and no H_2O inhibition was observed at higher temperatures. However, during addition of 20 vol.% H_2O, the inhibiting effect could be observed up to 600 °C, in qualitative agreement with Eriksson *et al.* [41].

Table 1. Effect of H$_2$O addition on T_{30} (temperature at 30% CH$_4$ conversion) during temperature-programmed CH$_4$ oxidation over Pd catalysts.

	Reference [16]			Reference [39]	
	1.1% Pd/Al$_2$O$_3$	1.1% Pd/SnO$_2$	1.1% Pd/Al$_2$O$_3$-36NiO	0.9% Pd/ZrO$_2$ [a]	0.9% Pd/ZrO$_2$ [b]
GHSV, h^{-1}	48,000			50,000	
Dry feed gas composition, vol.%	1% CH$_4$/20% O$_2$ in N$_2$			0.4% CH$_4$/0.05% CO/5% CO$_2$/10% O$_2$ in N$_2$	
	T_{30}, °C			T_{30}, °C	
Added H$_2$O, vol.%					
0	345	290	372	360	300
1	400	315	372	-	-
5	430	335	420	-	-
10	460	360	425	410	350
20	510	365	445	-	-

[a] Calcined at 873K for 6 h; [b] Calcined at 1273 K for 6 h.

Table 2. H$_2$O inhibition over Pd catalysts during CH$_4$ oxidation.

	Reaction conditions					Water addition				
							CH$_4$ Conversion, %			
Catalyst	Temp °C	GHSV h^{-1}	Feed Gas mol %	Conc. vol %	Period [a] min	Before H$_2$O addition	During H$_2$O addition	After H$_2$O addition	Comments	Refs.
0.1 wt.% Pd/H-beta	400	120,000	0.2% CH$_4$/10%O$_2$ in N$_2$	10	100	75	15	58	Conversion after 400 min TOS with periodic water addition	[38]

Table 2. *Cont.*

Catalyst	Reaction conditions					Water addition CH$_4$ Conversion, %			Comments	Refs
	Temp °C	GHSV h^{-1}	Feed Gas mol %	Conc. vol %	Period [a]	Before H$_2$O addition	During H$_2$O addition	After H$_2$O addition		
1.3 wt.% Pd/HTNU-10[c]	400	120,000	1% CH$_4$/4%O$_2$ in N$_2$	5	900	43	8	40	Conversion after 35 h TOS with periodic water addition	[33]
5 wt.% Pd/Al$_2$O$_3$	500	100,000	1.5% CH$_4$ in air	5	300	95	13	30	Initial activity Conversion after 400 min	[35]
2 wt.% Pd/Al$_2$O$_3$	550	160,000	0.4% CH$_4$ in air	10	60	95	79	92	TOS with periodic water addition Conversion after 300 min	[32]
1 wt.% Pd/Al$_2$O$_3$	600	160,000	0.4% CH$_4$ in air	8	60	95	90	93	TOS with periodic water addition	[36]

[a] Period: Time length of water addition period. [b] TOS: Time-on-stream. [c] HTNU-10 is the H-form of a medium pore zeolite with Si/Al = 7.1.

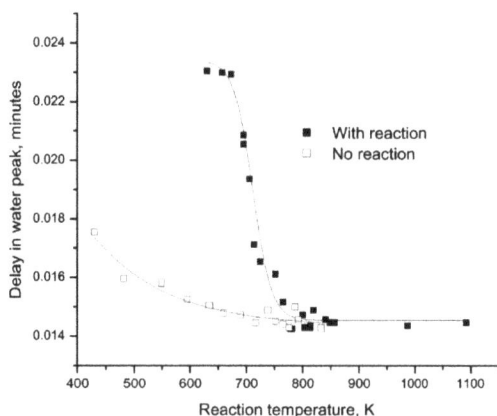

Figure 2. Delay in the H_2O peak with respect to other products obtained by passing pulses of $CH_4/O_2/He$ (closed square) and 1 vol.% O_2/3.45 vol.% H_2O/He (open square) over Pd/ZrO_2 at different temperatures. Reproduced with permission from [42]. Copyright © 2001 Elsevier.

Further insight into the H_2O adsorption/desorption phenomena on Pd/ZrO_2 catalysts has been obtained using pulsed-flow experiments [42,43]. Accordingly, pulses of $CH_4/O_2/He$ (1:4:95 vol %) were passed over a Pd/ZrO_2 catalyst at various temperatures and the products monitored by mass spectrometer. The time at which the peak maximum for H_2O appeared in each spectrum, compared to other products, was reported as the delay in the H_2O peak. The data (Figure 2) show that the H_2O generated during CH_4 oxidation lags other products, suggesting a slow H_2O adsorption/desorption equilibrium which might include spillover to the support. As the temperature increases above 450 °C (723 K), the desorption rate of H_2O increases and the delay in the H_2O peak compared to the other products is insignificant. This behavior is in agreement with observations from other studies [30,31,44] that the desorption rate of H_2O produced during CH_4 oxidation is slow and on the order of seconds below 450 °C, even though CO_2, the other product of reaction, desorbs very quickly. Increasing temperature above 450 °C removes the desorption time gap between CO_2 and H_2O, and thus, no inhibition by H_2O occurs. Ciuparu *et al.* [42] also pulsed gas containing 3.45 vol.% $H_2O/O_2/He$ but no CH_4 (and hence no reaction) through the same catalyst bed (Figure 2), showing that the H_2O generated from CH_4 oxidation lags the H_2O added to the feed. These data demonstrate that the adsorption/desorption of H_2O from the Pd catalyst surface is temperature dependent and reaches equilibrium at temperatures above ~450 °C (723 K), even for H_2O added in the gas phase.

Figure 3 compares temperature-programmed-reaction (TPR) profiles for CH_4 oxidation obtained over a Pd/ZrO_2 catalyst, from both pulsed and continuous flow experiments with or without H_2O added [42,43]. The pulsed flow TPR profile was obtained by injecting pulses of the reaction mixture (1/4/95:$CH_4/O_2/He$ for the "dry" feed and 1/4/95:$CH_4/O_2/He$ saturated with ~2% H_2O for the "wet" feed) into a He stream every 3 min while ramping the temperature at 0.5 K min^{-1}. Between consecutive pulses the catalyst was purged in flowing He. The pulsed flow data of Figure

3 show that at temperatures above 450 °C (723 K), there is no H_2O inhibition, since the conversions of "dry" and "wet" reaction mixtures are essentially the same. At <450 °C, inhibition is observed due to a low H_2O desorption rate. When H_2O is added to the gas phase, the H_2O adsorption rate is enhanced and the rate of desorption is further decreased. With continuous flow of reactants and a higher H_2O concentration, H_2O inhibition occurs at high temperatures due to re-adsorption. The addition of H_2O to the feed directs the equilibrium towards more H_2O adsorption on the surface and hence a greater decrease in catalyst activity during CH_4 oxidation.

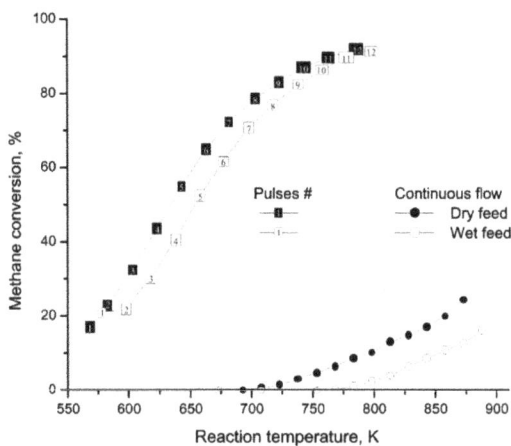

Figure 3. Temperature-programmed reactions during pulsed or continuous flow of reactants over Pd/ZrO_2 with or without H_2O in the feed. Reproduced with permission from [42]. Copyright © 2001 Elsevier.

The above observations are consistent with the following hypotheses: (1) product inhibition of CH_4 oxidation by H_2O on PdO catalysts occurs at temperatures below 450 °C; (2) product inhibition by H_2O is enhanced by its slow rate of desorption from the PdO catalyst relative to a higher rate of CH_4 oxidation; (3) PdO and H_2O may interact via the reversible reaction: $PdO + H_2O \leftrightarrow Pd(OH)_2$ yielding inactive $Pd(OH)_2$ and thus reversibly deactivating PdO as first proposed by Cullis *et al.* [40]; and (4) the extent of the CH_4 oxidation reaction increases with increasing temperature but is reduced with increasing H_2O concentration in the gas phase.

2.2. Catalyst Sintering by H_2O

The possibility that addition of H_2O may degrade Pd catalysts through a sintering mechanism [26] has also been investigated. According to Hansen *et al.* [45], the sintering rate of metal nanoparticles depends on their size. For nanoparticles <3 nm in diameter, Ostwald ripening is the most likely sintering mechanism. For larger particles (3–10 nm), both Ostwald ripening and particle migration and coalescence may occur, but the sintering rate is much slower than for the smaller particles [45]. The particle sintering rate has also been shown to correlate with the vapor pressure of the surface species [4]. Pd is unique among the PGMs in that the oxide (PdO) has a much lower vapor pressure than the metal (Pd), and consequently, one would expect a very low sintering rate of PdO by Ostwald ripening [4]. The rate of sintering is also dependent on the support. Lamber *et al.* [46] suggested that on SiO$_2$ in the presence of H$_2$O, the formation of silanol (Si-OH) groups favors the migration and coalescence of Pd, whereas in the absence of H$_2$O, Ostwald ripening is favored. Sintering suppression has been demonstrated for Pt catalysts using supports that enhance metal-support interactions [28]. Nagai *et al.* [47] demonstrated a correlation between the O electron density of the support, the strength of the Pt-O interaction and the resulting crystallite size. Thus, supports with a stronger metal-support interaction have a higher O electron density and yield smaller Pt crystallites in the order SiO$_2$ < Al$_2$O$_3$ < ZrO$_2$ < TiO$_2$ < CeO$_2$ [28,47].

Xu *et al.* [48] reported that the main deactivation mechanism of Pd/Al$_2$O$_3$ catalysts following exposure to 10 (*v/v*)% H$_2$O/N$_2$ at 900 °C for up to 200 h is Pd sintering. A substantial decrease in Pd dispersion from 3.7% to 0.9% over 7 wt.% Pd/Al$_2$O$_3$ and similar decreases at other Pd loadings after 96 h hydrothermal aging, were observed. As noted by Xu *et al.* [48], aging the catalyst at 900 °C ensures that PdO decomposition to Pd^0 is complete and consequently the more rapid sintering observed is relevant to the behavior of Pd^0 rather than PdO.

Escandon *et al.* [49] examined effects of hydrothermal aging at lower temperatures, where PdO is thermodynamically stable [6]. A 1 wt.% Pd/ZrO$_2$-Ce catalyst was hydrothermally aged at 300, 425, and 550 °C in 2% H$_2$O/Air for 30 h, before being evaluated for CH$_4$ oxidation under lean-burn conditions (5000 ppmv CH$_4$ in dry air). The results, shown in Figure 4, are compared with the same catalyst, thermally aged at 550 °C in dry air for 30 h (identified as Pd/ZrO$_2$-Ce-550 in Figure 4) [49]. A significant irreversible decrease in CH$_4$ conversion occurs and the extent of catalyst deactivation increases with aging temperature (Figure 4). The $T_{50\%}$ increases from 375 °C for the fresh oxidized catalyst (identified as Pd/ZrO$_2$-Ce in Figure 4), to 450 °C for the air-aged catalyst and to > 550 °C for the hydrothermally aged catalyst. Pd dispersion and BET surface area of the aged catalysts did not change [49]. Comparing the activity results of the catalyst thermally aged in air (Pd/ZrO$_2$-Ce-550) with that aged in 2% H$_2$O/air at 550 °C (Pd/ZrO$_2$-Ce-550h), confirms that catalyst deactivation rate increases in the presence of H$_2$O. The stability of the hydrothermally aged catalysts during reaction was also evaluated, using both isothermal deactivation experiments at 500 °C and light-off measurements made after 50 h reaction with 5000 ppmv CH$_4$ in air. The catalysts aged in the presence of H$_2$O at 300 °C underwent a significant deactivation whereas the catalyst aged in the presence of H$_2$O at 425 °C was much more resistant to deactivation, and after 25 h time-on-stream was the most active of all the catalysts examined. XRD analysis of the catalysts showed that the more stable catalysts are associated with the most stable supports [49].

In another study of CH$_4$ oxidation at low temperature (250–450 °C), a change in PdO dispersion was suggested as the main cause of deactivation of 0.5% Pd/Al$_2$O$_3$ and 0.5% Pd/SiO$_2$ catalysts [50]. Dispersion decreased from 10% for the unused 0.5% Pd/SiO$_2$ catalyst to 5.6% for the catalyst reacted in 1% CH$_4$/air feed at 450 °C for 7 h, whereas for the 0.5% Pd/Al$_2$O$_3$ catalyst the corresponding changes in dispersion were 67% to 6.3%, respectively. These observations are in good agreement with that of Narui *et al.* [51], in which the PdO dispersion of a 0.5% Pd/Al$_2$O$_3$ catalyst decreased from 14% to 11% after 6 h reaction at 350 °C. Zhang *et al.* [52] investigated Pd catalysts supported on ZSM-5 and reported that catalyst stability is improved when CH$_4$ oxidation is carried out in the presence of H_2O at 430–480 °C, compared to the reaction in a dry feed. In both cases, the loss in catalyst activity could be related to reduced PdO dispersion, as determined by the Pd/Si ratio measured by XPS, but the loss in dispersion is smaller in the presence of H_2O [52]. By contrast, Araya *et al.* [53] reported an insignificant drop in PdO dispersion (from 31.7% to 28.2%) of a Pd/SiO$_2$ catalyst after 96 h of reaction at 325 °C in 1.5% CH$_4$/6% O$_2$ in He, despite a significant decrease in CH$_4$ conversion from 32% to 22%. The extent of catalyst deactivation was found to further increase in the presence of 3% H_2O added to the feed.

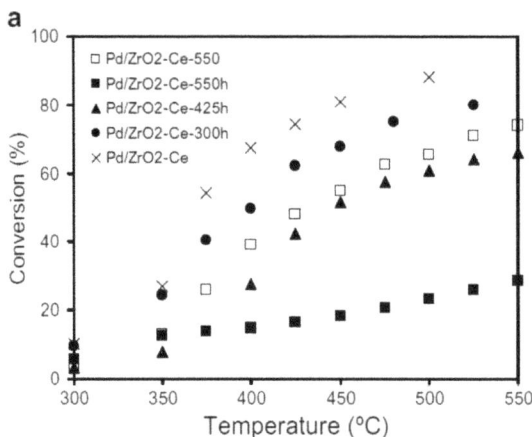

Figure 4. CH$_4$ conversion over fresh 1 wt.% Pd/ZrO$_2$-Ce catalyst compared to 1 wt.% Pd/ZrO$_2$-Ce thermally aged in air at 550 °C (Pd/ZrO$_2$-Ce-550) and hydrothermally aged at different temperatures in 2% H$_2$O/air (identified as Pd/ZrO$_2$-Ce-TTTh where TTT is the aging temperature in °C). Reproduced with permission from [49]. Copyright© 2008 Elsevier.

Several studies have demonstrated that catalyst sintering can be reduced by encapsulating Pd/PdO nanoparticles in support materials. Sinter-resistant Pd catalysts have been prepared by atomic layer deposition of Al$_2$O$_3$ overlayers on Pd [54], as well as by the synthesis of Pd/SiO$_2$ core-shell structures [55,56]. Cargnello *et al.* [22] reported a Pd/CeO$_2$ core-shell catalyst supported on Al$_2$O$_3$ for CH$_4$ oxidation that is about 200xs more active than an equivalent Pd-CeO$_2$/Al$_2$O$_3$ catalyst prepared by wet impregnation. The authors demonstrated that the Pd cores remain isolated even after heating the catalyst to 850 °C and that the CH$_4$ light-off curves (measured at GHSV of 200,000 h^{-1} in a feed gas of 0.5% CH$_4$, 2% O$_2$ in Ar) are the same for the fresh catalyst and one that has been aged at 850 °C for 12 hours. The Pd nanoparticles encapsulated by CeO$_2$ enhance the metal-support interaction that leads to exceptionally high CH$_4$ oxidation activity and good thermal stability [22].

2.3. Effects of Support

The data of Table 1 show that the inhibition of CH$_4$ oxidation by H$_2$O on Pd catalysts is dependent upon the support. Pd/Al$_2$O$_3$ shows significantly more inhibition with 10% H$_2$O added to the feed than either the Pd/SnO$_2$ or Pd/ZrO$_2$ catalysts. More detailed data from Kikuchi *et al.* [16] comparing CH$_4$ light-off curves for a 1.1 wt.% Pd/Al$_2$O$_3$ catalyst and a 1.1 wt.% Pd/SnO$_2$ catalyst with H$_2$O added to the feed over a range of concentrations (1–20 vol.%), are shown in Figures 5 and 6. By increasing the H$_2$O concentration, the CH$_4$ light-off curves for both catalysts shift to higher temperatures. However, the temperature shift is larger over the Pd/Al$_2$O$_3$ catalyst than the Pd/SnO$_2$. The authors completed a simplified kinetic analysis of the CH$_4$ oxidation rate data to show that the enthalpy of adsorption of H$_2$O is strongest on the Pd/Al$_2$O$_3$ catalyst ($\Delta H_{ad} \sim -49$ kJ/mol), from which they concluded that the significant loss in activity of the Pd/Al$_2$O$_3$ in the presence of H$_2$O is due to a high coverage of the active sites by H$_2$O [16]. These results could also be interpreted according to the more recent proposals by Schwartz *et al.* [44,57], that hydroxyl accumulation on the support hinders oxygen migration and exchange, and hence CH$_4$ oxidation. The strong adsorption of H$_2$O determined by kinetic analysis on the Pd/Al$_2$O$_3$ catalyst [16] is consistent with a large hydroxyl accumulation on the catalyst surface that could inhibit the O exchange.

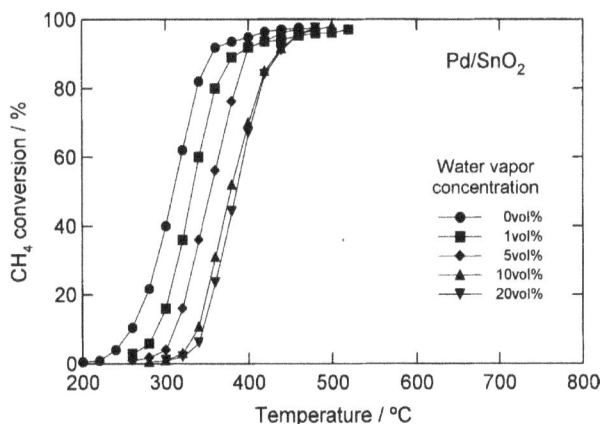

Figure 5. Catalytic combustion of CH$_4$ over 1.1 wt.% Pd/SnO$_2$ with different amounts of water added (vol.%). Reaction conditions: CH$_4$, 1 vol.%; O$_2$, 20 vol.%; H$_2$O, 0–20 vol.%; N$_2$, balance; GHSV 48,000 h^{-1}. Reprinted with permission from [16]. Copyright© 2002 Elsevier.

The rate of deactivation during CH$_4$ oxidation in the presence of H$_2$O has been shown to be reduced by using a support with high oxygen surface mobility. At temperatures below 450 °C, Ciuparu *et al.* [30] reported the inhibition effect of H$_2$O to be dependent upon the oxygen mobility of the support. Comparing PdO supported on oxides with increasing surface oxygen mobility: Al$_2$O$_3$ < ZrO$_2$ < Ce$_{0.1}$Zr$_{0.9}$O$_2$, they show that the resistance to H$_2$O inhibition during CH$_4$ oxidation increases in the same order. The deactivation rate of PdO was also compared over Al$_2$O$_3$, MgO, and TiO$_2$ supports by Schwartz *et al.* [44,57] at temperatures <450 °C. Deactivation is shown to be a consequence of reduced oxygen mobility due to hydroxyl adsorption. They also reported that PdO/MgO catalyst has a slower deactivation rate compared with Al$_2$O$_3$ and TiO$_2$ supports because of the higher oxygen surface mobility on the MgO [44,57]. However, Pd catalysts dispersed on other supports such as MCM-41, which have high surface area (1113 m^2/g) and lower oxygen mobility than MgO and Al_2O_3, did not deactivate either, suggesting that other factors also play a role, depending on the catalyst and the support.

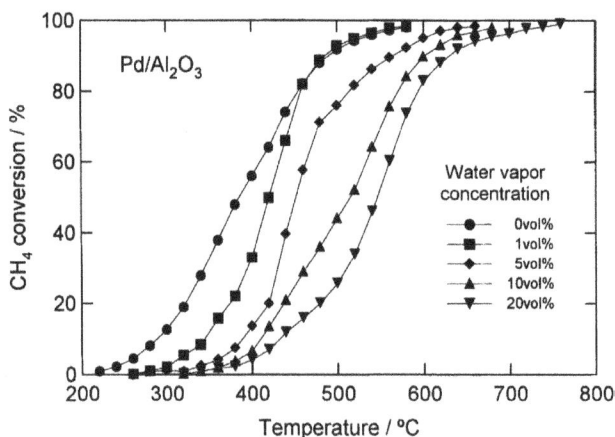

Figure 6. Catalytic combustion of CH_4 over 1.1 wt.% Pd/Al_2O_3 with different amounts of water added (vol.%). Reaction conditions: CH_4, 1 vol.%; O_2, 20 vol.%; H_2O, 0–20 vol.%; N_2, balance; GHSV 48,000 h^{-1}. Reprinted with permission from [16]. Copyright © 2002 Elsevier.

Another study compared the stability of Pd/SiO_2 and Pd/ZrO_2 during CH_4 oxidation using a dry feed gas [53]. The data (Figure 7) show that the Pd/ZrO_2 is stable after 40 h time-on-stream, while the CH_4 conversion over the Pd/SiO_2 catalyst increases from 13% to 32% in the first 3 h, and then decreases to 22% after 96 h (see Figure 7). Although the Pd/ZrO_2 catalyst is more stable than the Pd/SiO_2 catalyst, its conversion is lower than for the Pd/SiO_2 catalyst. The lower deactivation rate observed on the Pd/ZrO_2 is consistent with the higher oxygen mobility of this catalyst compared to Pd/SiO_2, as noted above.

Metal-support interactions, support stability and the tendency of the support to encapsulate Pd, may also play a role in the deactivation of Pd catalysts during CH_4 oxidation. Gannouni *et al.* [58] compared Pd catalysts supported on silica and mesoporous aluminosilicas and showed that, according to the light-off curves measured with 1% CH_4, 4% O_2 in He, CH_4 oxidation activity is enhanced on the pure silica support, whereas on the aluminosilica, the beneficial effect of Al^{3+} on metal dispersion and catalytic activity is counterbalanced by partial metal encapsulation. Above 500 °C in the presence of H_2O, the structural collapse of the support, metal sintering, and metal encapsulation by the support all occur [58]. Similar effects were reported with SiO_2 supports by Zhu *et al.* [59]. SiO_2 desorbs chemisorbed H_2O (silanol groups –Si-OH) at ~397 °C [46] and the formation of hydroxides according to the reaction: $SiO_{2 (s)} + 2H_2O_{(g)} \leftrightarrow Si(OH)_{4 (g)}$ is feasible at temperatures above 700 °C [60,61]. Hydroxyl mobility can change the extent of metal-support interactions [45,46]. Zhu *et al.* [59] reported the encapsulation of PdO by SiO_2 during CH_4 oxidation at only 325 °C. The authors suggested that silica migration by (i) formation of a palladium silicide during H_2 reduction at 650 °C that is subsequently oxidized during CH_4 oxidation and (ii) migration of SiO_2 during CH_4 oxidation caused by the water formed during reaction, are important related factors facilitating the encapsulation of PdO by the SiO_2. Migration of SiO_2 onto the metal crystallites in other catalyst systems containing H_2O has also been reported in the literature [46,62].

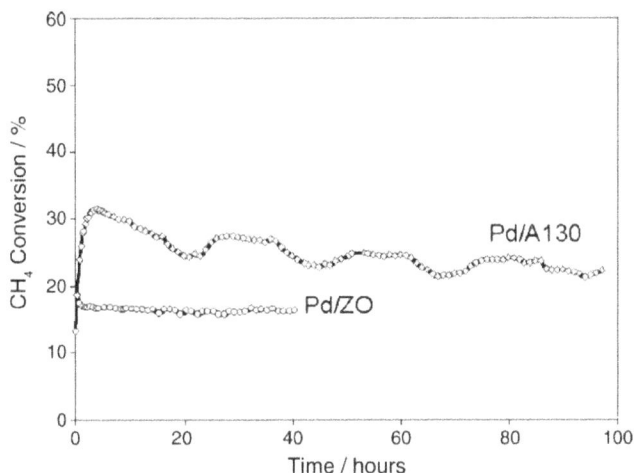

Figure 7. Methane conversion over time of Pd/ZrO_2 and Pd/Aerosil130 catalysts. Reaction conditions: 1.5% CH_4; 6% O_2; total flow = 90 $cm^3 min^{-1}$, balanced in He; temperature = 325 °C; catalyst mass = 0.2 g. Reprinted with permission from [53]. Copyright © 2005 Elsevier.

Yoshida *et al.* [63] also examined the effects of various metal oxide supports of Pd on the low temperature oxidation of CH_4 as summarized in Table 3. The catalytic activity varies with the support, but the support oxides with moderate acid strength (Al_2O_3 and SiO_2) give maximum CH_4 conversion. For these catalysts higher activity corresponds to a higher oxidation state of Pd (bulk PdO). The lower activity of Pd on basic supports is attributed to the formation of binary oxides from PdO and the support (such as Pd/MgO_x), in spite of a high Pd oxidation state.

The effect of metal oxides added to Pd/Al_2O_3 to improve the hydrothermal stability has been reported by Liu *et al.* [36] who showed in particular, that the addition of NiO or MgO improved the hydrothermal stability of Pd/Al_2O_3 through the formation of $NiAl_2O_4$ and $MgAl_3O_4$ spinel structures. According to the authors, the spinel results in weakened support acidity that suppresses the formation of $Pd(OH)_2$ during hydrothermal aging.

Table 3. Effect of support on properties of 5 wt.% Pd catalysts and their CH_4 oxidation conversion. Data adapted from [63].

Support	Support Acid Strength	Pd Dispersion		CH_4 conversion [a], %
	(H_o)	Fresh	Used	
MgO	22.3	0.21	0.20	12
ZrO_2	9.3	0.41	0.12	3
Al_2O_3	3.3	0.35	0.20	59
SiO_2	-5.6	0.09	0.11	58
SiO_2-ZrO_2	-8.2	0.16	0.13	20
SiO_2-Al_2O_3	-11.9	0.12	0.06	10
SO_4^{2-}-ZrO_2	-13.6	-	0.02	11

[a] measured at 350 °C in 0.25% CH_4/3%O_2 in He at GHSV of 1,200,000 h^{-1}.

A comparison of initial CH_4 oxidation activity as a function of temperature for Pd-Pt catalysts on Al_2O_3, ZrO_2, $LaMnAl_{11}O_{19}$, Ce-ZrO_2, and Y-ZrO_2 was reported by Persson et al. [64]. Monolith catalysts were tested in a tubular quartz flow reactor at atmospheric pressure in 1.5 vol.% CH_4 in dry air and at a space velocity of 250,000 h^{-1}. In steady-state experiments, reaction temperature was set initially at 470 °C and then increased to 720 °C stepwise in 50 °C increments, with 1-h holds at each temperature. The Pd-Pt/Al_2O_3 catalyst had the highest activity at lower temperatures (470–570 °C), while the Pd-Pt/Ce-ZrO_2 catalyst had the highest activity between 620 °C and 800 °C [64]. The authors suggested that the higher surface area of the Al_2O_3 compared to the other supports (e.g., 90 m^2/g for Al_2O_3 versus 10 m^2/g for Ce-ZrO_2) accounts for the higher activity of Pd-Pt/Al_2O_3 at lower temperatures, due to higher dispersion of Pd-Pt oxides, while at higher reaction temperatures the Pd-Pt catalyst probably undergoes reduction to the metal. A combination of lower activity for Pd metal and its propensity for rapid sintering probably explain the lower activity. The authors also suggested that the Ce-ZrO_2 likely enhances the stability of the PdO, similar to the enhanced stability observed on CeO_2 [30]. In addition, ZrO_2 has high oxygen mobility [30] and the ability to re-oxidize metallic Pd into PdO should be higher. Indeed, Pd/alumina is re-oxidized very slowly, whereas Pd supported on ceria-stabilized ZrO_2 is re-oxidized more rapidly.

Since H_2O adsorption on the Pd and/or the support is an important step in inhibiting CH_4 oxidation over Pd, support hydrophobicity may be expected to impact the inhibition effect of H_2O. Araya et al. [53] studied this effect on the deactivation of Pd-based catalysts by preparing 1 wt.% Pd on two different commercial silicas, Aerosil130 and Aerosil R972. The Aerosil R972 is hydrophobic since the OH groups have been replaced by methyl groups. Both 1% Pd/A130 and 1% Pd/R972 were tested at 325 °C in a gaseous mixture of 1.5% CH_4 and 6% O_2 in He at a total flow rate of 90 cm^3 min^{-1} with addition of 3% H_2O after 2 h As shown in Figure 8, the effect of H_2O addition to the feed gas is approximately the same for the hydrophobic silica, Pd/R972, and the hydrophilic Pd/A130. In both cases, a large decrease in CH_4 conversion is observed with the introduction of H_2O to the reactor. The authors reported a reaction order with respect to H_2O of −0.25 for both Pd/A130 and Pd/R972, emphasizing that the hydrophobicity of the support does not affect the extent of H_2O inhibition observed on either catalyst.

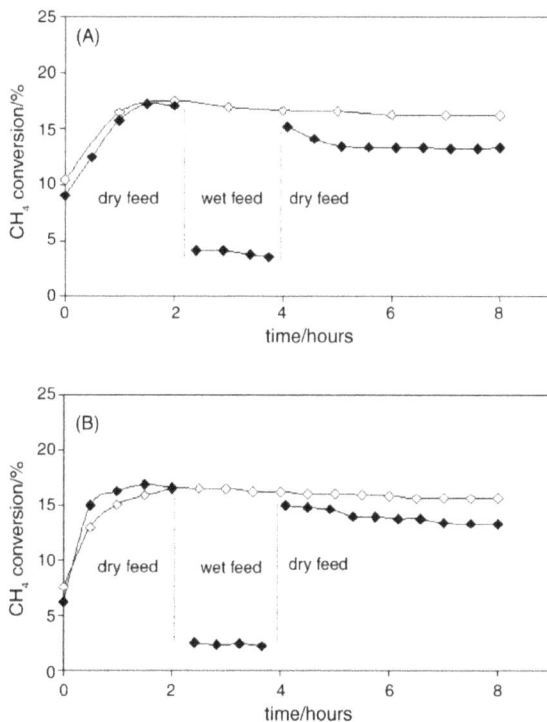

Figure 8. (A) Pd/Aerosil130 catalyst, **(B)** Pd/R972 catalyst. Reaction conditions: total flow $= 90$ cm^3 (STP) min^{-1}, temperature $= 325$ °C; catalyst mass $= 0.2$ g. Open symbols: dry feed 1.5% CH$_4$; 6% O$_2$;balance He; closed symbol: wet feed 1.5% CH$_4$; 6% O$_2$ with 3% H$_2$O, balance He. Reproduced with permission from [53]. Copyright © 2005 Elsevier.

2.4. H₂O Inhibition and Hydroxyl Formation

Although Pd(OH)$_2$ has been postulated as a cause for deactivation of PdO catalysts in the presence of H$_2$O [18,31,32,40], and while this mechanism is consistent with many of the observations discussed above, recent evidence obtained from FTIR and isotopic labeling experiments that monitor the formation and conversion of hydroxyls on the catalyst surface during reaction, suggest an alternative mechanism of deactivation.

Using DRIFTS, Persson *et al.* [35] reported an increase in signal intensity from surface hydroxyls weakly H-bonded to the support (3200–3800 cm^{-1}) [65] after introducing 1.5% CH$_4$ in air to a PdO/Al$_2$O$_3$ catalyst at low temperature (200 °C; Figure 9). The peak at 3016 cm^{-1} in Figure 9a, assigned to gas phase CH$_4$, increases with time-on-stream because of catalyst deactivation. The hydroxyls have characteristic adsorptions at 3733, 3697, 3556 and 3500 cm^{-1}, with the hydroxyls at 3697 and 3733 cm^{-1} assigned to bridged and terminal isolated hydroxyl species, respectively. Upon CH$_4$ removal from the feed (Figure 9b), the peaks associated with OH species remain, highly consistent with a slow desorption of OH species produced during CH$_4$ oxidation. Hence, Persson *et al.* [35] suggested that catalyst deactivation on PdO/Al$_2$O$_3$ might be due to the formation and accumulation of hydroxyls on the catalyst surface, bound either to the PdO, Al$_2$O$_3$ or the interface between the two [30]. Gao *et al.* [32] reported similar hydroxyl bands at 3733, 3697, 3556 and 3500 cm^{-1} during lean-burn CH$_4$ oxidation (0.4% CH$_4$ in air) at 250 °C. The FTIR spectra from reaction with 2 vol.% H$_2$O added to the CH$_4$-O$_2$ feed also yield a broad band at 3445 cm^{-1} that is associated with OH species on Al$_2$O$_3$ [32].

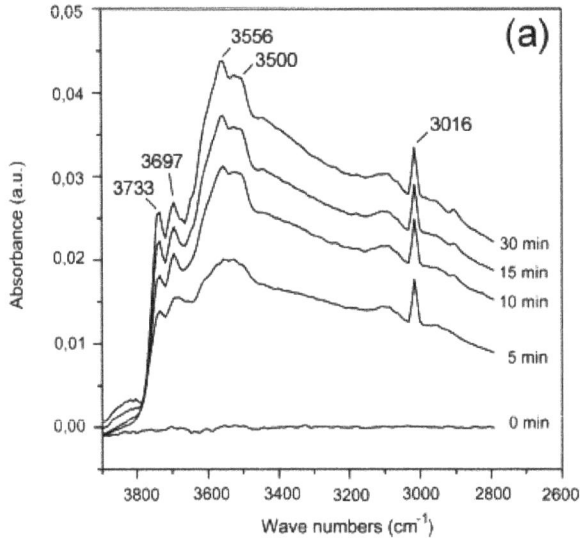

Figure 9. *Cont.*

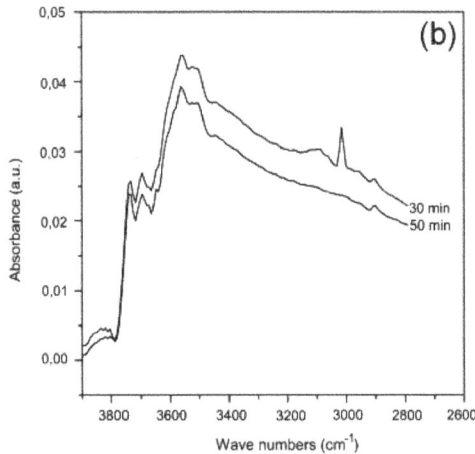

Figure 9. FTIR spectra of 5 wt.% Pd/Al$_2$O$_3$ at 200 °C (**a**) during the CH$_4$-O$_2$ reaction; (**b**) desorption when CH$_4$ was removed. Reproduced with permission from [35]. Copyright© 2007 Elsevier.

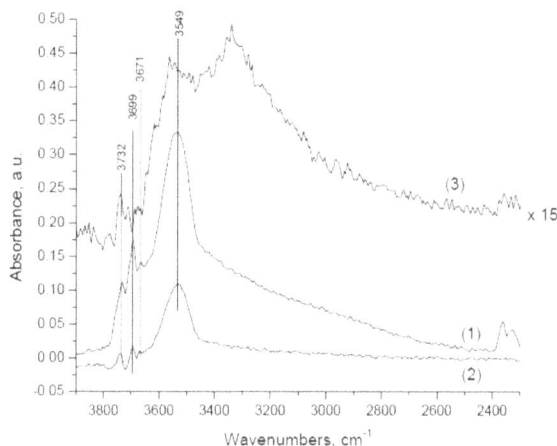

Figure 10. FTIR spectra at highest surface coverage and 350 °C on (1) PdO/Al$_2$O$_3$ during CH$_4$-O$_2$ reaction, (2) PdO/Al$_2$O$_3$ and (3) Al$_2$O$_3$ when injecting H$_2$O pulses. Reproduced with permission from [30]. Copyright © 2004 Elsevier.

Ciuparu *et al.* [30] also identified three well-defined peaks at 3732 (OH$_I$), 3699 (OH$_{II}$), and 3549 (OH$_{III}$) cm^{-1} associated with surface hydroxyls generated during CH$_4$ oxidation on a PdO/Al$_2$O$_3$ catalyst (3.5 wt % Pd) at 350 °C using a feed gas of 0.128% CH$_4$ and 17.3% O$_2$ in He/N$_2$ (Figure 10). The spectrum was compared to that measured at the same temperature when injecting pulses of ~3% H$_2$O into an air flow over the PdO/Al$_2$O$_3$ catalyst and the Al$_2$O$_3$ support (see Figure 10). Since Al$_2$O$_3$ has been shown to have a significantly lower hydroxyl coverage compared to PdO/Al$_2$O$_3$ when injecting H$_2$O pulses at 350 °C (the spectrum of Al$_2$O$_3$ is magnified by a factor of 15 in Figure 10), they concluded that the three peaks are associated with the presence of OH adsorbed on the PdO catalyst surface. The higher hydroxyl coverage during CH$_4$ oxidation compared to pulse injection of H$_2$O onto the PdO/Al$_2$O$_3$ catalyst, indicates that (1) adsorbed H$_2$O is dissociated on the surface of PdO/Al$_2$O$_3$ and (2) hydroxyls formed from H$_2$O pulses are less strongly bound to the surface than hydroxyls produced by the CH$_4$ oxidation reaction.

Since the frequencies of the OH$_I$ and OH$_{II}$ species are shifted to higher wave numbers for OH species more weakly bound to Pd, Ciuparu *et al.* [30] suggested that the high frequency peaks (OH$_I$, OH$_{II}$) can be assigned to terminal and bridged hydroxyl species, respectively, and the low frequency peak at ~3549 cm^{-1} with broad maximum values can be associated with OH species bound to different sites (multi-bound OHs; OH$_{III}$) (Figure 10). Transient temperature experiments show that the hydroxyl binding energy increases in the order OH$_I$ < OH$_{II}$ < OH$_{III}$ [30].

The peak areas of the terminal, bridged, and multi-bound hydroxyls were monitored with time-on-stream at different temperatures during reaction, as illustrated by Figure 11 for reaction at 175 °C [30]. Upon removal of CH$_4$ from the feed, the peak areas for the bridged and multi-bound OH species continue to increase, whereas the area of the terminal OH species decreases (Figure 11). This decrease is attributed to the conversion of terminal OH species to bridged or multi-bound OH species. Based on the intensities of the various hydroxyl species at different temperatures, the authors proposed the inter-conversion among the OH species as: $OH_{III} \leftrightarrow OH_{II} \leftrightarrow OH_I \rightarrow H_2O_{(g)}$ where only terminal OH species recombine and desorb as H_2O and the transformation of bridged OH species to terminal OH species is the rate determining step (RDS) for hydroxyl desorption and hence low temperature CH_4 oxidation [30]. Importantly the authors show that the surface coverage by the hydroxyls (Figure 11) correlates with the activity loss at low temperature, meaning that the activity loss and surface coverage have similar timescales, from which they conclude that the former is likely an effect of the latter [30].

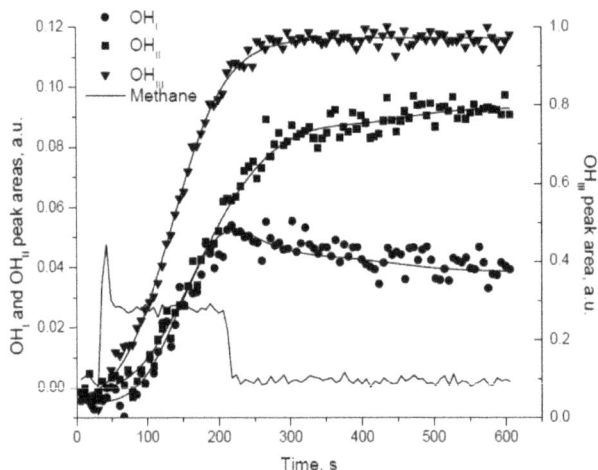

Figure 11. The normalized peak areas of different surface OH species generated during lean-CH_4-O_2 reaction at 175 °C. Reproduced with permission from [30]. Copyright © 2004 Elsevier.

FTIR spectra measured during CH_4 oxidation at 325 °C with 0.1% CH_4/4%O_2 in He over a series of 3 wt.% PdO catalysts supported on Al_2O_3, MgO, TiO_2 and MCM-41 [44] show that the hydroxyl coverage is dependent on the support. On Al_2O_3, well defined peaks similar to those identified by Ciuparu et al. [30] are observed, but no common peak among all catalysts that would provide evidence for Pd-OH bond formation, are present. Furthermore the large contribution from OH bonding on the supports makes it impossible to directly identify the presence of $Pd(OH)_2$ on these supports [32,44]. However, by using ^{18}O isotopic labeling and FTIR, the authors demonstrate that peaks associated with the accumulation of hydroxyls on PdO are not present at 325 °C. Hence, the more recent evidence suggests that deactivation by $Pd(OH)_2$ formation is unlikely, in agreement with the experimental observation that $Pd(OH)_2$/C decomposes in N_2 at about 250 °C [66]. In addition, evidence from temperature-programmed desorption studies of H_2O adsorbed on PdO(101) thin films, suggests the formation of an OH-H_2O complex at low temperature (<127 °C) and low coverage (< ½ monolayer), whereas H_2O preferentially chemisorbs in molecular form at higher coverages [67].

Schwartz et al. [44] showed, however, that catalyst deactivation during CH_4 oxidation correlates with hydroxyl accumulation on the oxide support. The redox mechanism for CH_4 combustion on

Pd/PdO generally assumes dissociation of a CH_4 molecule to yield a methyl fragment and a hydroxyl group (CH_4 + Pd-O + Pd-* →Pd-OH + Pd-CH_3, where Pd-* represents a vacancy) [68,69]. H atoms are abstracted sequentially from the methyl group by neighboring Pd-O to form surface hydroxyl groups (Pd-OH). Recombination of surface hydroxyls yields water and a surface vacancy (2Pd-OH→H_2O + Pd-O +Pd-*), that is regenerated by oxygen (2Pd-* + O_2→2Pd-O) [68,69]. Based on their experimental studies, Schwartz et al. [44,57], proposed that during lean-burn CH_4 oxidation, O_2 molecules dissociate on Pd-* sites and exchange with oxygen on the support so that Pd active sites are re-oxidized with oxygen atoms from the support during the catalytic reaction as follows:

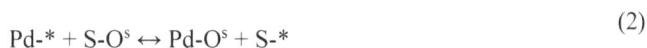

$$Pd\text{-}O + S\text{-}* \leftrightarrow Pd\text{-}* + S\text{-}O \tag{1}$$

$$Pd\text{-}* + S\text{-}O^s \leftrightarrow Pd\text{-}O^s + S\text{-}* \tag{2}$$

and overall:

$$Pd\text{-}O + S\text{-}O^s \leftrightarrow Pd\text{-}O^s + S\text{-}O \tag{3}$$

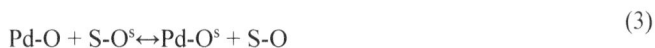

where S represents the support, S-* is an O vacancy on the support and O^s represents an O atom associated with the solid oxide. This proposed mechanism suggests the possibility that a primary cause for catalyst deactivation is hydroxyl accumulation on the support, which hinders oxygen migration and exchange processes.

Evidence for O exchange with the support is provided by the isotopic labeling experiments summarized in Figure 12, during which $Pd^{18}O/Al_2{}^{16}O_3$ and $Pd^{18}O/Mg^{16}O$ were exposed to $^{18}O_2$/He flow at 400 °C [57]. An increase in $^{16}O^{18}O$ signal intensity with time is proposed to arise from oxygen exchange with the catalyst support [44]. The $^{16}O^{18}O$ signal (see lower, separate dashed line in Figure 12) is reduced when $H_2{}^{16}O$ is injected to the feed and is recovered when $H_2{}^{16}O$ is removed. Apparently, hydroxyl groups tend to migrate to the oxide support rather than desorb. By increasing the concentration of hydroxyl groups, through addition and dissociation of H_2O, oxygen exchange of Pd-* active sites with the oxide support (S-O^s) is interrupted. Thus, the number of PdO sites participating in the CH_4 oxidation reaction decreases with time, as H_2O dissociates and OH coverage of the support increases, with a consequent decrease in CH_4 conversion [44]. This proposed mechanism of catalyst deactivation is believed to occur at temperatures below 450 °C. Finally, the authors note that the rate of deactivation on Pd/Al_2O_3 catalysts, with higher concentrations of hydroxyl during reaction, is higher than on catalysts containing a support with higher oxygen mobility (Pd/MgO) [44,57].

Ciuparu et al. [70] also reported on pulsed experiments with $^{18}O_2$ over pure Pd and Pd/ZrO_2 catalysts, oxidized before reaction, to clarify the effect of hydroxyls on the surface oxygen exchange. They determined that due to the slow recombination of hydroxyls and hence H_2O desorption from the Pd catalyst surface during CH_4 oxidation (2Pd-OH→H_2O + Pd-O +Pd-*), the isotopic exchange of oxygen with the Pd sites (see Figure 13) occurs before H_2O desorption from the surface. The oxygen vacancies on the PdO surface resulting from H_2O desorption are thus rapidly filled by oxygen from the PdO bulk or oxide support (Pd-* + S-O^s↔Pd-O^s + S-*). In fact, in this unsteady-state experiment, the labeled oxygen pulsed through the catalyst bed, is purged from the reactor before H_2O is desorbed [70]. These observations are in agreement with the studies of Schwartz et al. [44,57] already discussed and confirm that the accumulation of hydroxyls on the Pd catalyst surface impedes the oxygen exchange and limits Pd catalyst activity.

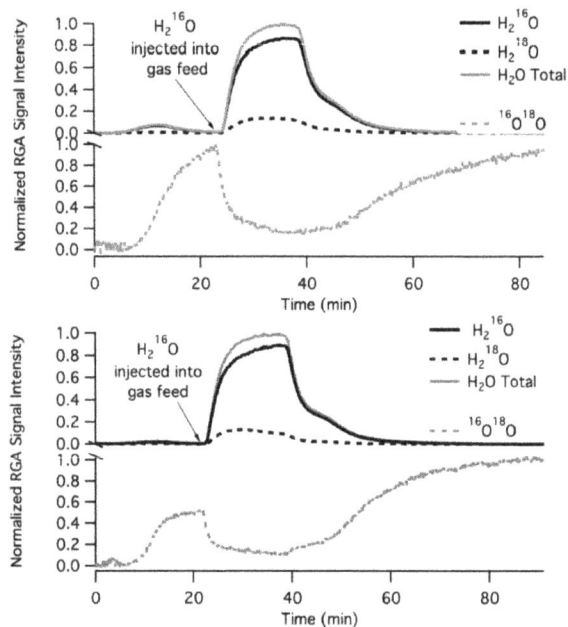

Figure 12. Oxygen exchange of (a) 3 wt.% Pd^{18}O/Al$_2{}^{16}$O$_3$ (top) and (b) 3 wt.% Pd^{18}O/Mg^{16}O (bottom) with catalyst supports in a flow of ^{18}O$_2$/He at 400 °C. H$_2{}^{16}$O was injected at some time to probe its effect on oxygen exchange. Reproduced with permission from [44]. Copyright ©2012 American Chemical Society.

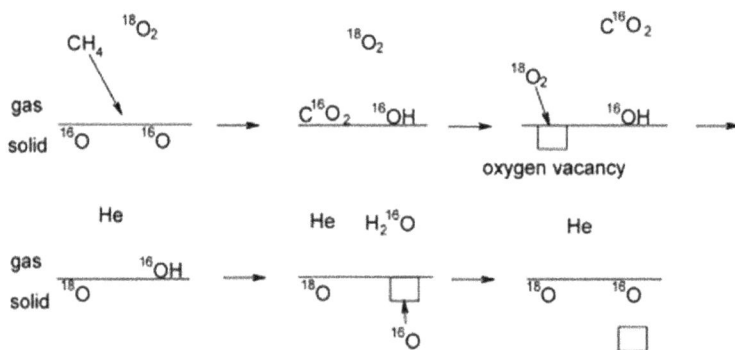

Figure 13. Schematic of oxygen exchange during CH$_4$ oxidation using labeled pulsed experiments. Reproduced with permission from [70]. Copyright © 2002 Elsevier.

3. The Use of Pd-Bimetallic Catalysts for CH4 Oxidation

Pd-bimetallic catalysts have been studied to improve stability of Pd catalysts for CH4 oxidation [19,51,71,72]. Pd-bimetallic catalysts are usually less active than Pd alone [64,73–75] simply because they contain less Pd, the most active metal for CH4 oxidation [20,25]. The lower activity of the bimetallic compared to Pd alone may also be due to the presence of smaller amounts of PdO as a result of alloy formation between Pd and Pt [64], or the transformation of PdO to Pd metal [76]. According to Ozawa *et al.* [77] the addition of Pt improves PdO/Al2O3 catalyst stability by preventing the growth of PdO and Pd–Pt particles during CH4 oxidation at high temperature (800 °C) [77].

Several studies have reported higher initial activity of Pd-bimetallic catalysts compared to Pd alone [17,19,51,78]. These researchers suggest that the second metal added to Pd dissociates O2 and the resulting O atoms are adsorbed by Pd, helping to maintain PdO active sites. Ishihara *et al.* [78] reported a T_{50} of 533 °C for a 1 wt.% Pd/Al2O3 catalyst, whereas for a Pd-Ni/Al2O3 catalyst (Pd:Ni = 9:1) T_{50} was 380 °C. In another study, it was reported that a higher dispersion of PdO on PdO-Pt/α-Al2O3 catalyst (27%) compared to PdO/α-Al2O3 (14%) results in higher initial activity and higher stability of the bimetallic catalyst [51] . After exposing the PdO/α-Al2O3 catalyst to the reaction feed stream for 6 h at 350 °C, an increase in average particle size from 8 to 11 nm is observed, whereas the average particle size does not change significantly for the PdO-Pt/α-Al2O3 catalyst [51].

Persson *et al.* [73] examined a series of Pd-bimetallics supported on Al2O3 finding that the metallic phase structure has a significant influence on the catalyst stability. For example, in several bimetallic systems (PdAg, PdCu, PdRh, and PdIr) separate phases of each metal oxide are formed after calcination (at 1000 °C for 1h followed by 1000 °C for 2h after loading the supported metal oxide powders onto a cordierite monolith) and this enhances catalyst stability in the case of the PdCu and PdAg (as measured stepwise at temperatures from 400–800 °C in 1.4% CH4 in dry air at a space velocity of 250,000 h^{-1}). Formation of a Co or Ni aluminate spinel in PdCo and PdNi bimetallics, however, does not improve catalyst stability, whereas alloy formation in PdPt and PdAu on Al2O3 increases hydrothermal stability in the presence of 15% H2O/air at 1000 °C for 10 h. In another study by Persson *et al.* [64], Pd-Pt bimetallic catalysts on various supports (alumina, zirconia) were shown to have higher thermal stability than monometallic Pd during CH4 oxidation in dry air (1.5% CH4 in air at a GHSV 250,000 h^{-1}). The stability of the Pd-Pt catalysts improved at lower temperatures (up to 620 °C). At temperatures of 520 °C and 570 °C CH4 conversion on Pd-Pt catalysts increased with time-on-stream. Above 620 °C (especially at 670 °C and 720 °C) conversion decreased with time-on-stream. Those catalysts with higher initial activity also had higher deactivation rates. The deactivation cannot be attributed to PdO decomposition because the initial activity test showed that PdO decomposition started at higher temperature (770 °C with 1.5 vol.% CH4 in air). According to XRD results, no PdO decomposition was observed at temperatures below 800 °C for the Pd/Al2O3, although PdO decomposition at ~700 °C may have yielded Pd that was not detectable by XRD (due to low concentration or high dispersion).

The amount of second metal added to the Pd can also affect the stability of the bimetallic catalyst. Persson *et al.* [74] reported that Pd-Pt bimetallic catalysts with Pd:Pt ratios of 2:1 and 1:1 are stable. Time-on-stream CH_4 oxidation experiments (in 1.5% CH_4 in air at a space velocity of 250,000 h^{-1}) for both a 5 wt.% Pd/Al_2O_3 and a 2:1 $Pd:Pt/Al_2O_3$ bimetallic with total metal loading of 5 wt.% were studied over a wide range of temperatures (470–720 °C) [64]. The temperature was increased from 470 °C to 720 °C stepwise by 50 °C and held for 1 h at each temperature. CH_4 conversion over the Pd/Al_2O_3 and $Pd-Pt/Al_2O_3$ catalyst decreased during the 1 h reaction time at each temperature. However, the decrease in conversion was lower for the bimetallic catalyst compared to the Pd catalyst. The decrease in activity was higher at higher temperatures (670 °C and 720 °C), especially for the Pd catalyst. *In situ* XRD spectra of the Pd-Pt bimetallic catalysts are shown in Figure 14. At room temperature, a sharp peak corresponding to Pd-Pt (111) and a small peak corresponding to PdO (101) are observed for the $PdPt-Al_2O_3$ catalyst. By increasing the temperature to 300 °C, the PdO peak disappears and then reappears at 500 °C. The Pd-Pt peak intensity reaches a maximum at 700 °C while the PdO peak disappears at this temperature. The formation of Pd-Pt instead of PdO is consistent with deactivation of the bimetallic catalyst at high temperature (700 °C).

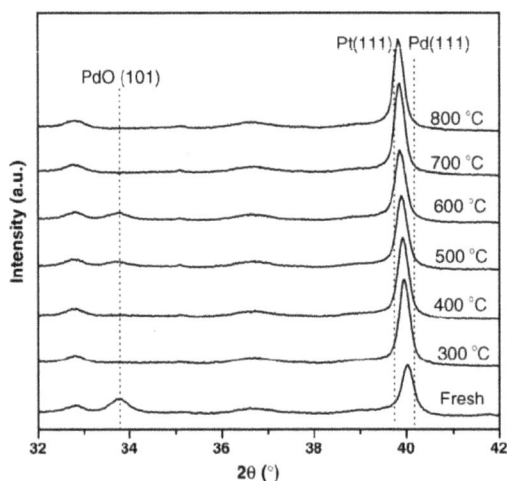

Figure 14. High-temperature *in situ* XRD profiles of $PdPt-Al_2O_3$ during heating. Reprinted with permission from [64]. Copyright © 2006 Elsevier.

Steady-state experiments using a 18.7 wt.% Pd/Al_2O_3 catalyst with different loadings of Pt (1.6, 3.1 and 3.9 wt.%) (Figure 15) reported by Ozawa *et al.* [77], also provide some insight into the improved stability of bimetallic catalysts as Pt content increases. In this study, reaction temperature was held at 800 °C and CH_4 combustion rate was measured over a 10 h period using a 1% CH_4 in air feed gas at a GHSV of 1,500,000 mL/(g_{cat}-h). Deactivation rate is shown to decrease as the Pt loading of the Pd-Pt bimetallics increases. For example, the combustion rate for the 18 wt.% Pd-3.9 wt.% Pt/Al_2O_3 decreases from 710 µmol s^{-1} g^{-1} to 460 µmol s^{-1} g^{-1} after 10 h, whereas it decreases to 400 µmol s^{-1} g^{-1} for the 18.4 wt.% Pd-1.6 wt.% Pt/Al_2O_3 catalyst.

XRD analysis of the catalysts studied by Ozawa *et al.* [77] after 10 h reaction indicates PdO to be present in the Pt-doped catalysts while no Pd^0 is observed. However, Pd^0 is present in the Pd monometallic catalyst, likely because of the decomposition of PdO at the high temperature of the reaction (800 °C). In addition, the crystallite size of the PdO (101) in the Pd catalyst is larger than for the Pd-Pt catalysts. Table 4 compares changes in PdO particle size and BET surface area before and after 10 h reaction for the same Pd and Pd-Pt catalysts. From these data it is clear that the extent of sintering of the Pd catalyst is greater than for the Pd-Pt catalysts. The time-on-stream conversion data reported by Ozawa *et al.* [77] (Figure 15) were fitted to a deactivation equation with two terms, the first representing rapid transformation of PdO to Pd^0 of the Pd-Pt alloy phase, and the second associated with the slow growth of the PdO crystallite [77]. The deactivation is affected more by the second term suggesting that particle growth of the PdO is the main cause of catalyst deactivation at the chosen reaction conditions [77].

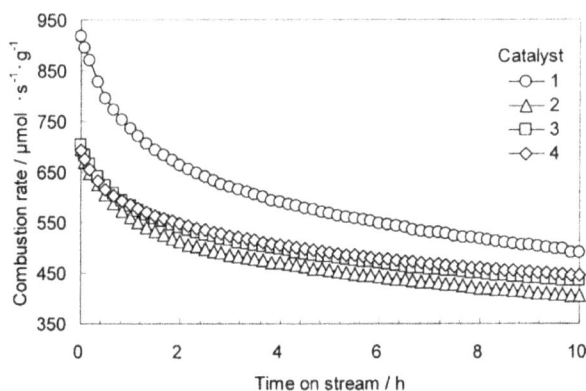

Figure 15. CH₄ combustion rate at 800 °C with time on stream. Combustion conditions: 1 vol.% CH₄, 99 vol.% air, CH₄/air flow = 450 L.h⁻¹, catalyst weight = 0.3 g. Catalyst 1, 2, 3, and 4 represent 18.7 wt.% Pd, 18.4 wt.% Pd-1.6 wt.% Pt, 18.1 wt.% Pd-3.4 wt.% Pt, and 18.0 wt.% Pd-3.9 wt.% Pt over Al₂O₃ catalysts. Reprinted with permission from [77]. Copyright © 2004 Elsevier.

Table 4. Changes in Pd and Pt-Pd catalyst properties before and after aging. Adapted from [77].

Catalyst, wt.% on Al₂O₃		18.7% Pd	18.4% Pd-1.6% Pt	18.1% Pd-3.1% Pt	18.0% Pd-3.9% Pt
BET area, m²/g	Fresh	56	51	51	46
	Aged	46	46	46	52
PdO size, nm	Fresh	12.5	15.3	15.2	14.7
	Aged	17.9	18.0	16.7	16.2

These results are in a good agreement with the results reported by Yamamoto *et al.* [72] in which a Pd-Pt bimetallic catalyst was more active for CH₄ conversion than Pd (as measured by the temperature required for 50% CH₄ conversion) and the conversion was maintained following 2500 h time-on-stream at 385 °C. XRD analyses showed that the crystallite growth as a function of time for both Pd (111) and PdO (101) was faster on the Pd (10 g/L)/Al₂O₃ catalyst than the Pd(10 g/L)-Pt (10 g/L)/Al₂O₃ catalyst. Hence one concludes that the presence of Pt retards the sintering of PdO.

Effects of H_2O on deactivation of Pt *versus* Pt-Pd catalysts have also been reported, at both thermal and hydrothermal aging conditions [17,19,71]. Pieck *et al.* [17] reported that the T_{50} of a 0.4% Pt-0.8% Pd/Al₂O₃ catalyst after thermal treatment at 600 °C for 4 h in wet air (60 cm³ min⁻¹ air flow with 0.356 cm³ h⁻¹ water), is ~50 °C lower than that obtained over a Pd catalyst. Lapisardi *et al.* [19] reported that a fresh Pd₀.₉₃-Pt₀.₀₇/Al₂O₃ catalyst (total metal loading 2.12 wt.% with Pd:Pt molar ratio of 0.93:0.07) is as active as a fresh Pd/Al₂O₃ catalyst in a dry feed [19]. Interestingly, the Pd₀.₉₃-Pt₀.₀₇/Al₂O₃ catalyst is less affected by addition of 10 vol.% steam to the feed stream than the 2.2 wt.% Pd/Al₂O₃ catalyst. The T_{50} for the Pd-Pt bimetallic increases from 320 °C to 400 °C when 10 vol.% steam is added to the feed stream, whereas the corresponding increase in T_{50} for the Pd/Al₂O₃ catalyst is from 320 °C to 425 °C. Thus, the Pd-Pt bimetallic, containing only 0.26 wt % Pt is more active and stable than the Pd catalyst for CH₄ oxidation in the presence of steam.

The stabilities of Pt and Pt-Pd catalysts each loaded on a wash coated monolith have also been reported [71]. A feed stream with 4067 ppmv CH₄ in air was reacted over these catalysts as reaction temperature increased from 300 to 700 °C stepwise in 50 °C increments. CH₄ conversion was monitored for a period of 1 h at each temperature. Subsequently the temperature was decreased to 300 °C also in 50 °C steps, again holding at each temperature for 1 h. The conversion of CH₄ was compared for both heating and cooling cycles. The results show that the Pt-Pd catalyst is more active than the Pt catalyst. The comparison between the heating and cooling cycles was also done for steam-aged catalysts, in which the catalysts were exposed to the feed stream at 650 °C with 5 vol.% water for 20 h. Table 5 lists the T_{50} for both fresh Pt and Pd-Pt catalysts, the steam-aged catalysts during tests in a dry feed and the steam-aged catalysts tested in a wet feed, containing 5 wt % H_2O. The data show that the fresh Pd-Pt catalyst is more active than the fresh Pt catalyst. Higher activities are also observed for steam-aged Pd-Pt catalysts tested in dry or wet feed gas.

Table 5. T_{50} for fresh and steam aged Pd and Pt-Pd catalysts operated in dry and wet feed. Combustion conditions: 4067 vol. ppm CH₄; total flow rate of 234.5 cm³/min; 500 mg catalyst; 5 vol.% water in wet feed. Adapted from [71].

	Temperature at 50% CH₄ conversion (T_{50}), °C		
Catalyst	Fresh	Steam-aged	Steam-aged
	Dry feed	Dry feed	Wet feed
Pt	540	610	610
4:1 Pt-Pd	400	470	535

4. Kinetic Consequences of H_2O on CH_4 Oxidation over Pd Catalysts

The rate of CH_4 oxidation over Pd catalysts is influenced by temperature, reactant partial pressures, the state of the Pd at reaction conditions (Pd^0, PdO or a sub-oxide), possibly Pd crystallite size (*i.e.*, may be structure-sensitive), and inhibition by products H_2O and CO_2. Consequently, kinetic parameters reported in the literature vary over wide ranges; this is especially true of the apparent activation energy for CH_4 oxidation [20]. As noted by Carstens *et al.* [79], rate data must account for the inhibition effect of H_2O when determining the activation barrier, but Ciuparu *et al.* [43] has shown that the correction is complicated by the fact that the effect of H_2O inhibition is temperature dependent. For example, the apparent activation energy for CH_4 oxidation over a Pd/ZrO_2 catalyst is estimated to be 180 kJ/mol from data measured at temperatures below 192 °C, whereas a value of 87 kJ/mol is obtained at temperatures above 192 °C [42]. The higher value of the apparent activation energy at lower temperatures is attributed to the strong inhibiting effect of H_2O on the Pd catalyst.

Zhu *et al.* [80] reported kinetic parameters for CH_4 oxidation over a series of model Pd and PdO surfaces and foils, and compared the values to literature data on supported Pd catalysts (Table 6). From Table 6 the reaction orders for CH_4 and O_2 are probably not sensitive to the structure of the Pd catalyst, although on the supported catalysts the reaction orders for H_2O vary from −0.25 to −1.3. Taking account of the error in the E_a estimates (±20 kJ/mol), Zhu *et al.* [80] concluded that on the large single-crystal model catalysts, the activation energies are similar and the combustion of CH_4 over Pd or PdO is not sensitive to the structure of the catalyst. Larger E_a values are reported for the Pd/oxide-supports (150–185 kJ/mol) corrected for the effect of H_2O (assuming an order of −1) [79], whereas the much smaller E_a for the Pd/zeolite catalysts (72–77 kJ/mol) are possibly associated with the high acidity and high OH surface concentration of zeolites, in obvious contrast to the observed inhibition by OH groups for PdO supported on conventional supports. The negative orders of reaction for H_2O are indicative of the varying degrees of inhibition of CH_4 oxidation by H_2O on Pd and PdO surfaces and catalysts.

The role of H_2O in the inhibition of PdO catalysts during CH_4 oxidation has been documented in this review to relate to the adsorption and slow desorption of H_2O on active sites during reaction. Kikuchi *et al.* [16] proposed a kinetic model assuming competitive adsorption between H_2O and CH_4 on PdO sites, where dissociative CH_4 adsorption was assumed to be the rate determining step (RDS) and the coverage by C-species was assumed to be negligible. The main elementary steps of the reaction are postulated as follows:

$$H_2O_{(g)} + s \rightarrow H_2O - s \tag{4}$$

$$CH_{4(g)} + 2s \rightarrow CH_3 - s + H - s \tag{5}$$

from which the following rate expression is derived [16]:

$$r = k_r \frac{P_{CH_4}}{1 + K_{H_2O}P_{H_2O}} \tag{6}$$

where r is the reaction rate, k_r is the rate constant for H abstraction, K_{H_2O} is the H_2O adsorption equilibrium constant, and P_{CH_4} and P_{H_2O} are the partial pressures of CH_4 and H_2O, respectively. K_{H_2O} is exponentially dependent upon the H_2O adsorption enthalpy (ΔH_{ads}). To increase the activity and durability of the Pd catalysts in the presence of H_2O, K_{H_2O} should be small according to the above reaction model. Based on the measured ΔH_{ads} values for water on supported Pd catalysts, water adsorbed on Pd/Al_2O_3 has the highest negative adsorption enthalpy ($\Delta H_{ads} = -49$ kJ mol^{-1}) compared to Pd/SnO_2 (-31 kJ mol^{-1}) and Pd/Al_2O_3-36NiO (-30 kJ mol^{-1}) (Table 7) despite the lower activation energy calculated for Pd/Al_2O_3 (see Table 7) [16]. A higher $|\Delta H_{ads}|$ implies stronger H_2O adsorption on the surface and is evidence of a higher coverage of active sites by H_2O molecules on Pd/Al_2O_3 catalysts and consequently lower catalyst activity. However, the larger negative enthalpy also predicts a more rapid decrease in K_{H2O} with increasing temperature for Pd/Al_2O_3.

Table 6. Kinetic parameters for CH_4 oxidation over Pd catalysts.

Catalyst	E_a, kJ/mol	Reaction order			T Range °C	Refs
		CH$_4$	O$_2$	H$_2$O		
Model Catalysts						
Pd foil	125	0.7	−0.1	0.05	296–360	[81]
Pd (111)	140	0.7	−0.1	0.05	296–360	[80]
Pd (100)	130	0.9	0.01	0.07	296–360	[80]
PdO foil	125	0.7	0.2	−0.9	296–360	[80]
PdO(111)	140	0.8	−0.1	−0.9	296–360	[80]
PdO(100)	125	0.8	0.1	−1.0	296–360	[80]
Supported Catalysts						
Pd black	135	0.7	0.1	−0.8	296–360	[81]
8.5% Pd/Al$_2$O$_3$	150	1	0	-1	232–360	[80,82]
0.5% Pd/Al$_2$O$_3$^{a}	60	0.90	0.08	−1.3 to −0.9	240–400	[83]
10% Pd/ZrO$_2$	185	1	0	−1	232–360	[80,82]
5% Pd/ZrO$_2$	185	1.1	0.1	−1.0	250–280	[68]
1% Pd/ZrO$_2$	172	1	0	−1.0	227–441	[53]
1% Pd/SiO$_2$	-	1	0	−0.25	227–441	[53]
2.8% Pd/H-Mord.	77	0.7	-0.1	−0.4	342–417	[33]
2.5% Pd-H-beta	72	0.5	0.2	−0.5	342–417	[33]

^{a} E_a determined under dry reaction conditions, correction for H_2O inhibition.

The larger negative value in the order of H_2O for the 1% Pd/ZrO_2 catalyst, compared to the Pd/SiO_2 catalyst, as reported by Araya et al. [53] (Table 6), reflects stronger H_2O adsorption on ZrO_2 than on the SiO_2 [53]. Hurtado et al. [83] observed a change in the power-law reaction order of H_2O from −1.3 to −0.9 as temperature increased from 300 °C to 350 °C using a H_2O-CH_4-O_2 reactant mixture and a commercial 0.5 wt.% Pd/γ-Al_2O_3 catalyst. Considering the equation proposed by Kikuchi et al. [16], with $K_{H_2O}P_{H_2O} \gg 1$, the H_2O reaction order will reduce to −1 but if $K_{H_2O}P_{H_2O}$ is small, the H_2O reaction order reduces to a value approaching zero.

Table 7. Estimated kinetic parameters for CH_4 oxidation using the rate equation $r = k_r \dfrac{P_{CH_4}}{1+K_{H_2O}P_{H_2O}}$ [16].

Catalyst	Pd loading (wt.%)	E_a kJ/mol	ΔH_{ads} for H_2O kJ/mol
Pd/Al₂O₃	1.1	81	−49
Pd/SnO₂	1.1	111	−31
Pd/Al₂O₃-36NiO	1.1	90	−30

Hurtado et al. [83] also attributed the inhibition effect of H_2O during reaction to the adsorption of H_2O on Pd catalysts. Based on this assumption the authors examined several Eley-Rideal, Langmuir-Hinshelwood and Mars-van Krevelen kinetic models finding that by considering competitive adsorption between H_2O and CH_4 on Pd oxide sites and slow desorption of products, the following kinetic model could be derived:

$$r = \frac{k_1 k_2 P_{CH_4} P_{O_2}}{k_1 P_{O_2}(1+K_{H_2O}P_{H_2O})+2k_2 P_{CH_4}+(k_1 k_2/k_3)P_{O_2}P_{CH_4}} \tag{7}$$

where k_1, k_2, and k_3 are the rate constants for (1) irreversible oxygen adsorption, (2) surface reaction with CH_4, and (3) product desorption steps in the mechanistic sequence, respectively. This model provides the best fit of their measured rate data. The ΔH_{ads} for water estimated from equation (7) is −54.5 kJ/mol, in agreement with the data of Table 7. The inhibiting effects of H_2O are assumed to be a consequence of a competitive adsorption between CH_4 and H_2O on PdO sites. Deactivation by H_2O was previously thought to be due to formation of inactive $Pd(OH)_2$ that does not participate in the CH_4 oxidation reaction and is reversible at temperatures above 250 °C [66]. Hurtado et al. [83] also note that the formation of $Pd(OH)_2$ is thermodynamically favored from PdO sites rather than from Pd^0. However, the more recent mechanism involving H_2O inhibition of the O exchange between Pd sites and oxide supports, proposed by Schwartz et al. [44,57] (see earlier discussion) appears to be supported by more definitive data.

5. Conclusions

Studies of the past decade provide new insights into the effects of H_2O on Pd catalysts during CH_4 oxidation, especially at lower temperatures. The principal effects of H_2O are:

(a) reaction inhibition by H_2O adsorption
(b) deactivation due to formation of $Pd(OH)_2$ and
(c) H_2O-assisted sintering at high reaction temperatures (>500 °C)

Reaction inhibition by H_2O increases with (a) decreasing reaction temperature at <500 °C and (b) higher H_2O concentrations, while this effect is generally negligible at >500 °C. O surface mobility of supports apparently influences H_2O inhibition, i.e., high O mobility (on CeO_2 and ZrO_2) results in less inhibition by H_2O than for Al_2O_3.

The main cause of partially reversible deactivation has been related to hydroxyl adsorption on the support and PdO. Although earlier studies suggested that formation of inactive $Pd(OH)_2$ could be the cause of deactivation, recent studies provide definitive evidence that adsorbed hydroxyls suppress O exchange between the support and Pd active sites causing suppression of catalyst activity.

H_2O-assisted sintering of supported Pd catalysts is observed at >500 °C. Catalysts with stabilized supports or core-shell structures have higher resistance to hydrothermal sintering. Several studies show that Pd bimetallic catalysts also improve catalyst stability, although explanations for the role of the second metal are not well-defined. Suppression of PdO sintering, enhanced oxygen mobility and suppression of hydroxide formation are postulated to play a key role in higher stability of Pd bimetallic catalysts.

Rate expressions from kinetic studies of CH_4 oxidation at conditions relevant to natural gas vehicles are based on the assumptions of (a) product inhibition by H_2O is a consequence of a competitive adsorption mechanism between CH_4 and H_2O on PdO sites; and (b) deactivation by H_2O is due to the formation of inactive $Pd(OH)_2$. None of the previous kinetic studies have linked the observed kinetic effects of H_2O to O mobility that recent studies show is critical during CH_4 oxidation.

Acknowledgements

The financial support of Westport Innovations Inc. and Natural Sciences and Engineering Research Council of Canada (NSERC) is gratefully acknowledged.

References

1. US Government. The World Factbook 2013–14. Washington, DC: Central Intelligence Agency, 2013. Available online: https://www.cia.gov/library/publications/the-world-factbook/geos/xx.html (accessed on 1 September 2013).
2. Natural Gas Vehicle Knowledge Base, Statistics. Available online: http://www.iangv.org/category/stats/2014 (Accessed on 25 March 2015).
3. Bartholomew, C.H.; Farrauto, R.J. *Fundamentals of Industrial Catalytic Processes*, 2nd ed.; J. Wiley and Sons Inc: Hoboken, NJ, USA, 2006.
4. Barbier, J., Jr.; Duprez, D. Steam Effects in Three-Way Catalysis. *Appl. Catal. B* **1994**, *4*, 105–140.
5. Lox, E.S.J.; Engler, B.H. In *Handbook of Heterogeneous Catalysis*; Ertl, G., Knozinger, H., Weitkamp, J., Eds.; Wiley-VCH: Weinheim, Germany, 1997; Volume 4.
6. Ciuparu, D.; Lyubovsky, M.R.; Altman, E.; Pfefferle, L.D.; Datye, A. Catalytic Combustion of Methane over Palladium-Based Catalysts. *Catal. Rev. Sci. Eng.* **2002**, *44*, doi:10.1081/CR-120015482.
7. Silver, R.G.; Summers, J.C. In *Catalysis and Automotive Pollution Control III Studies in Surface Science and Catalysis*; Frennet, A., Bastin, J.M., Eds.; Elsevier Science: Amsterdam, The Netherlands, 1995; pp. 871–884.

8. Klingstedt, F.; Neyestanaki, A.K.; Byggningsbacka, R.; Lindfors, L.; Lundén, M.; Petersson, M.; Tengström, P.; Ollonqvist, T.; Väyrynen, J. Palladium Based Catalysts for Exhaust Aftertreatment of Natural Gas Powered Vehicles and Biofuel Combustion. *Appl. Catal. A* **2001**, *209*, 301–316.

9. Farrauto, R.J.; Hobson, M.C.; Kennelly, T.; Waterman, E.M. Catalytic Chemistry of Supported Palladium for Combustion of Methane. *Appl. Catal. A* **1992**, *81*, 227–237.

10. Lampert, J.K.; Kazi, M.S.; Farrauto, R.J. Palladium Catalyst Performance for Methane Emissions Abatement from Lean Burn Natural Gas Vehicles. *Appl. Catal. B* **1997**, *14*, 211–223.

11. Bounechada, D.; Groppi, G.; Forzatti, P.; Kallinen, K.; Kinnunen, T. Effect of Periodic lean/rich Switch on Methane Conversion over a Ce–Zr Promoted Pd-Rh/Al$_2$O$_3$ catalyst in the Exhausts of Natural Gas Vehicles. *Appl. Catal. B* **2012**, *119–120*, 91–99.

12. Gelin, P.; Primet, M. Complete Oxidation of Methane at Low Temperature over Noble Metal Based Catalysts: A Review. *Appl. Catal. B* **2002**, *39*, 1–37.

13. Choudhary, T.V.; Banerjee, S.; Choudhary, V.R. Catalysts for Combustion of Methane and Lower Alkanes. *Appl. Catal. A* **2002**, *234*, 1–23.

14. Centi, G. Supported Palladium Catalysts in Environmental Catalytic Technologies for Gaseous Emissions. *J. Mol. Catal. A* **2001**, *173*, 287–312.

15. Forzatti, P.; Groppi, G. Catalytic Combustion for the Production of Energy. *Catal. Today* **1999**, *54*, 165–180.

16. Kikuchi, R.; Maeda, S.; Sasaki, K.; Wennerström, S.; Eguchi, K. Low-Temperature Methane Oxidation over Oxide-Supported Pd Catalysts: Inhibitory Effect of Water Vapor. *Appl. Catal. A* **2002**, *232*, 23–28.

17. Pieck, C.L.; Vera, C.R.; Peirotti, E.M.; Yori, J.C. Effect of Water Vapor on the Activity of Pt-Pd/Al$_2$O$_3$ Catalysts for Methane Combustion. *Appl. Catal. A* **2002**, *226*, 281–291.

18. Roth, D.; Gélin, P.; Primet, M.; Tena, E. Catalytic Behaviour of Cl-Free and Cl-Containing Pd/Al$_2$O$_3$ Catalysts in the Total Oxidation of Methane at Low Temperature. *Appl. Catal. A* **2000**, *203*, 37–45.

19. Lapisardi, G.; Urfels, L.; Gélin, P.; Primet, M.; Kaddouri, A.; Garbowski, E.; Toppi, S.; Tena, E. Superior Catalytic Behaviour of Pt-Doped Pd Catalysts in the Complete Oxidation of Methane at Low Temperature. *Catal. Today* **2006**, *117*, 564–568.

20. Li, Z.; Hoflund, G.B. A Review on Complete Oxidation of Methane at Low Tempertaure. *J. Nat. Gas Chem.* **2003**, *12*, 153–160.

21. Colussi, S.; Gayen, A.; Camellone, M.F.; Boaro, M.; Llorca, J.; Fabris, S.; Trovarelli, A. Nanofaceted Pd-O Sites in Pd-Ce Surface Superstructures: Enhanced Activity in Catalytic Combustion of Methane. *Angew. Chem. Int. Ed.* **2009**, *48*, 8633–8636.

22. Cargnello, M.; Jaén, J.J.D.; Garrido, J.C.H.; Bakhmutsky, K.; Montini, T.; Gámez, J.J.C.; Gorte, R.J.; Fornasiero, P. Exceptional Activity for Methane Combustion over Modular Pd@CeO$_2$ Subunits on Functionalized Al$_2$O$_3$. *Science* **2012**, *337*, 713–717.

23. Hellman, A.; Resta, A.; Martin, N.M.; Gustafson, J.; Trinchero, A.; Carlsson, P.-A.; Balmes, O.; Felici, R.; van Rijn, R.; Frenken, J.W.M.; Andersen, J.N.; Lundgren, E.; Gronbeck, H. The Active Phase of Palladium during Methane Oxidation. *J. Phys. Chem. Lett.* **2012**, *3*, 678–682.

24. Briot, P.; Primet, M. Catalytic Oxidation of Methane over Palladium Supported on Alumina: Effect of Aging Under Reactants. *Appl. Catal.* **1991**, *68*, 301–314.

25. Gélin, P.; Urfels, L.; Primet, M.; Tena, E. Complete Oxidation of Methane at Low Temperature over Pt and Pd Catalysts for the Abatement of Lean-Burn Natural Gas Fuelled Vehicles Emissions: Influence of Water and Sulphur Containing Compounds. *Catal. Today* **2003**, *83*, 45–57.

26. Bartholomew, C.H. Mechanisms of Catalyst Deactivation. *Appl. Catal. A* **2001**, *212*, 17–60.

27. Kang, S.; Han, S.; Nam, S.; Nam, I.; Cho, B.; Kim, C.; Oh, S. Effect of Aging Atmosphere on Thermal Sintering of Modern Commercial TWCs. *Top. Catal.* **2013**, *56*, 298–305.

28. Shinjoh, H. Noble Metal Sintering Suppression Technology in Three-Way Catalyst: Automotive Three-Way Catalysts with the Noble Metal Sintering Suppression Technology Based on the Support Anchoring Effect. *Catal. Surv. Asia* **2009**, *13*, 184–190.

29. Fathali, A.; Olsson, L.; Ekstrom, F.; Laurell, M.; Andersson, B. Hydrothermal Aging-Induced Changes in Washcoats of Commercial Three-Way Catalysts. *Top. Catal.* **2013**, *56*, 323–328.

30. Ciuparu, D.; Perkins, E.; Pfefferle, L. *In Situ* DR-FTIR Investigation of Surface Hydroxyls on γ-Al_2O_3 Supported PdO Catalysts during Methane Combustion. *Appl. Catal. A* **2004**, *263*, 145–153.

31. Burch, R.; Urbano, F.J.; Loader, P.K. Methane Combustion Over Palladium Catalysts: The Effect of Carbon Dioxide and Water on Activity. *Appl. Catal. A* **1995**, *123*, 173–184.

32. Gao, D.; Wang, S.; Zhang, C.; Yuan, Z.; Wang, S. Methane Combustion over Pd/Al_2O_3 Catalyst: Effects of Chlorine Ions and Water on Catalytic Activity. *Chin. J. Catal.* **2008**, *29*, 1221–1225.

33. Park, J.; Kim, B.; Shin, C.; Seo, G.; Kim, S.; Hong, S. Methane Combustion over Pd Catalysts Loaded on Medium and Large Pore Zeolites. *Top. Catal.* **2009**, *52*, 27–34.

34. Stasinska, B.; Machocki, A.; Antoniak, K.; Rotko, M.; Figueiredo, J.L.; Gonçalves, F. Importance of Palladium Dispersion in Pd/Al_2O_3 Catalysts for Complete Oxidation of Humid Low-methane–air Mixtures. *Catal. Today* **2008**, *137*, 329–334.

35. Persson, K.; Pfefferle, L.D.; Schwartz, W.; Ersson, A.; Järås, S.G. Stability of Palladium-Based Catalysts during Catalytic Combustion of Methane: The Influence of Water. *Appl. Catal. B* **2007**, *74*, 242–250.

36. Liu, Y.; Wang, S.; Gao, D.; Sun, T.; Zhang, C.; Wang, S. Influence of Metal Oxides on the Performance of Pd/Al_2O_3 Catalysts for Methane Combustion Under Lean-Fuel Conditions. *Fuel Process. Technol.* **2013**, *111*, 55–61.

37. Zhu, G.; Han, J.; Zemlyanov, D.Y.; Ribeiro, F.H. Temperature Dependence of the Kinetics for the Complete Oxidation of Methane on Palladium and Palladium Oxide. *J. Phys. Chem. B* **2005**, *109*, 2331–2337.

38. Okumura, K.; Shinohara, E.; Niwa, M. Pd Loaded on High Silica Beta Support Active for the Total Oxidation of Diluted Methane in the Presence of Water Vapor. *Catal. Today* **2006**, *117*, 577–583.

39. Nomura, K.; Noro, K.; Nakamura, Y.; Yoshida, H.; Satsuma, A.; Hattori, T. Combustion of a Trace Amount of CH_4 in the Presence of Water Vapor over ZrO_2-Supported Pd Catalysts. *Catal. Lett.* **1999**, *58*, 127–130.

40. Cullis, C.F.; Nevell, T.G.; Trimm, D.L. Role of the Catalyst Support in the Oxidation of Methane over Palladium. *Faraday Trans.* **1972**, *68*, 1406–1412.

41. Eriksson, S.; Boutonnet, M.; Järås, S. Catalytic Combustion of Methane in Steam and Carbon Dioxide-Diluted Reaction Mixtures. *Appl. Catal. A* **2006**, *312*, 95–101.

42. Ciuparu, D.; Katsikis, N.; Pfefferle, L. Temperature and Time Dependence of the Water Inhibition Effect on Supported Palladium Catalyst for Methane Combustion. *Appl. Catal. A* **2001**, *216*, 209–215.

43. Ciuparu, D.; Pfefferle, L. Support and Water Effects on Palladium Based Methane Combustion Catalysts. *Appl. Catal. A* **2001**, *209*, 415–428.

44. Schwartz, W.R.; Ciuparu, D.; Pfefferle, L.D. Combustion of Methane over Palladium-Based Catalysts: Catalytic Deactivation and Role of the Support. *J. Phys. Chem. C* **2012**, *116*, 8587–8593.

45. Hansen, T.W.; DeLaRiva, A.T.; Challa, S.R.; Datye, A.K. Sintering of Catalytic Nanoparticles: Particle Migration Or Ostwald Ripening? *Acc. Chem. Res.* **2013**, *46*, 1720–1730.

46. Lamber, R.; Jaeger, N.; Schulz-Ekloff, G. Metal-Support Interaction in the Pd/SiO_2 System: Influence of the Support Pretreatment. *J. Catal.* **1990**, *123*, 285–297.

47. Nagai, Y.; Hirabayashi, T.; Dohmae, K.; Takagi, N.; Minami, T.; Shinjoh, H.; Matsumoto, S. Sintering Inhibition Mechanism of Platinum Supported on Ceria-Based Oxide and Pt-oxide–support Interaction. *J. Catal.* **2006**, *242*, 103–109.

48. Xu, Q.; Kharas, K.C.; Croley, B.J.; Datye, A.K. The Sintering of Supported Pd Automotive Catalysts. *ChemCatChem* **2011**, *3*, 1004–1014.

49. Escandón, L.S.; Niño, D.; Díaz, E.; Ordóñez, S.; Díez, F.V. Effect of Hydrothermal Ageing on the Performance of Ce-Promoted PdO/ZrO_2 for Methane Combustion. *Catal. Commun.* **2008**, *9*, 2291–2296.

50. Muto, K.; Katada, N.; Niwa, M. Complete Oxidation of Methane on Supported Palladium Catalyst: Support Effect. *Appl. Catal. A* **1996**, *134*, 203–215.

51. Narui, K.; Yata, H.; Furuta, K.; Nishida, A.; Kohtoku, Y.; Matsuzaki, T. Effects of Addition of Pt to PdO/Al_2O_3 Catalyst on Catalytic Activity for Methane Combustion and TEM Observations of Supported Particles. *Appl. Catal. A* **1999**, *179*, 165–173.

52. Zhang, B.; Wang, X.; M'Ramadj, O.; Li, D.; Zhang, H.; Lu, G. Effect of Water on the Performance of Pd-ZSM-5 Catalysts for the Combustion of Methane. *J. Nat. Gas Chem.* **2008**, *17*, 87–92.

53. Araya, P.; Guerrero, S.; Robertson, J.; Gracia, F.J. Methane Combustion over Pd/SiO_2 Catalysts with Different Degrees of Hydrophobicity. *Appl. Catal. A* **2005**, *283*, 225–233.

54. Lu, J.; Fu, B.; Kung, M.C.; Xiao, G.; Elam, J.W.; Kung, H.H.; Stair, P.C. Coking- and Sintering-Resistant Palladium Catalysts Achieved through Atomic Layer Deposition. *Science* **2012**, *335*, 1205–1208.

55. Forman, A.J.; Park, J.; Tang, W.; Hu, Y.; Stucky, G.D.; McFarland, E.W. Silica-Encapsulated Pd Nanoparticles as a Regenerable and Sintering-Resistant Catalyst. *ChemCatChem* **2010**, *2*, 1318–1324.

56. Adijanto, L.; Bennett, D.A.; Chen, C.; Yu, A.S.; Cargnello, M.; Fornasiero, P.; Gorte, R.J.; Vohs, J.M. Exceptional Thermal Stability of Pd@CeO$_2$ Core-Shell Catalyst Nanostructures Grafted Onto an Oxide Surface. *Nano Lett.* **2013**, *13*, 2252–2257.

57. Schwartz, W.R.; Pfefferle, L.D. Combustion of Methane over Palladium-Based Catalysts: Support Interactions. *J. Phys. Chem. C* **2012**, *116*, 8571–8578.

58. Gannouni, A.; Albela, B.; Zina, M.S.; Bonneviot, L. Metal Dispersion, Accessibility and Catalytic Activity in Methane Oxidation of Mesoporous Templated Aluminosilica Supported Palladium. *Appl. Catal. A* **2013**, *464–465*, 116–127.

59. Zhu, G.; Fujimoto, K.; Zemlyanov, D.Y.; Datye, A.K.; Ribeiro, F.H. Coverage of Palladium by Silicon Oxide during Reduction in H$_2$ and Complete Oxidation of Methane. *J. Catal.* **2004**, *225*, 170–178.

60. Jacobson, N.S.; Opila, E.J.; Myers, D.L.; Copland, E.H. Thermodynamics of Gas Phase Species in the Si–O–H System. *J. Chem. Thermodyn.* **2005**, *37*, 1130–1137.

61. Opila, E.; Jacobson, N.; Myers, D.; Copland, E. Predicting Oxide Stability in High-Temperature Water Vapor. *JOM* **2006**, *58*, 22–28.

62. Lund, C.R.F.; Dumesic, J.A. Strong Oxide-Oxide Interactions in Silica-Supported Fe$_3$O$_4$: III. Water-Induced Migration of Silica on Geometrically Designed Catalysts. *J. Catal.* **1981**, *72*, 21–30.

63. Yoshida, H.; Nakajima, T.; Yazawa, Y.; Hattori, T. Support Effect on Methane Combustion over Palladium Catalysts. *Appl. Catal. B* **2007**, *71*, 70–79.

64. Persson, K.; Ersson, A.; Colussi, S.; Trovarelli, A.; Järås, S.G. Catalytic Combustion of Methane over Bimetallic Pd–Pt Catalysts: The Influence of Support Materials. *Appl. Catal. B* **2006**, *66*, 175–185.

65. Guerrero, S.; Araya, P.; Wolf, E.E. Methane Oxidation on Pd Supported on High Area Zirconia Catalysts. *Appl. Catal. A* **2006**, *298*, 243–253.

66. Card, R.J.; Schmitt, J.L.; Simpson, J.M. Palladium-Carbon Hydrogenolysis Catalysts: The Effect of Preparation Variables on Catalytic Activity. *J. Catal.* **1983**, *79*, 13–20.

67. Kan, H.H.; Colmyer, R.J.; Asthagiri, A.; Weaver, J.F. Adsorption of Water on a PdO(101) Thin Film: Evidence of an Adsorbed HO-H$_2$O Complex. *J. Phys. Chem. C* **2009**, *113*, 1495–1506.

68. Fujimoto, K.; Ribeiro, F.H.; Avalos-Borja, M.; Iglesia, E. Structure and Reactivity of PdOx/ZrO$_2$ Catalysts for Methane Oxidation at Low Temperatures. *J. Catal.* **1998**, *179*, 431–442.

69. Chin, Y.; Iglesia, E. Elementary Steps the Role of Chemisorbed Oxygen, and the Effects of Cluster Size in Catalytic CH$_4$-O$_2$ Reactions on Palladium. *J. Phys. Chem. C* **2011**, *115*, 17845–17855.

70. Ciuparu, D.; Pfefferle, L. Contributions of Lattice Oxygen to the overall Oxygen Balance during Methane Combustion over PdO-Based Catalysts. *Catal. Today* **2002**, *77*, 167–179.

71. Abbasi, R.; Wu, L.; Wanke, S.E.; Hayes, R.E. Kinetics of Methane Combustion over Pt and Pt–Pd Catalysts. *Chem. Eng. Res. Des.* **2012**, *90*, 1930–1942.

72. Yamamoto, H.; Uchida, H. Oxidation of Methane over Pt and Pd Supported on Alumina in Lean-Burn Natural-Gas Engine Exhaust. *Catal. Today* **1998**, *45*, 147–151.

73. Persson, K.; Ersson, A.; Jansson, K.; Iverlund, N.; Järås, S. Influence of Co-Metals on Bimetallic Palladium Catalysts for Methane Combustion. *J. Catal.* **2005**, *231*, 139–150.

74. Persson, K.; Ersson, A.; Jansson, K.; Fierro, J.L.G.; Järås, S.G. Influence of Molar Ratio on Pd–Pt Catalysts for Methane Combustion. *J. Catal.* **2006**, *243*, 14–24.

75. Persson, K.; Jansson, K.; Järås, S.G. Characterisation and Microstructure of Pd and Bimetallic Pd–Pt Catalysts during Methane Oxidation. *J. Catal.* **2007**, *245*, 401–414.

76. Kuper, W.J.; Blaauw, M.; Berg, F.V.; Graaf, G.H. Catalytic Combustion Concept for Gas Turbines. *Catal. Today* **1999**, *47*, 377–389.

77. Ozawa, Y.; Tochihara, Y.; Watanabe, A.; Nagai, M.; Omi, S. Deactivation of Pt·PdO/Al$_2$O$_3$ in Catalytic Combustion of Methane. *Appl. Catal. A* **2004**, *259*, 1–7.

78. Ishihara, T. Effects of Additives on the Activity of Palladium Catalysts for Methane Combustion. *Chem. Lett.* **1993**, *22*, 407–410.

79. Carstens, J.N.; Su, S.C.; Bell, A.T. Factors Affecting the Catalytic Activity of Pd/ZrO$_2$ for the Combustion of Methane. *J. Catal.* **1998**, *176*, 136.

80. Zhu, G.; Han, J.; Zemlyanov, D.Y.; Ribeiro, F.H. The Turnover Rate for the Catalytic Combustion of Methane over Palladium is Not Sensitive to the Structure of the Catalyst. *J. Am. Chem. Soc.* **2004**, *126*, 9896–9897.

81. Monteiro, R.S.; Zemlyanov, D.; Storey, J.M.; Ribeiro, F.H. Turnover Rate and Reaction Orders for the Complete Oxidation of Methane on a Palladium Foil in Excess Dioxygen. *J. Catal.* **2001**, *199*, 291–301.

82. Ribeiro, F.H.; Chow, M.; Dallabetta, R.A. Kinetics of the Complete Oxidation of Methane over Supported Palladium Catalysts. *J. Catal.* **1994**, *146*, 537–544.

83. Hurtado, P.; Ordóñez, S.; Sastre, H.; Díez, F.V. Development of a Kinetic Model for the Oxidation of Methane over Pd/Al$_2$O$_3$ at Dry and Wet Conditions. *Appl. Catal. B* **2004**, *51*, 229–238.

Effect of Ce and Zr Addition to Ni/SiO$_2$ Catalysts for Hydrogen Production through Ethanol Steam Reforming

Jose Antonio Calles, Alicia Carrero, Arturo Javier Vizcaíno and Montaña Lindo

Abstract: A series of Ni/Ce$_x$Zr$_{1-x}$O$_2$/SiO$_2$ catalysts with different Zr/Ce mass ratios were prepared by incipient wetness impregnation. Ni/SiO$_2$, Ni/CeO$_2$ and Ni/ZrO$_2$ were also prepared as reference materials to compare. Catalysts' performances were tested in ethanol steam reforming for hydrogen production and characterized by XRD, H$_2$-temperature programmed reduction (TPR), NH$_3$-temperature programmed desorption (TPD), TEM, ICP-AES and N$_2$-sorption measurements. The Ni/SiO$_2$ catalyst led to a higher hydrogen selectivity than Ni/CeO$_2$ and Ni/ZrO$_2$, but it could not maintain complete ethanol conversion due to deactivation. The incorporation of Ce or Zr prior to Ni on the silica support resulted in catalysts with better performance for steam reforming, keeping complete ethanol conversion over time. When both Zr and Ce were incorporated into the catalyst, Ce$_x$Zr$_{1-x}$O$_2$ solid solution was formed, as confirmed by XRD analyses. TPR results revealed stronger Ni-support interaction in the Ce$_x$Zr$_{1-x}$O$_2$-modified catalysts than in Ni/SiO$_2$ one, which can be attributed to an increase of the dispersion of Ni species. All of the Ni/Ce$_x$Zr$_{1-x}$O$_2$/SiO$_2$ catalysts exhibited good catalytic activity and stability after 8 h of time on stream at 600 °C. The best catalytic performance in terms of hydrogen selectivity was achieved when the Zr/Ce mass ratio was three.

Reprinted from *Catalysts*. Cite as: Calles, J.A.; Carrero, A.; Vizcaíno, A.J.; Lindo, M. Effect of Ce and Zr Addition to Ni/SiO$_2$ Catalysts for Hydrogen Production through Ethanol Steam Reforming. *Catalysts* **2015**, *5*, 58–76.

1. Introduction

Energy sustainability and reduction of CO$_2$ emissions will be joined with a decrease in fossil fuel use and the development of green energies. In this sense, hydrogen could be the energy carrier of the future due to its clean and non-polluting nature [1–4]. However, the current hydrogen production routes imply the use of fossil fuel-derived products, like methane, as feedstock and, for this reason, the search of new alternatives for hydrogen production based on renewable resources is essential [5]. In line with this, hydrogen production from ethanol steam reforming is an attractive option, since ethanol can be obtained from a wide variety of biomass feedstocks, and therefore, it can minimize CO$_2$ emissions. In addition, ethanol has a high H/C atomic ratio, low toxicity and can be easily and safely manipulated and transported [6,7].

The ethanol steam reforming process (ESR) can be represented by the following equation:

$$CH_3CH_2OH + 3H_2O \rightleftarrows 6H_2 + 2CO_2 \tag{1}$$

which involves several steps. The main reaction mechanism comprises dehydrogenation or dehydration routes, which implies the formation of intermediates, like acetylene or ethylene, respectively. Coke formation reactions also participate in the reaction pathway leading to catalyst deactivation [8–10].

Ethanol steam reforming has been investigated using different catalysts [10,11]. In fact, Ni/SiO_2 has been widely used, because it provides good activity and, at the same time, high selectivity to H_2 and CO_x at a relatively low cost [12–19]. This catalytic behavior cannot only be ascribed to the nickel active phase, since the support also affects catalysts activity and product distribution [11,15–18]. In this sense, the use of silica in Ni-Cu-supported catalysts gave better results in comparison to other carrier materials, like alumina and MCM-41 [20]. Despite the good catalytic performance, the literature also describes that Ni/SiO_2 can be deactivated along time on stream, by metal sintering and coke deposition.

Trying to get over this drawback, a number of attempts have been made to prevent deactivation by modifying catalysts through the addition of different elements [16–18]. In this sense, increasing attention has been given to CeO_2 or ZrO_2-CeO_2 mixed oxide, because the oxygen mobility improves catalytic stability through coke gasification [9,11,21–24]. However, CeO_2 and ZrO_2-CeO_2 oxides have a very low surface area to be used as catalytic supports. To increase the textural properties, silica has been incorporated to ZrO_2-CeO_2 [25,26], and the literature also reports how CeO_2 or ZrO_2-CeO_2 were supported on silica to be used in catalytic applications [27,28]. Other researchers describe the addition of ZrO_2 to ceria and silica to increase their thermal resistance and prevent the growth of Ni crystallites [29–31].

Based on the above premises, this work proposes the use of Ce- and Zr-oxides as promoters in the synthesis of Ni/SiO_2 catalyst to reduce catalyst deactivation, allowing nickel particles to accommodate on a porous support in order to avoid or minimize the sinterization of the metallic phase.

2. Results and Discussion

2.1. Ethanol Steam Reforming over Reference Catalysts

Commercial silica, ceria and zirconia with BET surface areas of 276, 64 and 20 m^2/g, respectively, were used as supports of the reference catalysts. The physicochemical properties of the catalyst obtained after the incorporation of nickel into these supports are summarized in Table 1.

Table 1. Physicochemical properties of the reference catalysts.

Catalyst	Ni [a] wt%	S_{BET} m^2/g	Acidity [b] meq-NH_3/g	D^c_{NiO} nm	D^d_{Ni} nm
Ni/S	6.9	263	-	16.4	17.6
Ni/C	6.7	53	-	-	-
Ni/Z	6.8	15	0.275	-	-

[a] ICP-AES; [b] NH_3-temperature programmed desorption (TPD) analysis; [c] calculated from the (200) reflection of NiO in XRD; [d] calculated from the (111) reflection of Ni in XRD.

The actual nickel content measured by ICP-AES was near the nominal loading. Compared to the supports, the BET surface area of the catalysts decreased, although the trend is maintained, Ni/S >Ni/C >Ni/Z. Since the catalyst acidity can influence the products' distribution in the steam reforming reaction, it was measured by NH_3-TPD, verifying that only the Ni/Z catalyst exhibited measurable acidity with a desorption peak around 220 °C , corresponding to 0.275 meq-NH_3/g.

Figure 1 shows the XRD diffractograms of the calcined and reduced catalysts, where peaks corresponding to the carrier materials and those associated with the Ni phase can be observed. For calcined samples, reflections corresponding to the planes (111), (200) and (220) of cubic NiO can be observed at 2θ = 37.3°, 43.3° and 62.9° (JCPDS 78-0643), while peaks attributed to the planes (111) and (200) of metallic Ni appear at 2θ = 44.4° and 51.8° (JCPDS 70-1849) for the reduced catalysts.   The intensity of these reflections is hardly noticeable in the case of the Ni/C sample and overlap with ZrO_2 peaks in the Ni/Z sample.   Apart from the diffraction of the Ni phase, the pattern of Ni/S shows only one diffuse reflection around 2θ values of 22.5°, typical of certain amorphous materials, while diffraction peaks corresponding to cubic cerium (IV) oxide (JCPDS 89-8436; 2θ = 28.5°, 33.0°, 47.5°, 56.3°, 59.1° and 69.3°) can be clearly observed for Ni/C, and those corresponding to a mixture of monoclinic and tetragonal zirconium (IV) oxide (JCPDS 74-1200 and 80-0784, respectively; main peaks at 2θ = 28.3°, 31.5° and 50.2°) are seen for the Ni/Z catalyst. These analyses were used to calculate the Ni phase crystallites size of the Ni/S sample from the NiO(200) reflection at 2θ = 43.3° and the Ni(111) reflection at 2θ = 44.4° (Table 1). However, accurate crystallite diameters could not be calculated, neither for the Ni/C sample, due to the low intensity of the diffraction lines, nor for the Ni/Z sample, where Ni species peaks overlapped with those of ZrO_2.

The TPR analysis of calcined reference catalysts is shown in Figure 2. The profiles show a main peak between 270 and 350 °C , attributed to the reduction of NiO particles to Ni. In the case of the Ni/C and Ni/Z samples, the maximum of the reduction peak is placed at lower temperatures, and also, an additional shoulder can be observed, which is attributed to a fraction of the NiO particles strongly interacting with the support. Finally, in the case of the Ni/C sample, another peak arises around 800 °C , due to the partial reduction of Ce^{4+} to Ce^{3+} [21]. TPR analysis of the supports (not shown) demonstrated that, under our experimental conditions, only ceria may give rise to hydrogen consumption with maxima at 800 °C , corresponding to the diffusion-limited partial reduction of bulk CeO_2 particles. The low reduction temperature of Ni/C and Ni/Z catalysts suggests also a weaker metal support interaction in these two catalysts.

Table 2 summarizes the results obtained for the ethanol steam reforming reaction after 8 h of time on stream (TOS). While the Ni/C sample kept ethanol conversion above 99%, it decreased to 95.0% and 97.1% after 8 h TOS for the Ni/Z and Ni/S samples, respectively. Regarding intermediate products, acetaldehyde can be detected when using Ni/S and Ni/C, while ethylene is found with the Ni/Z sample due to the presence of acid sites, favoring the ethanol dehydration reaction. Methane selectivity decreases as hydrogen selectivity increases, indicating that methane steam reforming is one of the main steps in the ethanol steam reforming reaction mechanism. Thus, a slight hydrogen selectivity is reached by the Ni/S catalyst in comparison with the rest of samples.

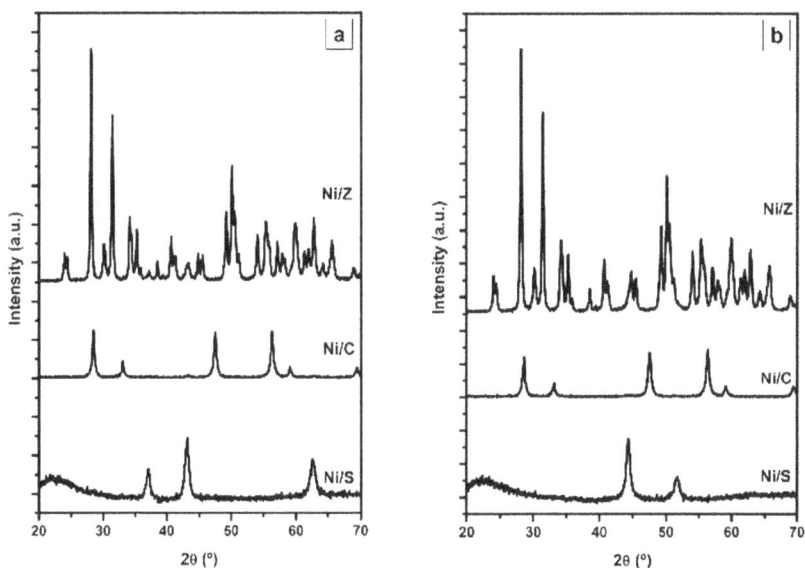

Figure 1. XRD patterns of the calcined (**a**) and reduced (**b**) reference catalysts.

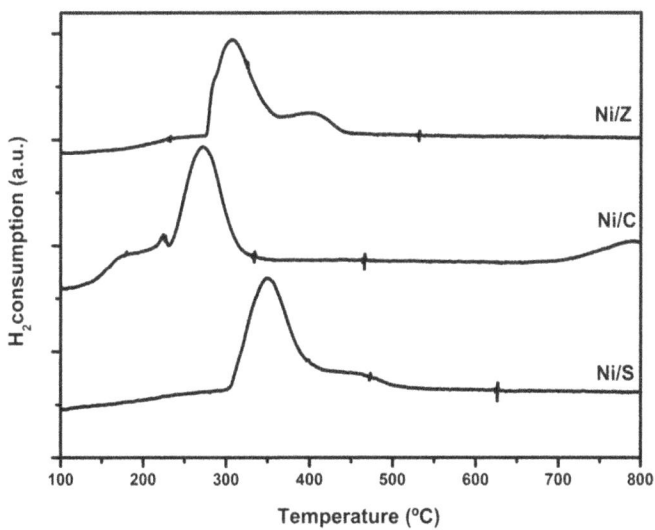

Figure 2. Temperature programmed reduction (TPR) profiles of the calcined reference catalysts.

Table 2. Results of ethanol steam reforming obtained with the reference catalysts (atmospheric pressure; T: 600 °C ; weight hourly space velocity; WHSV: 12.7 h^{-1}; $R_{H2O/EtOH}$: 3.7 mol/mol; time on stream (TOS): 8 h).

Catalyst	X_{EtOH}	Selectivity (mol%)					
	mol%	H_2	CO_2	CO	CH_4	C_2H_4	CH_3CHO
Ni/S	97.1	84.2	49.4	31.5	17.4	0	1.7
Ni/C	99.1	82.5	48.0	31.6	19.6	0	0.7
Ni/Z	95.0	80.7	45.3	31.8	20.4	2.4	0

Based on the above results and taking into account that silica is a porous support that could accommodate Ce an Zr oxides, the next study was the modification of Ni/SiO$_2$ catalysts by Ce or Zr addition.

2.2. Ni/SiO$_2$ Catalysts Modified by Ce or Zr

2.2.1. Characterization of Ce- or Zr-Modified Silica

For this study, four silica-supported Ni catalysts were prepared and modified by Ce or Zr addition, both before and after the incorporation of Ni. Thus, amorphous silica was used as the support of the Ce and Zr phases, and the XRD patterns of the resulting materials are shown in Figure 3, together with that of the Ni/S sample (described in Section 2.1), where Ce and Zr were later incorporated.

In the case of the CeS sample, the main diffraction peaks characteristic of the planes (111), (200), (220) and (311) of cubic CeO$_2$ (JCPDS 89-8436) are observed at 2θ = 28.5°, 33.0°, 47.5° and 56.3°. On the other hand, in the diffraction pattern corresponding to the ZrS sample, no clear peaks of Zr species can be seen, although a small broad slope appears around 2θ = 30.2°, which may be attributed to the plane (101) of highly dispersed ZrO$_2$ in the tetragonal phase (JCPDS 80-0784). This suggests that while Ce trends to aggregate into CeO$_2$ particles [16], Zr trends to form ZrO$_2$ crystallites too small to be detected by XRD. This fact agrees with the results published by Sánchez-Sánchez *et al.* [15] with Zr/Al$_2$O$_3$, where for Zr loadings up to 6.6 μmol Zr/m^2, Zr ions exists in atomic dispersion. The BET surface area of CeS and ZrS was 267 and 227 m^2/g, respectively, slightly lower in comparison to the silica used as the support, due to the metals' incorporation. Finally, TPR analysis of CeS and ZrS materials (not shown) evidenced that no detectable hydrogen consumption occurs at the metal loading used in this study (around 10 wt%), in spite of the partial reduction observed for the bulk CeO$_2$ used as a support in Section 2.1.

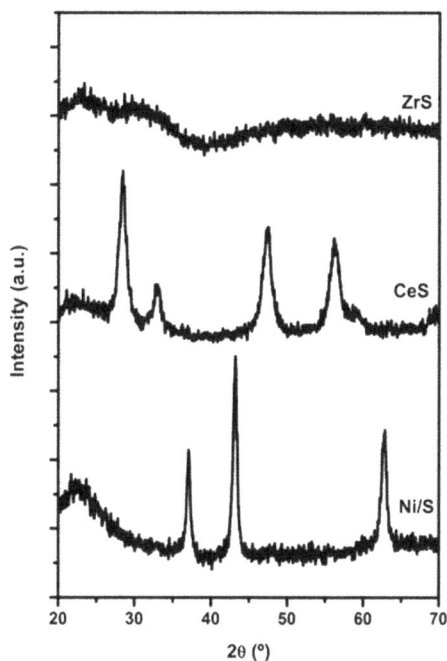

Figure 3. XRD patterns of the Ce- and Zr-modified silica supports.

2.2.2. Characterization of Ce- or Zr-Modified Ni/SiO$_2$ Catalysts

The physicochemical properties of the prepared Ni catalysts are summarized in Table 3. ICP-AES analysis revealed that silica support was loaded with near the desired amounts of Ni, Ce and Zr. Although the incorporation of a second metal into the previously described materials slightly decreased the value of their textural properties, all of the catalysts exhibited a high BET surface area. The acidity of the materials was measured by NH$_3$-TPD and is summarized in Table 3. Ce/NiS and Ni/CeS catalysts show negligible acidity, while Zr incorporation into silica support results in the presence of acid sites, which is more pronounced when Zr is incorporated after Ni, probably due to easier accessibility to these acid sites (not covered by Ni).

The XRD spectra of the calcined and reduced catalysts are shown in Figure 4. All of the calcined samples exhibit the peaks characteristic of cubic NiO (JCPDS 78-0643) at $2\theta = 37.3°$, $43.3°$ and $62.9°$, while the reduced samples show the reflections corresponding to metallic Ni (JCPDS 70-1849) at $2\theta = 44.4°$ and $51.8°$. As observed for the corresponding modified support, the main diffraction peaks characteristic of cubic CeO$_2$ appear in the case of the Ni/CeS and Ce/NiS catalysts, and the broad slope attributed to highly-dispersed ZrO$_2$ can be seen for the Zr/NiS and Ni/ZrS catalysts. The mean size of the CeO$_2$ particles, calculated by means of the Scherrer equation, resulted in being higher for the Ce/NiS sample, which indicates that CeO$_2$ particles are better dispersed on silica than on Ni/S. Regarding the Ni phase, the mean crystallite sizes are summarized in Table 3. Concerning the calcined samples, both Ce- and Zr-modified catalysts have slightly larger NiO crystallites than the unmodified Ni/S sample (Table 1), mainly when Ni is added to catalyst after Ce or Zr. A similar trend is found for the reduced samples. However, while the Ni phase crystallite size increases after the reduction process for the Ce-modified samples, attributed to the thermal effect, the addition of Zr leads even to a slight reduction of the crystallite size, which may be attributed to higher thermal stability together with lattice contraction from NiO to Ni due to the different molar volumes of these phases.

Table 3. Physicochemical properties of the Ni/SiO$_2$ catalysts modified by Ce or Zr.

Catalyst	Ni [a] wt%	Ce [a] wt%	Zr [a] wt%	S_{BET} m^2/g	Acidity [b] meq-NH$_3$/g	D^c_{CeO2} nm	D^d_{NiO} nm	D^e_{Ni} nm
Ce/NiS	6.7	9.7	-	228	-	8.1	19.8	19.9
Zr/NiS	6.7	-	9.4	236	0.144	-	20.0	19.9
Ni/CeS	6.8	9.2	-	247	-	7.0	21.8	23.2
Ni/ZrS	6.9	-	9.9	213	0.133	-	21.6	19.7

[a] ICP-AES measurements; [b] NH$_3$-TPD analysis; [c] calculated from the (220) reflection of CeO$_2$ in XRD; [d] calculated from the (200) reflection of NiO in XRD; [e] calculated from the (111) reflection of Ni in XRD.

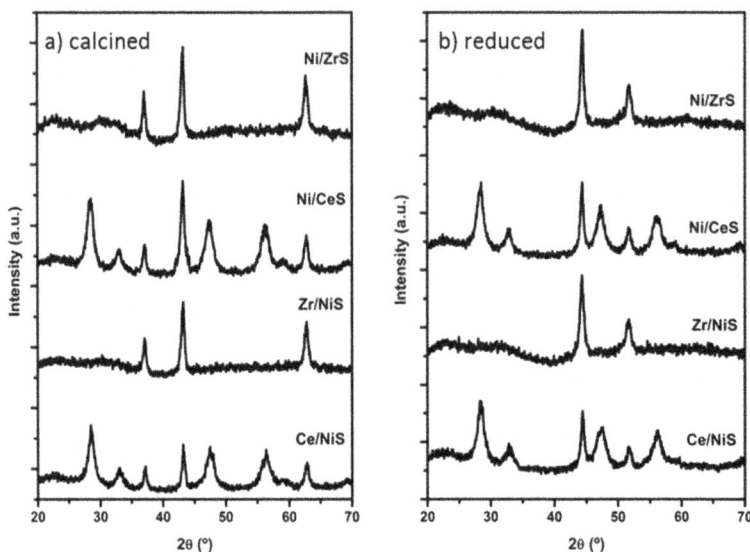

Figure 4. XRD patterns of the Ce- and Zr-modified Ni/SiO$_2$ catalysts: **(a)** calcined; **(b)** reduced.

The reducibility of these catalysts was determined by H$_2$-TPR experiments. The profiles shown in Figure 5 exhibit more than one peak attributed to the direct reduction of Ni^{2+} species to Ni, which means the presence of different reducible nickel species depending on the interaction with the support. Quantitative analysis of the TPR profiles revealed total reduction of the Ni^{2+} species for all of the samples. As in the case of the Ni/S sample (Figure 2), the peak at the lower temperature is attributed to the typical reduction of NiO to Ni, which may correspond to relatively large NiO particles, while the shoulder at the higher temperature is ascribed to nickel species in close contact with the support [22,23]. The Ce/NiS sample presents a lower reduction temperature than Ni/S catalyst, indicating the promotion of Ni reducibility by CeO$_2$ incorporation, an effect typically found for lanthanide elements [16]. On the contrary, reduction of the Zr/NiS sample takes place at higher temperatures, which may indicate an intimate contact between the Ni phase and the promoter due to partial covering of the NiO particles by the highly-dispersed Zr phase added to the Ni/S sample. On the other hand, when Ni was impregnated after Ce or Zr, reduction occurs in a broad temperature range, similarly to the reduction of Ni/S (Figure 2). A shoulder arises in the profile of the Ni/CeS catalyst at slightly lower temperatures than the beginning of the reduction of the Ni/S sample, probably due to the presence of oxygen vacancies in CeO$_2$, promoting the reduction of NiO [24,32], an effect facilitated by the small dimensions of the CeO$_2$ crystallites (Table 3). The profile corresponding to the Ni/ZrS sample reaches higher temperatures, probably due to the easier contact between NiO and ZrO$_2$, which is well dispersed over the support, unlike the CeO$_2$ particles, as the XRD patterns evidenced. The shift of the Zr-modified catalysts profiles to higher reduction temperatures in comparison to the rest of the samples indicates higher thermal stability and accounts for the decrease in the Ni phase crystallite size after the reduction of the catalyst, as previously observed from the XRD results.

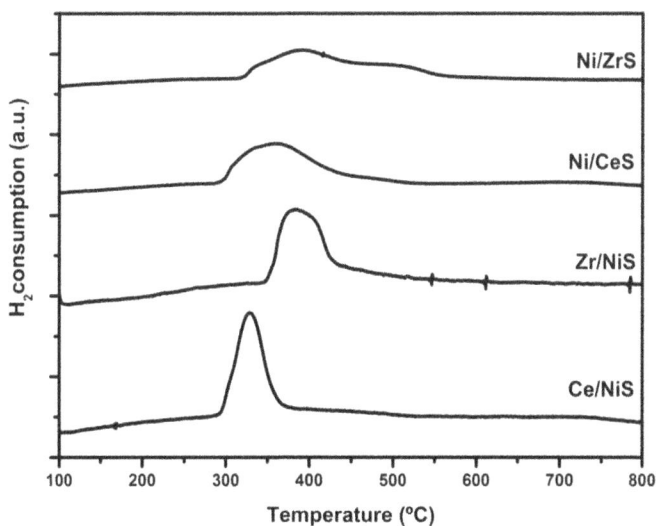

Figure 5. Temperature programmed reduction (TPR) profiles of the calcined Ce- or Zr-modified Ni/SiO$_2$ catalysts.

2.2.3. Catalytic Results Using Ni/SiO$_2$ Catalysts Modified by Ce or Zr

The catalytic results obtained by these catalysts after 8 h of TOS on ethanol steam reforming at 600 °C are shown in Table 4. Complete ethanol conversion was maintained over all of the catalysts, except in the case of the Ce/NiS sample, whereas it slightly decreased up to 98.5 after 8 h. On the other hand, hydrogen selectivity is lower for the samples where incorporation of Ni was carried out before the addition of the Ce or Zr, which may be due to some Ni particles covered by the promoters. This effect is more pronounced in the case of the Zr/NiS sample, which reached the lowest hydrogen selectivity, due to the high dispersion of the Zr phase, probably in the form of an overlayer over the Ni particles. However, total ethanol conversion is achieved with this sample, because the acidic nature of this catalyst favors ethanol dehydration to ethylene, as observed from the relatively high C_2H_4 amount obtained among the products. Ethylene selectivity is smaller for the Ni/ZrS sample, as expected from its lower acidity (Table 3). Therefore, regarding the impregnation order of the different elements, the incorporation of Ce or Zr before Ni on SiO_2 leads to lower selectivity to intermediate products and, consequently, higher selectivity towards the main products (H_2 and CO_2). This may be ascribed to the easier accessibility of reactants to the Ni sites. Finally, the CO_2/CO ratio among the gaseous products is increased by the presence of CeO_2, since it favors the water-gas shift reaction [33,34].

Table 4. Results on ethanol steam reforming obtained with the Ni/SiO_2 catalyst modified by Ce or Zr (atmospheric pressure; T: 600 °C ; weight hourly space velocity; WHSV: 12.7 h^{-1}; $R_{H2O/EtOH}$; 3.7 mol/mol; TOS: 8 h).

Catalyst	X_{EtOH}	Selectivity (mol%)						Coke
	mol%	H_2	CO_2	CO	CH_4	C_2H_4	CH_3CHO	g/g$_{cat}$ h
NiS	97.1	84.2	49.4	31.5	17.4	0	1.7	0.369
Ce/NiS	98.5	82.0	49.6	29.9	19.5	0	1.0	0.230
Zr/NiS	100	79.2	43.5	29.1	21.2	6.2	0	0.475
Ni/CeS	100	84.3	56.2	25.1	18.7	0.0	0	0.370
Ni/ZrS	100	88.3	56.5	28.0	12.9	2.6	0	0.518

The results shown in Table 4 indicate that coke formation occurs at a much higher rate on Zr-modified catalysts, because, as explained above, the acid sites joined with the presence of Zr promote ethanol dehydration to ethylene, which acts as a hard precursor for carbon deposition [35]. Although the highest coke amount was found on the Ni/ZrS sample, no loss of activity was observed for this sample, since carbon was deposited in the form of nanofibers (Figure 6). This type of coke does not embed metal particles [36], keeping the metal surface accessible for reactants, which, together with the high surface area of the silica support to accommodate large amounts of coke, has been described to have a minor effect on catalyst deactivation [37,38]. However, high coke formation is not desirable, because it would probably result in reactor plugging. On the other hand, when Ce was added over the Ni/S sample (Ce/NiS), coke formation diminished as a consequence of the enhancement of carbon gasification by CeO_2 [16,22]. However, in the case of the Ni/CeS sample, this reduction of the coke amount is not observed, due to the higher Ni particles of the catalyst that favors the growth of carbon nanofibers [39,40].

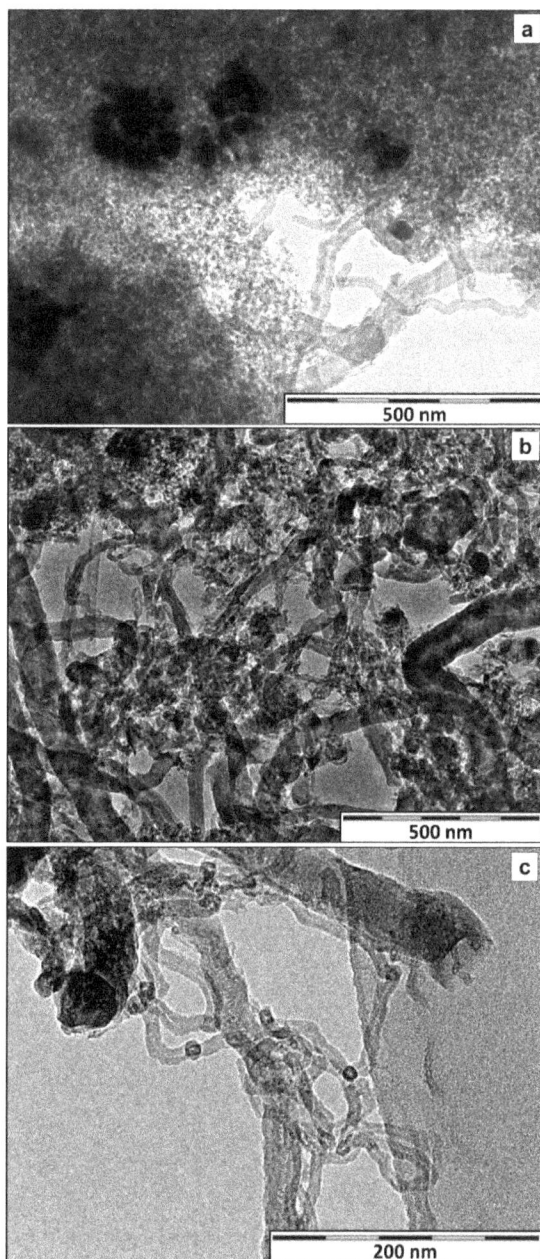

Figure 6. TEM image of coke formed on catalysts after being used for ethanol steam reforming for 8 h: (**a**) Ni/S; (**b**) Ce/NiS; (**c**) Ni/ZrS.

Consequently, taking into account that the highest hydrogen selectivity was reached with the Ni/ZrS catalyst at complete ethanol conversion and Ce/NiS led to a lesser amount of carbon deposited on the catalyst, the next study consisted of the modification of Ni/SiO$_2$ catalysts by both Ce and Zr incorporation before the Ni phase.

2.3. Ni/Ce$_x$Zr$_{1-x}$O$_2$/SiO$_2$ Catalysts: The Effect of the Zr/Ce Ratio

2.3.1. Characterization of the Ni/Ce$_x$Zr$_{1-x}$O$_2$/SiO$_2$ Catalysts

For this study, three Ni catalysts supported on Ce- and Zr-modified silica were prepared with Zr/Ce mass ratios of 1/3, 1 and 3. Their physicochemical properties are shown in Table 5, where it can be observed that there is a good agreement between the metal contents determined by ICP-AES and the theoretical values. All of the catalysts show a high surface area in comparison to the values corresponding to the Ni/C and Ni/Z catalysts (Table 1), as expected from the use of silica as a support. However, they exhibit lower BET surface areas than the Ni/S sample, with a slight decreasing trend as the Zr content increases (Table 3). As expected, the acidity of the materials measured by NH$_3$-TPD increased with the Zr loading, the Ni/ZrS catalyst having the highest number of acid sites.

Table 5. Physicochemical properties of the Ni/Ce$_x$Zr$_{1-x}$O$_2$/SiO$_2$ catalysts.

Catalyst	Ni [a] wt%	Ce [a] wt%	Zr [a] wt%	S_{BET} m^2/g	Acidity [b] meq$_{NH3}$/g	D^c_{CeO2} nm	D^d_{NiO} nm	D^e_{Ni} nm
Ni/CeZrS-1/3	6.8	6.0	2.3	243	0.062	6.3	19.1	19.7
Ni/CeZrS-1	7.0	4.2	4.5	242	0.077	5.0	18.1	16.6
Ni/CeZrS-3	6.6	2.4	6.6	238	0.088	3.4	17.6	15.7

[a] ICP-AES measurements; [b] NH$_3$-TPD analysis; [c] calculated from the (220) reflection of CeO$_2$ in XRD; [d] calculated from the (200) reflection of NiO in XRD; [e] calculated from the (111) reflection of Ni in XRD.

The XRD spectra of the calcined and reduced catalysts are shown in Figure 7, where samples containing both Ce and Zr have a pattern similar to that of Ni/CeS sample (Figure 4), but CeO$_2$ peaks become less intense and shift to higher 2θ values with increasing Zr loading. This fact has been ascribed to the formation of a Ce$_x$Zr$_{1-x}$O$_2$ solid solution, where the unit cell parameter changes with the Zr/Ce ratio [21,23,26]. The size of the CeO$_2$ crystallites on calcined catalysts were calculated from the width of the (220) characteristic peak. The values displayed in Table 5 show that crystallites found on these catalysts are smaller than those reported for unsupported Ce$_x$Zr$_{1-x}$O$_2$ [21], which evidences that the relatively high surface area of SiO$_2$ enhances their dispersion, and their sizes become smaller as the Zr/Ce ratio increases. Only a slight increment of these particles sizes ($\approx$ 0.1–0.3 nm) was detected after reduction treatment.

Additionally, the peaks characteristic of cubic NiO (JCPDS 78-0643) and those corresponding to metallic Ni (JCPDS 70-1849) can be observed for the calcined and reduced samples, respectively. Concerning the calcined samples, the mean NiO crystallites sizes, calculated by applying the Scherrer equation to the broadening of the (200), are summarized in Table 5. The NiO crystallite size decreases as the Zr content is increased, which has been previously reported for unsupported $Ce_xZr_{1-x}O_2$ catalysts [15,20]. In the case of the reduced samples, the Ni crystallite sizes follow a similar trend. It is noticeable that Ni crystallites of the reduced Ni/CeZrS-1 and Ni/CeZrS-3 catalysts are smaller than the Ni particles of the corresponding calcined samples and even smaller than Ni crystallites of Ni/S catalyst, Ni/CeZrS-3 being the catalyst with the smallest Ni crystallites.

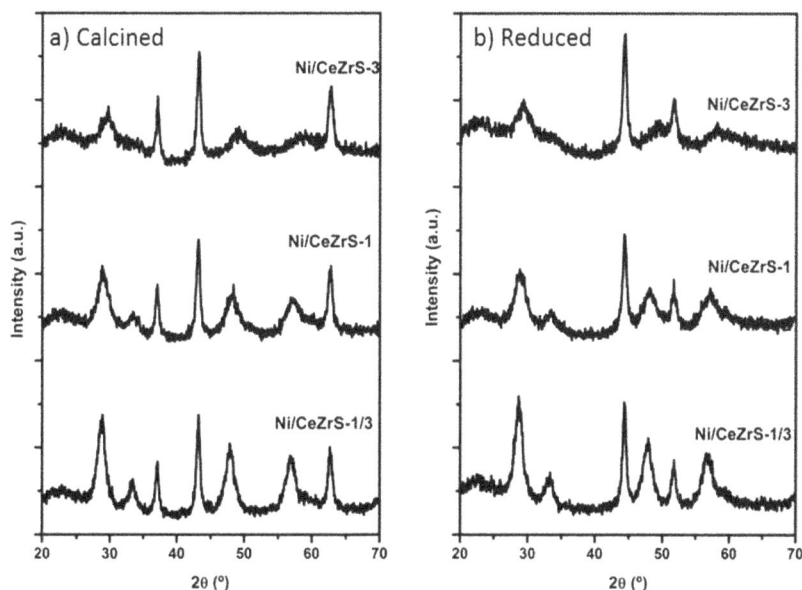

Figure 7. XRD patterns of the $Ni/Ce_xZr_{1-x}O_2/SiO_2$ catalysts: (**a**) calcined; (**b**) reduced.

The reducibility of these catalysts was determined by H_2-TPR experiments, and the corresponding profiles shown in Figure 8 evidenced a similarity with those of the Ni/CeS and Ni/ZrS samples. These profiles have a peak at a lower temperature due to the reduction of NiO particles and a shoulder at a higher temperature ascribed to the reduction of nickel species in close contact with the support. Quantitative analysis of the TPR profiles revealed total reduction of the Ni^{2+} species for all of the samples. By increasing the ZrO_2 content, the reduction profiles slightly shift towards higher temperatures, and this peak area increases while the area under the low temperature zone decreases. This fact indicates a stronger interaction of Ni^{2+} species with the support as ZrO_2 loading increases [15,18,19], which accounts for the hindered sintering of the Ni phase when Zr loading increases, as observed from the difference between the mean size of the nickel phase crystallites before and after reduction (see Table 5). The decrease of the $Ce_xZr_{1-x}O_2$ particle size as the Zr loading increases would make it possible for a higher contact between this phase and the NiO, which results in higher metal-support interaction [16].

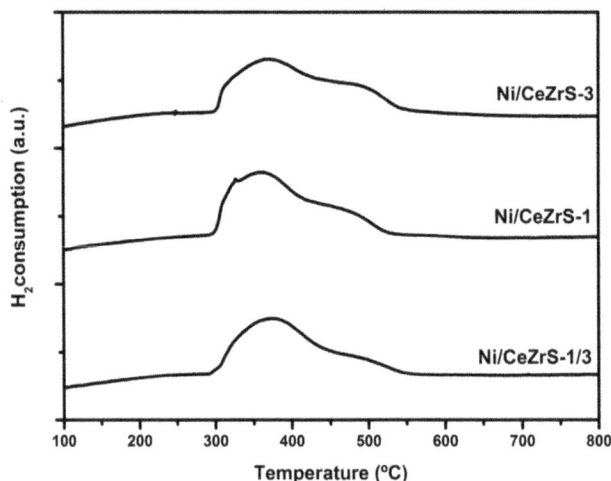

Figure 8. TPR profiles of the calcined $Ni/Ce_xZr_{1-x}O_2/SiO_2$ catalysts.

2.3.2. Catalytic Results Using the $Ni/Ce_xZr_{1-x}O_2/SiO_2$ Catalysts

These catalysts were tested at 600 °C in the ethanol steam reforming, and the obtained results are shown in Table 6. All of the catalysts showed complete ethanol conversion after 8 h of TOS. On the other hand, it can be observed how product selectivity depends on the Zr/Ce ratio. The only H-containing intermediate compound found among the reaction products was methane, while neither ethylene nor acetaldehyde could be detected. Besides, as the Zr/Ce ratio is increased, selectivity towards CH_4 decreases, and thus, hydrogen selectivity increases. This may be attributed to both a decrease of Ni crystallite size and an increase of thermal stability when the Zr content increases, as determined by XRD and TPR (Table 5 and Figure 8). Regarding the CO_2/CO ratio, it is higher for $Ni/Ce_xZr_{1-x}O_2/SiO_2$ catalysts than for the Ni/S sample and increases with the Zr/Ce ratio. An enhanced oxygen storage capacity due to the formation of the $Ce_xZr_{1-x}O_2$ solid solution has been reported to favor the water-gas shift reaction [23,26].

Table 6. Results on ethanol steam reforming obtained with the $Ni/Ce_xZr_{1-x}O_2/SiO_2$ catalysts (atmospheric pressure; T: 600 °C ; weight hourly space velocity (WHSV): 12.7 h^{-1}; $R_{H2O/EtOH}$: 3.7 mol/mol; TOS: 8 h).

Catalyst	X_{EtOH}	Selectivity (mol%)						Coke
	mol%	H_2	CO_2	CO	CH_4	C_2H_4	CH_3CHO	g/g_{cat} h
Ni/CeZrS-1/3	100	84.6	52.4	29.6	18.0	0	0	0.500
Ni/CeZrS-1	100	85.6	54.0	29.0	17.0	0	0	0.443
Ni/CeZrS-3	100	85.8	54.3	28.9	16.8	0	0	0.308

Finally, Table 6 compares catalysts in terms of coke formation, which decreases as the Zr/Ce mass ratio in the catalyst increases. However, since ceria and zirconia improve oxygen mobility [32,41,42], favoring carbon gasification, coke deposition on promoted catalysts decreased by increasing Zr content in the $Ce_xZr_{1-x}O_2$ solution, which agrees with the smaller size of NiO and Ni crystallites found in samples with a higher Zr content.

3. Experimental Section

3.1. Catalysts Preparation

A series of catalysts were prepared to study the effect of the addition of Ce and Zr to Ni/SiO_2 catalyst (Ni: 7 wt%). The modified catalysts contain a total loading of Ce and Zr of 10 wt% and Ce/Zr mass ratios of 0, 1/3, 1, 3 and ∞. Moreover, the order of the Ce or Zr incorporation with SiO_2, before or after Ni addition, was also studied. The addition of metals was accomplished by incipient wetness impregnation of amorphous silica (Ineos, Warrington, UK) with aqueous solutions of the metal precursors: $Ni(NO_3)_2 \cdot 6H_2O$ (Scharlab, Sentmenat, Barcelona, Spain), $Ce(NO_3)_3 \cdot 6H_2O$ (Scharlab, Sentmenat, Barcelona, Spain) and $ZrO(NO_3)_2 \cdot 6H_2O$ (Sigma–Aldrich, St. Louis, MO, USA). After the incorporation of each metal (Ni, Zr or Ce), the solid sample was air-dried overnight and further calcined at 500 °C for 5 h with a heating rate of 1.8 °C /min. Afterwards, the addition of the other metal, if needed, was carried out by following the same procedure. Ni/SiO_2, Ni/CeO_2 and Ni/ZrO_2 were prepared as reference materials by impregnation of Ni over commercial silica (Ineos, Warrington, UK), ceria (Riedel-de-Haën, Hanover, Germany) and zirconia (Sigma–Aldrich, St. Louis, MO, USA). Catalyst are denoted as A/BY-x, where A/B are the metals added to the support (A, secondly, and B, firstly added), Y is the support (S: silica, C: ceria and Z: zirconia), and x is the Zr/Ce mass ratio in the final sample.

3.2. Catalysts Characterization

X-ray powder diffraction (XRD) analysis was used to find out the crystalline phases and calculate mean crystallite size in the calcined and reduced catalysts. Measurements were carried out on a Phillips (Eindhoven, The Netherlands) X'Pert PRO diffractometer using Cu Kα radiation. The patterns were recorded with a 2θ increment step of 0.020° and a collection time of 2 s. Mean metallic crystallite diameters were calculated by applying the Scherrer equation.

The inductively-coupled plasma atomic emission spectroscopy (ICP-AES) technique was used to determine the total Ce, Zr and Ni content in the catalysts using a Varian (Palo Alto, CA, USA) VISTA-PRO AX CCD-Simultaneous ICP-AES spectrophotometer. Samples were previously dissolved with an acidic solution (HF and H_2SO_4).

Textural properties of the materials were calculated by nitrogen adsorption-desorption at 77 K in a Micromeritics (Norcross, GA, USA) TRISTAR 2050 sorptometer. Samples were first outgassed under vacuum at 200 °C for 4 h. Surface areas were determined using the BET method.

Acid properties of the catalysts and supports were obtained by ammonia temperature programmed desorption analysis (NH_3-TPD) in a Micromeritics (Norcross, GA, USA) Autochem 2910 equipment. Samples were previously outgassed under a He flow (50 NmL/min) at 560 °C for 30 min. After saturation of the sample with ammonia and removal of the physisorbed fraction by flowing He at 180 °C , NH_3 was desorbed from the sample by increasing the temperature up to 550 °C with a heating rate of 15 °C /min, keeping this temperature constant for 30 min.

The reducibility of supported metals was determined by hydrogen temperature programmed reduction analysis (H_2-TPR). Catalysts were analyzed with the same apparatus described for NH_3-TPD. Catalysts were previously degasified in flowing argon (35 mL/min) for 30 min at 110 °C with a heating rate of 15 °C /min. Afterwards, the H_2-TPR profile was obtained by flowing 10% H_2 in Ar (35 NmL/min) from 25 °C to 800 °C with a heating rate of 5 °C /min.

The morphology of catalyst particles and metal crystallites was analyzed by transmission electron microscopy (TEM). Samples were prepared by dispersion of the powdered material in acetone and following deposition on a copper grid with carbon support. Micrographs were attained on a Phillips (Eindhoven, The Netherlands) TECNAI 20 equipped with a LaB_6 filament and an accelerating voltage of 200 kV.

Carbon deposited on the used catalysts was evaluated by thermogravimetric analysis (TG). TG measurements were carried out on a TA instruments (New Castle, DE, USA) SDT 2960 thermobalance using an air flow of 100 mL/min and a heating rate of 5 °C /min up to 700 °C .

3.3. Catalytic Test

Catalysts activity and selectivity were measured in the ethanol steam reforming reaction on a MICRO ACTIVITY-PRO unit, as described elsewhere [19]. The catalyst was placed into the reactor and *in situ* reduced under flowing pure hydrogen (30 NmL/min) for 4.5 h at 550 °C with a heating rate of 2 °C /min. After activation, the catalytic tests were carried out at atmospheric pressure and 600 °C . A liquid water/ethanol (3.7 molar ratio) stream (0.075 mL/min) was fed into the system, vaporized at 150 °C and further diluted by N_2 (30 NmL/min). Weight hourly space velocity (WHSV), defined as the water-ethanol mass flow rate related to the mass of the catalyst, was fixed at 12.7 h^{-1}. The composition of the output gas stream was determined online by a gas chromatograph, Varian (Palo Alto, CA, USA) CP-3380, equipped with Hayesep Q and Molecular Sieve 13X columns, a thermal conductivity detector and using He as the carrier and reference gas. Condensable vapors (ethanol, water and acetaldehyde) were retained in a condenser and analyzed by chromatography.

4. Conclusions

Ethanol has been successfully converted into hydrogen though steam reforming over Ni catalysts supported on SiO_2, CeO_2, ZrO_2 and Ce- and/or Zr-modified SiO_2 with different Zr/Ce mass ratios. The Ni/SiO_2 catalyst led to higher hydrogen selectivity than Ni/CeO_2 and Ni/ZrO_2, but it could not maintain complete ethanol conversion. Incorporation of Zr to Ni/SiO_2 increased the Ni-support interaction, leading to complete ethanol conversion along time and higher hydrogen selectivity. However, large amounts of coke were formed as a consequence of ethylene formation induced by acid sites found on Zr-modified catalysts. On the contrary, lower coke formation was observed when Ce was incorporated into Ni/SiO_2, which can be ascribed to the presence of oxygen vacancies in

CeO_2 promoting coke gasification. Concerning the impregnation order, the incorporation of Ce and Zr prior to nickel on the silica support led to steam reforming catalysts with better performance. When both Zr and Ce were incorporated into Ni/SiO_2, a $Ce_xZr_{1-x}O_2$ solid solution was formed, increasing the Ni-support interaction and producing smaller Ni crystallites as the Zr/Ce mass ratio increased. As a result, complete ethanol conversion and hydrogen selectivity above 84 mol% were reached with the $Ni/Ce_xZr_{1-x}O_2/SiO_2$ catalysts. Concretely, using a Zr/Ce mass ratio of three, a trade-off between high selectivity to the main products, relatively low coke deposition and high thermal stability was achieved.

Acknowledgments

The authors acknowledge the financial support from the Ministerio de Economía y Competitividad through the project "Hydrogen and liquid fuels production from microalgae by thermochemical proceses" (Ref: CTQ2013-44447-R), and from the Comunidad de Madrid through the project "Residuals to Energy part 2" project (Ref: P2013/MAE-2882).

Author Contributions

This work was conceived by Jose Antonio Calles and Alicia Carrero. Montaña Lindo did the experimental work, assisted by Arturo Javier Vizcaíno who also draft the first version of manuscript. Furthermore, Arturo Javier Vizcaíno, Jose Antonio Calles and Alicia Carrero wrote and revised the final version of paper.

Conflicts of Interest

The authors declare no conflict of interest.

References

1. Züttel, A.; Remhof, A.; Borgschulte, A.; Friedrichs, O. Hydrogen: the future energy carrier. *Phil. Trans. R. Soc. A* **2012**, *368*, 3329–3342.
2. Guo, L.J.; Zhao, L.; Jing, D.W.; Lu, Y.J.; Yang, Y.J.; Bai, B.F.; Zhang, X.M.; Ma, L.J.; Wu, X.M. Solar hydrogen production and its development in China. *Energy* **2009**, *34*, 1073–1090.
3. Olateju, B.; Kumar, A. Hydrogen production from wind energy in Western Canada for upgrading bitumen from oil sands. *Energy* **2011**, *36*, 6326–6339.
4. Zahedi, A. Hydrogen as storage option for intermittent renewable technologies such as solar and wind. In Proceedings of the 2006 Australasian Universities Power Engineering Conference (AUPEC'06), Melbourne, Australia, 10–13 December 2006.
5. Goltsova, V.A.; Veziroglu, T.N.; Goltsova, L.F. Hydrogen civilization of the future—A new conception of the IAHE. *Int. J. Hydrogen Energy* **2006**, *31*, 153–159.

6. Grzegorczyk, W.; Denis, A.; Gac, W.; Ioannides, T.; Machocki, A. Hydrogen Formation via Steam Reforming of Ethanol Over Cu/ZnO Catalyst Modified with Nickel, Cobalt and Manganese. *Catal. Lett.* **2009**, *128*, 443–448.

7. Vizcaíno, A.J.; Carrero, A.; Calles, J.A. Hydrogen Production from Bioethanol. In *Hydrogen Production: Prospects and Processes*; Nova Science Publishers, Inc.: New York, NY, USA, 2012; pp. 247–294

8. Papadopoulou, E.; Delimaris, D.; Denis, A.; Machocki, A.; Ioannides, T. Alcohol reforming on cobalt-based catalysts prepared from organic salt precursors. *Int. J. Hydrogen Energy* **2007**, *37*, 16375–16381.

9. Navarro, R.M.; Peña, M.A.; Fierro, J.L.G. Hydrogen production reactions from crbon feedstocks: Fossil fuels and biomass. *Chem. Rev.* **2007**, *107*, 3952–3991.

10. Ni, M.; Leung, D.Y.C.; Leung, M.K.H. A review on reforming bio-ethanol for hydrogen production. *Int. J. Hydrogen Energy* **2007**, *32*, 3238–3247.

11. Auprête, F.; Descorme, C.; Duprez, D. Bio-ethanol catalytic steam reforming over supported metal catalysts. *Catal. Commun.* **2002**, *3*, 263–267.

12. Comas, J.; Mariño, F.; Laborde, M.; Amadeo, N. Bio-ethanol steam reforming on Ni/Al$_2$O$_3$ catalyst. *Chem. Eng. J.* **2004**, *98*, 61–68.

13. Fatsikostas, A.N.; Verykios, X.E. Reaction network of steam reforming of ethanol over Ni–based catalysts. *J. Catal.* **2004**, *225*, 439–452.

14. Frusteri, F.; Freni, S.; Spadaro, L.; Chiodo, V.; Bonura, G.; Donato, S.; Cavallaro, S. H$_2$ production for MC fuel cell by steam reforming of ethanol over MgO supported Pd, Rh, Ni and Co catalysts. *Catal. Commun.* **2004**, *5*, 611–615.

15. Sánchez-Sánchez, M.C.; Navarro, R.M.; Fierro, J.L.G. Ethanol steam reforming over Ni/M_xO_y-Al$_2$O$_3$ (M = Ce, La, Zr and Mg) catalysts: Influence of support on the hydrogen production. *Int. J. Hydrogen Energy* **2007**, *32*, 1462–1471.

16. Calles, J.A.; Carrero, A.; Vizcaíno, A.J. Ce and La modification of mesoporous Cu-Ni/SBA-15 catalysts for hydrogen production through ethanol steam reforming. *Micropor. Mesopor. Mater.* **2009**, *119*, 200–207.

17. Vizcaíno, A.J.; Carrero, A.; Calles, J.A. Ethanol steam reforming on Mg- and Ca-modified Cu-Ni/SBA-15 catalysts. *Catal. Today* **2009**, *146*, 63–70.

18. Lindo, M.; Vizcaíno, A.J.; Calles, J.A.; Carrero, A. Ethanol steam reforming on Ni/Al-SBA-15 catalysts: Effect of the aluminium content. *Int. J. Hydrogen Energy* **2010**, *35*, 5895–5901.

19. Vizcaíno, A.J.; Lindo, M.; Carrero, A.; Calles, J.A. Hydrogen production by steam reforming of ethanol using Ni catalysts based on ternary mixed oxides prepared by coprecipitation. *Int. J. Hydrogen Energy* **2012**, *37*, 1985–1992.

20. Vizcaíno, A.J.; Carrero, A.; Calles, J.A. Hydrogen production by ethanol steam reforming over Cu-Ni supported catalysts. *Int. J. Hydrogen Energy* **2007**, *32*, 1450-1461.

21. Biswas, P.; Kunzru, D. Steam reforming of ethanol for production of hydrogen over Ni/CeO$_2$-ZrO$_2$ catalysts: Effect of support and metal loading. *Int. J. Hydrogen Energy* **2007**, *32*, 969–980.

22. Biswas P.; Kunzru, D. Oxidative steam reforming of ethanol over Ni/CeO$_2$-ZrO$_2$ catalyst. *Chem. Eng. J.* **2008**, *136*, 41–49.

23. Youn, M.H.; Seo, J.G.; Cho, K.M.; Park, S.; Park, D.R.; Jung, J.C.; Song, I.K. Hydrogen production by auto-thermal reforming of ethanol over nickel catalysts supported on Ce-modified mesoporous zirconia: Effect of Ce/Zr molar ratio. *Int. J. Hydrogen Energy* **2008**, *33*, 5052–5059.

24. Wan, H.; Li, X.; Ji, S.; Huang, B.; Wang, K.; Li, C. Effect of Ni loading and Ce$_x$Zr$_{1-x}$O$_2$ promoter on Ni-based SBA-15 catalysts for steam reforming of methane. *J. Nat. Gas Chem.* **2007**, *16*, 139–147.

25. Rocchini, E.; Vicario, M.; Llorca, J.; Leitenburg, C.; Dolcetti, G.; Trovarelli, A. Reduction and oxygen storage behavior of noble metals supported on silica-doped ceria. *J. Catal.* **2002**, *211*, 407–421.

26. Raju, V.; Jaenicke, S.; Chuah, G.K. Effect of hydrothermal treatment and silica on thermal stability and oxygen storage capacity of ceria-zirconia. *Appl. Catal. B* **2009**, *191*, 92–100.

27. Reddy, B.M.; Thrimurthulu, G.; Saikia, P.; Bharali, P. Silica supported ceria and ceria-zirconia nanocomposite oxides for selective dehydration of 4-methylpentan-2-ol. *J. Mol. Catal. A* **2007**, *275*, 167–173.

28. Reddy, B.M.; Saikia, P.; Bharali, P.; Katta, L.; Thrimurthulu, G. Highly dispersed ceria and ceria-zirconia nanocomposites over silica surface for catalytic applications. *Catal. Today* **2009**, *141*, 109–114.

29. Takahashi, R.; Sato, S.; Sodesawa, T.; Yoshida, M.; Tomiyama, S. Addition of zirconia in Ni/SiO$_2$ catalyst for improvement of steam resistance. *Appl. Catal. A* **2004**, *273*, 211–215.

30. Seo, J.G.; Youn, M.H.; Song, I.K. Effect of SiO$_2$-ZrO$_2$ supports prepared by a grafting method on hydrogen production by steam reforming of liquefied natural gas over Ni/SiO$_2$-ZrO$_2$ catalysts. *J. Power Sources* **2007**, *168*, 251–257.

31. Roh, H.S.; Potdar, H.S.; Jun, K.W. Carbon dioxide reforming of methane over co-precipitated Ni-CeO$_2$, Ni-ZrO$_2$ and Ni-Ce-ZrO$_2$ catalysts. *Catal. Today* **2004**, *93–95*, 39–44.

32. Trovarelli, A. *Catalysis by Ceria and Related Materials*; Imperial College Press: London, UK, 2002.

33. Djinovic, P.; Batista, J.; Pintar, A. WGS reaction over nanostructures CuO-CeO$_2$ catalysts prepared by hard template method: Characterization, activity and deactivation. *Catal. Today* **2009**, *147*, 191–197.

34. Andreeva, D.; Ivanov, I.; Ilieva, L.; Sobczak, J.W.; Avdeev, G.; Tabakova, T. Nanosized gold catalysts supported on ceria and ceria-alumina for WGS reaction: Influence of preparation method. *Appl. Catal. A* **2007**, *333*, 153–160.

35. Trim, D.L. Coke formation and minimisation during steam reforming reactions. *Catal. Today* **1997**, *37*, 233–238.

36. Carrero, A.; Calles, J.A.; Vizcaíno, A.J. Hydrogen production by ethanol steam reforming over Cu-Ni/SBA-15 supported catalysts prepared by direct synthesis and impregnation. *Appl. Catal. A* **2007**, *327*, 82–94.

37. Wang, S.; Lu, G.Q. Reforming of methane with carbon dioxide over Ni/Al$_2$O$_2$ catalysts: Effect of nickel precursor. *Appl. Catal.* **1998**, *69*, 271–280.

38. Juan-Juan, J.; Román-Martínez, M.C.; Illán-Gómez, M.J. Effect of potassium content in the activity of K-promoted Ni/Al$_2$O$_2$ catalysts for the dry reforming of methane. *Appl. Catal.* **2006**, *301*, 9–15.

39. Carrero, A.; Calles, J.A.; Vizcaíno, A.J. Effect of Mg and Ca addition on coke deposition over Cu-Ni/SiO$_2$ catalysts for ethanol steam reforming. *Chem. Eng. J.* **2010**, *163*, 395–402.

40. Helveg, S.; Sehested, J.; Rostrup-Nielsen, J.R. Whisker carbon in perspective. *Catal. Today* **2011**, *178*, 42–46.

41. Natesakhawat, S.; Watson, R.B.; Wang, X.; Ozkan, U.S. Deactivation characteristics of lanthanide-promoted sol-gel Ni/Al$_2$O$_2$ catalysts in propane steam reforming. *J. Catal.* **2005**, *234*, 496–508.

42. Bellido, J.D.A.; Assaf, E.M. Nickel catalysts supported on ZrO$_2$, Y$_2$O$_3$-stabilized ZrO$_2$ and CaO-stabilized ZrO$_2$ for the steam reforming of ethanol: Effect of the support and nickel load. *J. Power Sources* **2008**, *177*, 24–32.

Deactivation Pattern of a "Model" Ni/MgO Catalyst in the Pre-Reforming of *n*-Hexane

Giuseppe Trunfio and Francesco Arena

Abstract: The deactivation pattern of a "model" Ni/MgO catalyst in the pre-reforming of *n*-hexane with steam (T, 450 °C; P, 5–15 bar) is reviewed. The influence of the steam-to-carbon ratio (S/C, 1.5–3.5) on the rate of catalyst fouling by coking is ascertained. Catalyst fouling leads to an exponential decay in activity, denoting 1st-order dependence of the coking process on active sites availability. Hydrogen hinders the coking process, though slight activity decay is due to sintering of the active Ni phase. Deactivation by thiophene causes a sharp, almost linear, drop to nearly zero activity within only 6 h; this deactivation is likely due to dissociative adsorption of thiophene with subsequent strong, irreversible chemical adsorption of S-atoms on active Ni sites, *i.e.*, irreversible poisoning. Modeling of activity decay curves (α, a_t/a_0) by proper kinetic equations allows assessing the effects of temperature, pressure, S/C, H_2 and thiophene feed on the deactivation pattern of the model Ni/MgO catalyst by coking, sintering, and poisoning phenomena.

Reprinted from *Catalysts*. Cite as: Trunfio, G.; Arena, F. Deactivation Pattern of a "Model" Ni/MgO Catalyst in the Pre-Reforming of *n*-Hexane. *Catalysts* **2014**, *4*, 196-214.

1. Introduction

There is a strong commitment worldwide to enhance the practical and economical feasibility of the steam reforming (SR) technology through efficiency improvement and feedstock versatility, which could result in significant improvements in fuel processing economy [1–7]. In this context, pre-reforming constitutes an established technology with economic and operational benefits on the overall syngas production, representing an important tool, especially for the revamping of older SR plants [2–4,8–10]. A pre-reforming unit consists in a tubular adiabatic reformer, allowing whatever hydrocarbon feed (NG, LPG, VN) be converted to CH_4 and CO_x at low temperature, typically in the 450–550 °C range, with many practical advantages consisting of [2–4]:

i.   an increased production capacity with smaller reformer furnaces;
ii. a higher feedstock flexibility;
iii. enhanced SR tube and catalyst lifetime;
iv. design of innovative process configuration for low energy consumption and investment costs.

Furthermore, the pre-reforming unit drastically limits the risk of carbon formation inside tubular reformers, ensuring a total conversion of higher hydrocarbons, and allowing, then, SR operations at lower steam-to-carbon (S/C) ratio, with a higher preheat temperature and heat flux and capacity [9–11]. Moreover, the pre-reforming catalyst ensures the total elimination of S-containing impurities, contributing to enhance SR catalysts' lifetime [2–5,8,11–14].

Considering the importance for the improvement of syngas producing plants, little previous literature deals with the basic aspects of the steam reforming of higher hydrocarbons reaction under low temperature and high pressure conditions [12,14–17]. In fact, only after the pioneering work of Rostrup-Nielsen on the steam reforming of C1-C7 alkanes on magnesia-supported metal catalysts, which led to the development of the SPARG process [2,10,14], has light been shed on the reaction mechanism and kinetics, leading to further improvements in catalysts' performance [11,14,16,17].

Although noble-metal-based systems feature a high reactivity in the reforming of superior hydrocarbons to methane and carbon oxides, typical formulations of pre-reforming catalysts include nickel as the active phase [6–8,16,17]. Due to different operating conditions, pre-reforming catalysts are generally richer in Ni loading (50–70 wt%), with the addition of rare earth elements (e.g., La, Sm, Ce, Y, Zr, etc.) as promoters, and are characterized by larger total (SA) and metal surface area (MSA) than SR counterparts [6,16]. Loading of the active phase and type of promoters, however, depend upon process characteristics and targets [6,16].

Generally, coking processes on Ni-based catalysts in steam reforming and methanation reactions proceed through common carbon intermediates leading to the formation of several types of deposits depending on operating conditions [2,8,12–15,18–24]. Moreover, effects of metal sintering on the active surface area of typical pre-reforming catalysts have been assessed [25]. Catalyst poisoning is generally caused by strong chemisorption of sulfur impurities present in the feed stream as inorganic and/or organic sulfides in most naturally occurring feedstock; sulfur poisoning depends on many variables including the kind of sulfur compound, nature of the catalyst and operating conditions [7,26,27].

In earlier papers, we documented that Ni/MgO catalysts obtained by an original non-aqueous preparation route exhibit a remarkable performance in both "steam" and "dry" reforming of methane [28–30]. Hindering the formation of the $NiMgO_x$ solid solution, indeed, such a preparation procedure allows tuning both physico-chemical and catalytic properties through a proper selection of Ni loading, calcination and reduction treatments [28,29,31].

This review provides an overview of reaction and deactivation networks occurring early (*i.e.*, during the first 17 h) on a "model" Ni/MgO catalyst in the steam reforming of *n*-hexane under pre-reforming conditions (T, 450 °C; P, 5–15 atm). Rates, selectivities, rate constants, and equilibrium constants of important reaction paths as a function of time and rate constants for deactivation by sintering, coking, and poisoning from the previous literature (with an emphasis on the authors' previous works) are presented, and the effects of catalyst deactivation on activity, selectivity, and stability are analyzed and discussed.

2. Results and Discussion

2.1. Effects of the Experimental Conditions on the Activity-Selectivity-Stability Pattern

Activity data of the "model" Ni/MgO catalyst in the pre-reforming of *n*-hexane at standard reaction conditions (T, 450 °C; P, 10 bar; P_{C6H14}, 0.18 bar; P_{H2O}, 3.0 bar; S/C, 2.8) are shown in Figure 1 in terms of n-hexane conversion (χ) and product selectivity *vs.* reaction time (t.o.s.). The only detected products are CH_4, CO_2, and CO [11–17] with a relative distribution which depends on the conversion level. After 30 min, taken as the initial reaction time (t_0, a_0), the hexane conversion is ≈70% with CH_4, CO_2 and CO selectivities of *ca.* 55, 43 and 2%, respectively. A drop in conversion of about 50% over a period of only 17 h involves a S_{CH4} decrease from 55 to 30%, counterbalanced by a rise in S_{CO2} from 45 to 70%, although no significant changes in S_{CO} (2–3%) are observed.

Figure 1. n-hexane conversion (χ) and product selectivity ($S\chi$) *vs.* t.o.s. at standard reaction conditions (adapted from ref. [11] with the kind permission of Elsevier, copyright 2004).

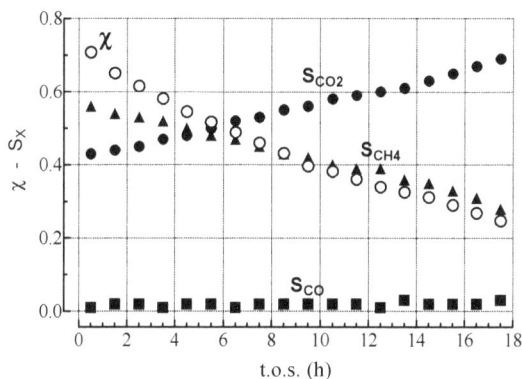

Activity data at 450 °C with varying space velocity (GHSV), feed stream composition (P_{H2O} and P_{H2}) and pressure are summarized in Table 1 in terms of initial and final hexane conversion (χ, %), reactor output (mol$_{C6H14}$·h^{-1}·g$_{cat}^{-1}$), product selectivity (S_χ, %) and weight gain of the "used" catalyst samples (W_C, g·g$_{cat}^{-1}$) due to carbon deposition and/or coke formation, determined by TGA-DSC measurements. In addition, the influence of experimental conditions on the "relative activity" ($\alpha = \chi/\chi_0$, the ratio between conversion at the time "t" and "t_0" respectively) is shown in Figure 2.

Table 1. Initial and final (17 h) activity data and amount of carbon on the "used" catalyst samples under different reaction conditions (T, 450 °C; p$_{C6H14}$/p^0, 0.018).

Run	P (atm)	S/C	GHSV [a] (h^{-1})	χ_{C6H14} (%)	Reactor output (mol$_{C6H14}$ h^{-1} g$_{cat}^{-1}$)	S_{CH4} (%)	S_{CO2} (%)	S_{CO} (%)	S_{C2-C5} (%)	W_C [b] (g g$_{cat}^{-1}$)
a	10	1.5	12,000	53–08	0.26 – 0.04	55–9	43–88	2–3	-	1.32
b	10	2.8	12,000	71–25	0.35–0.12	56–28	42–69	2–3	-	0.96
c	10	3.5	12,000	76–34	0.38–0.17	56–42	43–57	1–1	-	0.89
d	10	2.8	24,000	38–11	0.38–0.11	39–5	59–93	2–2	-	1.23
e [c]	10	2.8	24,000	77–73	0.77–0.73	98–97	1–2	0–0	1–1	0.03
f	15	2.8	12,000	69–37	0.51–0.27	60–50	39–49	1–1	-	0.75
g	5	2.8	12,000	67–21	0.17–0.05	49–25	49–73	1–2	-	1.02
h [d]	10	2.8	24,000	65–00	0.65–0.00	97–97	2–2	1–1	-	0.21

[a] F$_{C6H14}$/V$_{cat}$; [b] weight gain due to carbon deposition and/or coke formation; [c] hydrogen (H$_2$/C, 1) in the feed; [d] hydrogen (H$_2$/C, 1) and thiophene (50 ppm on C$_6$H$_{14}$) in the feed.

Data in Table 1 indicate that feed composition and operating conditions affect both hexane conversion and CH$_4$, CO$_2$ and CO selectivities, while trace amounts of C2-C5 hydrocarbons ($S_{C2-C5} \approx 1\%$) are detected only in presence of hydrogen (run e).

Figure 2. Relative activity *vs.* t.o.s. at different experimental conditions (see Table 1).

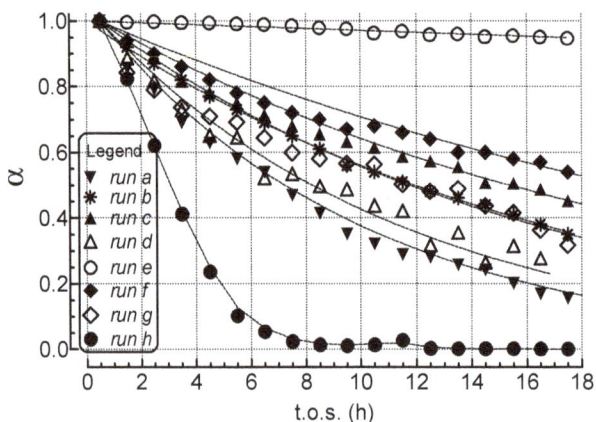

At a constant pressure of 10 atm, the most obvious effects of the S/C ratio increase from 1.5 to 3.5 (runs a–c) consist in a progressive increase of hexane conversion (from 53 to 76%), indicating a positive effect of water pressure on the reactor output, and a better catalyst stability (Figure 2) coupled to a relative influence on product selectivity during reaction time (Table 1). The pressure increase from 5 to 15 atm (S/C, 2.8) has no significant effect on the conversion level (runs b, f, g), as an increase from 0.17 to 0.51 mmol$_{C6H14} \cdot h^{-1} \cdot g_{cat}^{-1}$ (Table 1) mirrors a fairly constant pressure-normalized reactor output (0.035 mol$_{C6H14} \cdot h^{-1} \cdot g_{cat}^{-1}$/atm).

Figure 3. Overall conversion-selectivity (S_χ) data and H/O$_{out}$ ratio at different conditions (see Table 1) and reaction times (adapted from ref. [15] with the kind permission of Wiley, copyright 2006).

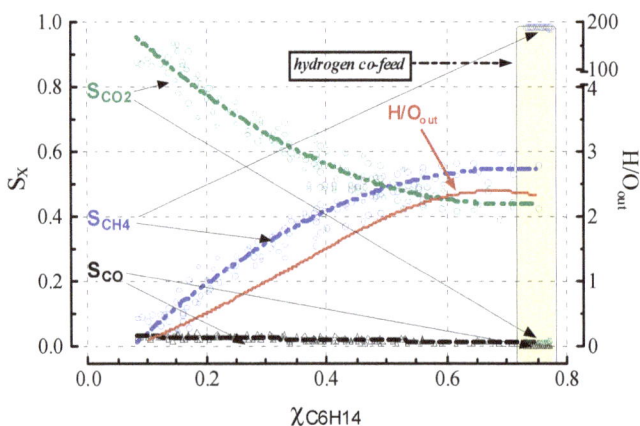

In addition, an enhancement of S_{CH4} (49–60%) and, even more so, of catalyst stability are evident, especially at 15 atm (run f).

Furthermore, a twofold increase of GHSV from 12,000 to 24,000 h^{-1} (runs b, d) causes an almost proportional decrease of the initial conversion (38%) according to an unchanging reactor output (0.35–0.38 $mol_{C6H14} \cdot h^{-1} \cdot g_{cat}^{-1}$), a drop in S_{CH4} from 55 to 39% with a corresponding growth of S_{CO2} (Table 1), and a considerably stronger activity decay (Figure 2).

In all the cases, a decreasing trend of the relative activity curves denotes noticeable deactivation phenomena, the extents of which depend on S/C ratio, P, and GHSV (Figure 2). The observed loss of activity can be largely accounted for by the weight loss recorded by TGA-DSC analysis of "used" catalyst samples (Table 1) during combustion of the carbonaceous deposits. This points to carbon formation and coking as the main causes of catalyst deactivation [11,15]. However, such experiments did not allow to differentiate between carbon and coke nor quantitatively measure CO_2 and H_2O.

Under simulated pre-reforming conditions (run e) with H_2 in the feed [3–6,16–18], a markedly higher initial conversion (77%) and a two-fold higher reactor output are observed relative to the standard tests (runs b, d). Moreover, the addition of hydrogen ($H_2/C = 1$) to the feed increases S_{CH4} to 97–98% and greatly improves catalyst stability, with only a minor (ca. 5%) activity loss during 17 h of t.o.s. (Figure 2, run e).

All the conversion-selectivity data at different conditions (see Table 1) and reaction times provide some general relationships showing that the selectivity pattern depends only on the activity level (Figure 3). Thus, S_{CH4} increases steadily with increasing conversion, reaching an asymptotic value of 55% in the conversion range of 60–80%, while S_{CO2} decreases from 90 to 45% and S_{CO} from 2 to 0.5% over the full range of measured conversion (8–75%). Changes in CO/CO_2 are consistent with the equilibrium composition predicted from the WGS reaction [11,15,16,18,19,32–37]. The overall result of these changes in product selectivities is a trend of increasing output hydrogen/oxygen atomic ratio (H/O_{out}) from zero at 8% conversion to a maximum value of 2.4 at 67% conversion (Figure 3), which is essentially the same as the H/C ratio of the hexane molecule of 2.33. The apparent increase in the "hydrogenation" functionality with increasing conversion level is consistent with a reaction network involving (1) cracking of hexane; (2) water dissociation; (3) primary formation of CO, which is further transformed to CO_2 via the water-gas-shift (WGS); and (4) CH_4 via methanation (MET) paths respectively [11,15], according to the following surface reaction network, which includes two irreversible (1 and 8) and six reversible (2–7) steps [11,13–16,19,21,33,35].

The above reaction scheme explains the inhibition of the methanation path (5) at low P_{CO} and P_{H2}, that is the case at low conversion, while the excess of steam favors the formation of CO_2 via the WGS path (6). Thus, hydrogen "availability", determined by the extent of hexane conversion, is the key-parameter controlling the CH_4/CO_2 distribution [38]. Therefore, enabling the gasification of coke-precursor species [11,15–17,21,22,35,36], the methanation step is the "competitor" process of coking, explaining the high catalyst stability and the negligible weight gain of the used catalysts under H_2 co-feeding (Table 1). In fact, the same carbon-intermediate (C_α) can undergo methanation at high H_2

concentrations or, at low H_2 concentrations, generate the following less reactive carbon species, precursors of C-deposit buildup [19–22,35,36], by ageing

- amorphous carbon (C_β),
- vermicular carbon (C_V),
- bulk Ni carbide (C_γ), and
- crystalline, graphitic carbon (C_C).

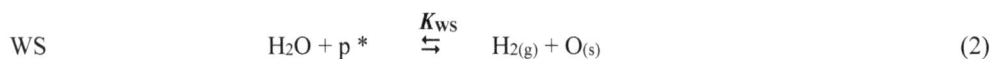

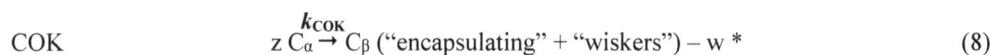

CRK $\qquad$ $C_6H_{14} + n *$ $\qquad \longrightarrow \qquad$ $(6 - m) C_{\alpha(s)} + m\, CH_{x(s)} + [(14 - x)/2]\, H_{2(g)}$ $\qquad$ (1)

WS $\qquad$ $H_2O + p *$ $\overset{K_{WS}}{\rightleftarrows}$ $H_{2(g)} + O_{(s)}$ $\qquad$ (2)

HC $\qquad$ $H_2 + 2 *$ $\overset{K_{H2}}{\rightleftarrows}$ $2\, H_{(s)}$ $\qquad$ (3)

GAS $\qquad$ $C_{\alpha(s)} + O_{(s)}$ $\overset{K_{GAS}}{\rightleftarrows}$ $CO_{(s)} + (n/6 + m - q) *$ $\qquad$ (4)

MET $\qquad$ $C_{\alpha(s)} + 4\, H_{(s)}$ $\overset{K_{MET}}{\rightleftarrows}$ $CH_{4(g)} + (n/6 + 4 - t) *$ $\qquad$ (5)

WGS $\qquad$ $CO_{(s)} + O_{(s)}$ $\overset{K_{WGS}}{\rightarrow}$ $CO_{2(s)} + (m + q - s) *$ $\qquad$ (6)

DES $\qquad$ $CO_{2(s)}$ $\overset{K_{CO2}}{\rightleftarrows}$ $CO_{2(g)} + s *$ $\qquad$ (7)

COK $\qquad$ $z\, C_\alpha \overset{k_{COK}}{\rightarrow} C_\beta$ ("encapsulating" + "wiskers") $-$ w $*$ $\qquad$ (8)

2.2. Modeling of Deactivation Phenomena

From a mechanistic point of view, the coking process can be viewed as a consequence of the surface accumulation of carbon moieties due to a lower gasification rate relative to the accumulation rates of C_{alpha}, polymeric carbon and coke, by the CO decomposition (scheme 1a) or hexane cracking (scheme 1b).

Scheme 1. Carbon formation by CO decomposition (**a**) or hexane cracking (**b**).

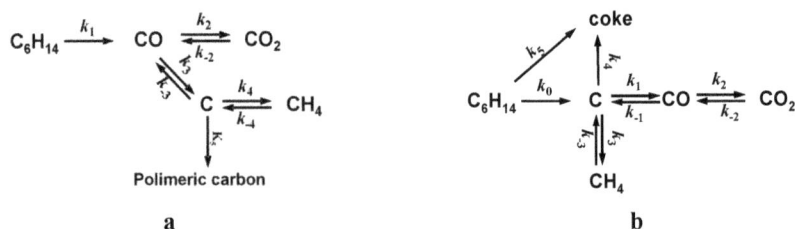

a $\qquad\qquad\qquad\qquad\qquad\qquad$ b

TEM pictures of the used catalyst samples in Figure 4 confirm that catalyst deactivation, feeding or not hydrogen, proceeds by different paths to form different inactive carbon or coke species. According to previous TGA results, under standard reaction conditions (run b) the catalyst grains (Figure 4B) appear embedded into an array of large carbon fibers (d_{fiber} = 20–50 nm), while no carbon deposits are apparent on the used catalyst in presence of H_2 (Figure 4C) [11,15,17,35], although halos surrounding several crystallites may be due to films of coke precursors.

Figure 4. TEM views (**A–C**) and PSD (**D**) of the "fresh" and "used" catalyst samples: (**A**) "fresh"; (**B**) "used" (run b); (**C**) "used" (run e); (**D**) CSD's of samples reproduced in panels A–C.

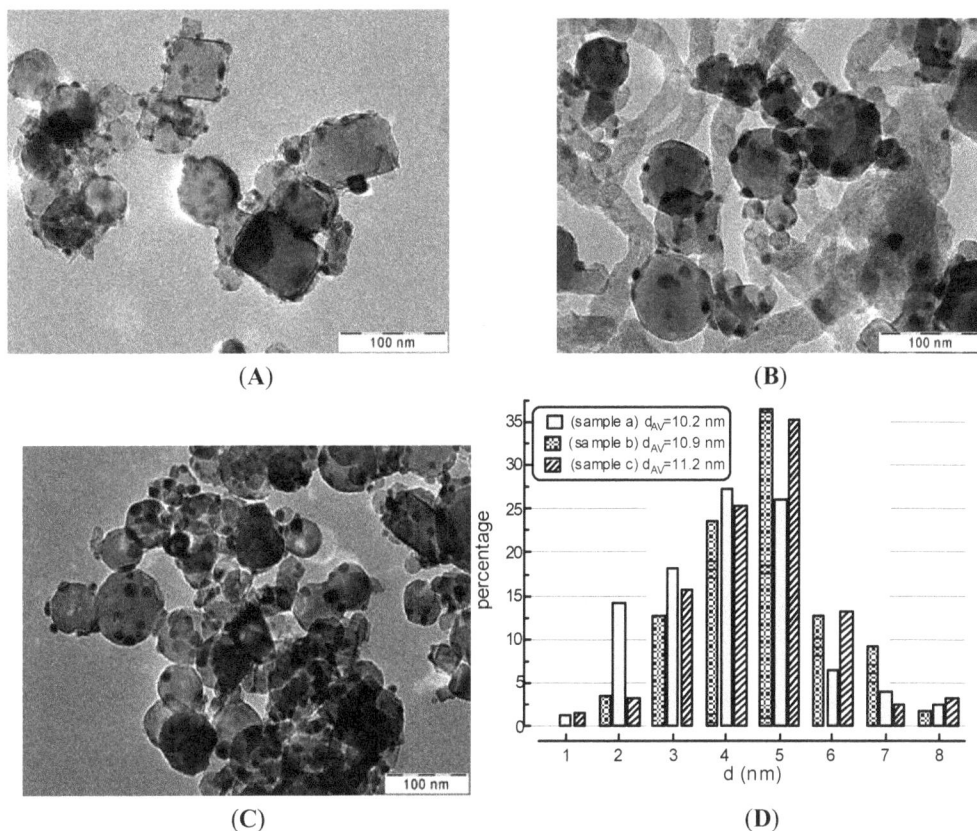

(A)

(B)

(C)

(D)

It should be emphasized that the principal cause of deactivation in prereforming or reforming of C_{2+} hydrocarbons at relatively low reaction temperatures is the formation of a hydrocarbonaceous film on the nickel surface [19] which poisons nickel sites. This work establishes that deactivation is minimized at high H_2 concentrations.

In both Runs b and e, an observed "smoothing" of the cubic habit of the magnesia carrier (Figure 4B,C) is consistent with moderate restructuring of the support, probably due to reaction with steam [15,16]; this is likely the origin of very modest metal sintering, apparent from minor changes in Ni crystallite size distributions (Figure 4D), accounting for *ca.* a 10% decrease in metal dispersion in agreement with H_2 uptake data [11].

As might be expected, thiophene co-feeding causes the most rapid decrease in reaction rate with a linear drop during the first 6 h, (Figure 2, run h) [7]. The amount of C_5H_5S fed during this time (0.05 g/kg hexane) corresponds to *ca.* 3 μmol (0.51 μmol/h), while the H_2 uptake of the fresh catalyst (120 µmol/g$_{cat}$) corresponds to a surface Ni atom concentration for a 0.025 g sample of 6.0 µmol (*i.e.*, 240 µmol/g$_{cat}$ of surface Ni sites assuming a H/Ni adsorption stoichiometry of 1 [7,15]). Thus, during pre-reforming with the above specified feed of thiophene (run h), the maximum S/Ni surface atomic ratio (*i.e.*, average chemisorption stoichiometry) would be 0.5, assuming complete and irreversible adsorption of sulfur atoms, which is likely under these conditions; in other words, each thiophene molecule deactivates two Ni atoms. However, due to the rapid and irreversible adsorption of S atoms on surface Ni atoms, the rate of poisoning could be influenced by pore diffusional resistances; thus the amount of adsorbed S in the catalyst sample would drop off sharply at the front of the bed and the sharp interface of sulfur saturated nickel would travel slowly, similar to a chromatographic wave, through the catalyst bed. Hence, the equations modeling thiophene adsorption are very complex and by necessity would include expressions for diffusion into the pores, adsorption, dissociation, and the slowly moving axial concentration gradient through the bed, as illustrated by the sophisticated two-dimensional model of H$_2$S adsorption on a Ni steam reforming catalyst described by Rostrup-Nielsen [39].

From the previous discussion, the deactivation pattern of the Ni/MgO catalyst can be assessed taking into account the overall effects of coking, sintering, and poisoning. However, other than Run h, the behavior of deactivation *versus* time observed in Figures 2 and 5 is due primarily to the formation of carbon and coke, since negligible sulfur was present and effects of sintering were negligibly small. With the exception of Run h, in the presence of thiophene, deactivation was log-linear (see Figure 5), consistent with an exponential activity decay

$$\alpha = \exp[-k_{deact} \cdot t] \tag{9}$$

Accordingly, the first order deactivation rate equation is

$$-\frac{da}{dt} = k_{deact} \cdot a \quad \text{or} \quad -ln\alpha = k_{deact} \cdot t \tag{10}$$

These can be also expressed in terms of active sites availability (C_{site}), assuming that activity is directly proportional to the concentration of surface sites [20],

$$\alpha = \frac{C_{site}}{C_{site}^0} \tag{11}$$

and, thus

$$-\frac{dC_{site}}{dt} = k_{deact} \cdot C_{site} \quad \text{or} \quad -ln\left(\frac{C_{site}}{C_{site}^0}\right) = k_{deact} \cdot t \tag{12}$$

where C_{site} and C_{site}° are the instantaneous and initial concentrations of active sites, respectively. Hence, the exponential decay of activity by coking is described by the equation

$$\alpha = e^{-(k_{cok})t} \tag{13}$$

which accounts for the general exponential-decay of catalytic activity for each of the runs in Figure 2. From the slopes of the linear fits of the deactivation rate data in Figure 5, the deactivation rate constants (k_{deact}) listed in Column 3 of Table 2 are obtained. It is further assumed that the rate of sintering (which depends mainly on temperature) is negligibly small for all runs given the very small change in crystallite diameter during reaction.

Figure 5. Plot of the relative activity at different experimental conditions (see Table 1) *vs.* t.o.s..

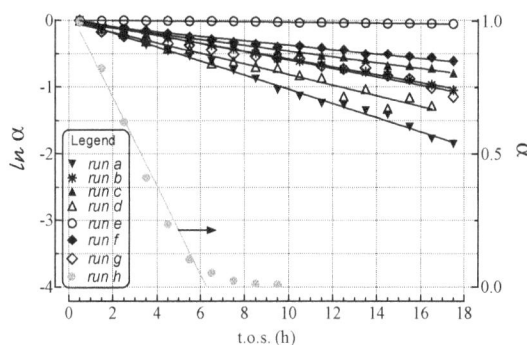

Table 2. Deactivation kinetic constants for coking.

Run	r^2	k_{cok}[a] (h^{-1})
a	1.00	$1.1 \times 10^{-1} \pm 2 \times 10^{-3}$
b	1.00	$5.5 \times 10^{-2} \pm 6 \times 10^{-4}$
c	0.99	$7.0 \times 10^{-2} \pm 1 \times 10^{-3}$
d	0.97	$7.7 \times 10^{-2} \pm 5 \times 10^{-3}$
e	0.96	-
f	0.98	$3.0 \times 10^{-2} \pm 1 \times 10^{-3}$
g	0.97	$6.3 \times 10^{-2} \pm 3 \times 10^{-3}$

a $k_{cok} = k_{deact}$.

It is emphasized that the calculation of rate constants in Table 2 (Run h involving sulfur poisoning excepted) relies on the following assumptions:

1 The rates of deactivation by coking and sintering are small relative to the rate of the main reaction. Hence the pseudo steady-state approximation can be made. This is a reasonable assumption in view of 2–3 orders of magnitude differences between reaction and coking rates at any condition and t.o.s. [15];

2 The rate of coking is slow enough that it is not influenced by a pore diffusional resistance for the deactivation process;

3 Pore diffusion resistance for the main reaction is neglected which, at high reaction rates and conversions, would lower the concentration of hydrocarbon coke precursors into the pores, then lowering coking rates along the pores; similarly concentration gradients through the bed due to high conversions should cause a decrease in adsorbed carbon concentration through the bed. This is probably a poor assumption in view of the high conversions of this study. Thus, the rate constants could be significantly in error due to differences in pore

diffusional resistance between runs and represent at best rough averages across pellets and bed.

Figure 6. Log-plot of the kinetic constant of coking *vs.* water partial pressure (**A**); and Relationship between the weight gain of the used samples (Table 1) and the kinetic constant of coking (Table 2) at different reaction conditions (**B**).

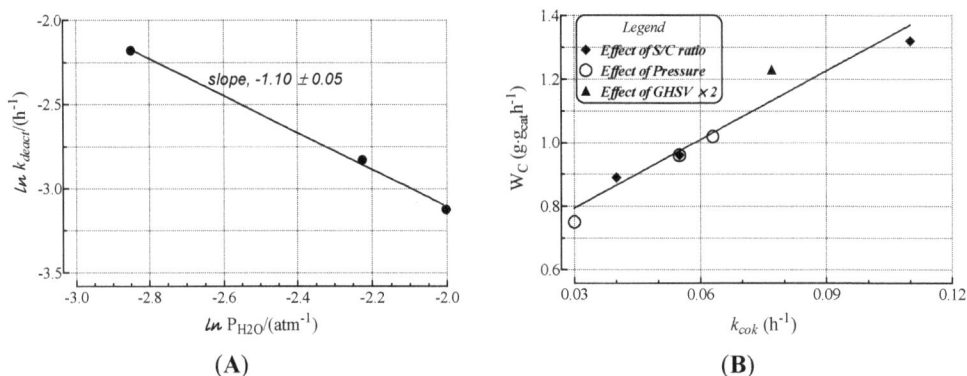

(**A**)   (**B**)

Values of k_{cok} in Table 2 are not constant as predicted by the simple first order deactivation model; indeed, they vary by a factor of 20. In fact, the log-plot of k_{cok} (Figure 6A) discloses an inverse dependence (-1.10 ± 0.05) on P_{H2O}, confirming that water concentration is a key-factor in controlling coking rate at low H_2 concentrations (typical of steady-state), a logical result given that water is also a gasifying agent for carbon. The straight-line relationship between k_{cok} and weight gain in the used samples in Figure 6B supports a link between deactivation at low H_2 concentrations to fouling by C-deposits for a range of typical prereforming conditions.

2.3. Effect of Deactivation on Product Selectivity and Surface Functionalities

Thermodynamic and kinetic evaluations of the outlet stream composition [2,9,15,21–23] shed light on driving force for carbon deposition and, in turn, how carbon deposits influence selectivity. In particular, assuming hexane cracking to be irreversible [9,15,22], a rough estimate of the outlet stream composition can be made from a thermodynamic analysis of gasification (GAS), water-gas-shift (WGS), and methanation (MET) reactions (Table 3) [9–11,15]. However, also the Boudouard reaction has been considered to evaluate the incidence of CO dismutation on carbon formation and coking [11]. Specifically, P_x values obtained from conversion-selectivity data were substituted in the equilibrium expressions to calculate experimental K_{exp} values of MET, GAS, WGS and DISP reactions as a function of t.o.s. [11].

Table 3. Thermodynamic constants of the various reactions of the steam reforming network (adapted from ref. [11] with the kind permission of Elsevier, copyright 2004).

Reaction	Stoichiometry			K_P [a]
CRK	C_6H_{14}	$\rightarrow$	$6\,C + 7\,H_2$	-
GAS	$C + H_2O$	$\leftrightarrows$	$CO + H_2$	5.09×10^{-4}
WGS	$CO + H_2O$	$\leftrightarrows$	$CO_2 + H_2$	7.47×10^{0}
MET	$C + 2\,H_2$	$\leftrightarrows$	CH_4	5.41×10^{1}
DISP	$2\,CO$	$\leftrightarrows$	$C + CO_2$	1.47×10^{4}

Equilibrium pressure constant values at 450 °C with P_x values in MPa.

The ratio (β) of K_{exp} to the corresponding calculated value of the equilibrium constant K_{eq} at 450 °C (from Table 3) provides a measure of the distance from equilibrium for each reaction, allowing a reliable assessment of the effects of the S/C ratio (Figure 7A–C) and H_2 feeding (Figure 7D) on activity *versus* time patterns. Figure 7 shows that irrespective of t.o.s. and S/C ratio, the WGS reaction is always near thermodynamic equilibrium conditions, given that β_{WGS} values vary between 0.6 and 1.0. In contrast, gasification ($\beta_{GAS} \approx 0.5$) and, even more so, methanation (β_{MET}, 0.2–0.05) reactions operate far from equilibrium and are thus kinetically controlled. It follows from the previous discussion that lower rates of GAS and MET relative to hexane cracking constitutes a major driving force to carbon deposits build-up [17,19,20,35]. Furthermore, the observation that β_{DISP} is always larger than one implies a $[P_{CO2}/(P_{CO})^2]$ ratio greater than that allowed by thermodynamics, indicating that under prereforming conditions on Ni/MgO, the forward reaction, which also contributes to carbon formation, is more kinetically favorable than the reverse gasification of carbon by CO_2. However, this deviation could also be an artifact of the approximate reaction model.

It is interesting that while β_{GAS} and β_{MET} decrease with trends similar to that of the relative activity (α), β_{DISP} increases progressively with t.o.s. (almost exponentially under the conditions of Figure 7A), suggesting that carbon and coke reactivity decrease accordingly by ageing (*i.e.*, by conversion of coke precursors to highly polymerized coke and by conversion of $C_\alpha \rightarrow C_\beta$, C_v, C_γ, C_C) [16,17,19,21,36]. In fact, the changes in product selectivity with the activity level (Figure 3) mirror the different influences of coke and carbon on different functionalities of Ni catalysts, consistent with the ^{14}CO and $^{14}CO_2$ isotopic labeling experiments of Jackson *et al.* who found that Ni/Al$_2$O$_3$ catalyst retains high activity towards the scrambling of C-atoms and surface carbon gasification, despite a considerable decrease in the availability of metal surface area due to growth of gum-like carbonaceous deposits [35].

In the present paper, the role of the methanation reaction in modeling of prereforming has been emphasized. Bartholomew emphasized that methanation and steam reforming functionalities of Ni catalysts are closely related on the basis of the activation energy values [22]. Furthermore, Rostrup-Nielsen stressed that hydrogenolysis and steam reforming reactions share many surface steps, requiring also analogous ensembles (*n*, 2.5–2.7) of active sites [17]. Hence, effects of deactivation on prereforming catalyst functionalities can be assessed by inspecting the trends of the relative rates of MET, GAS, and WGS reactions ($\alpha_x = \alpha \cdot S_x$), as shown in Figure 8. The semi-log plot of the relative rates of CH_4 (α_{CH4}), CO (α_{CO}), and CO_2 (α_{CO2}) formation provides satisfactory linear trends

(Figure 8), the slopes of which (k_{MET}, k_{GAS}, k_{WGS}) represent the decay constants for the above functionalities, listed in Table 4.

Figure 7. Pre-reforming of the *n*-hexane (T, 450 °C; P, 10 bar). Effect of S/C ratio and H_2 feeding on α and β values of GAS, WGS, MET and DISP reactions *vs.* t.o.s.: **(A)** S/C, 1.5; H_2/C, 0; **(B)** S/C, 2.8; H_2/C, 0; **(C)** S/C, 3.5; H_2/C, 0; **(D)** S/C, 2.8; H_2/C, 2. Legend: (●) α; (●) β_{WGS}; (○) β_{DISP}; (●) β_{GAS}; (●) β_{MET} (reproduced from ref. [11] with the kind permission of Elsevier, copyright 2004).

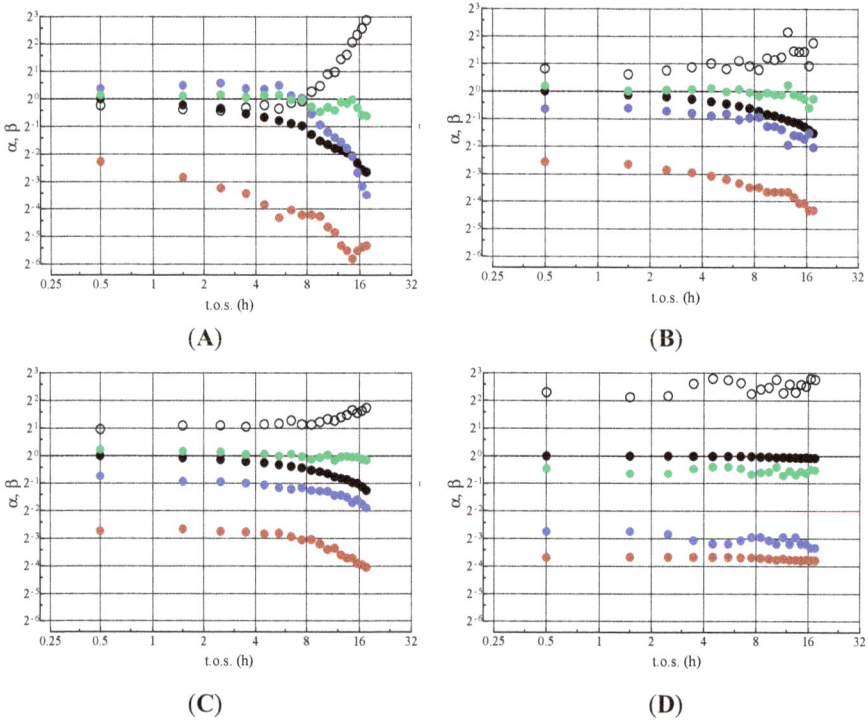

(A)

(B)

(C)

(D)

These figures reveal the following trends:

1. MET rate (α_{CH4}) decreases much more steeply than relative activity;
2. GAS rate (α_{CO}) decreases similar to relative activity;
3. WGS rate (α_{CO2}) decreases less than relative activity.

Figure 8. Relative rates of hexane conversion (α, ●), CH$_4$ (α_{CH4}, ▲), CO$_2$ (α_{CO2}, ◆) and CO (α_{CO}, ▼) formation *vs.* t.o.s. at different reaction conditions (see Table 1) (adapted from ref. [15] with the kind permission of Wiley, copyright 2006).

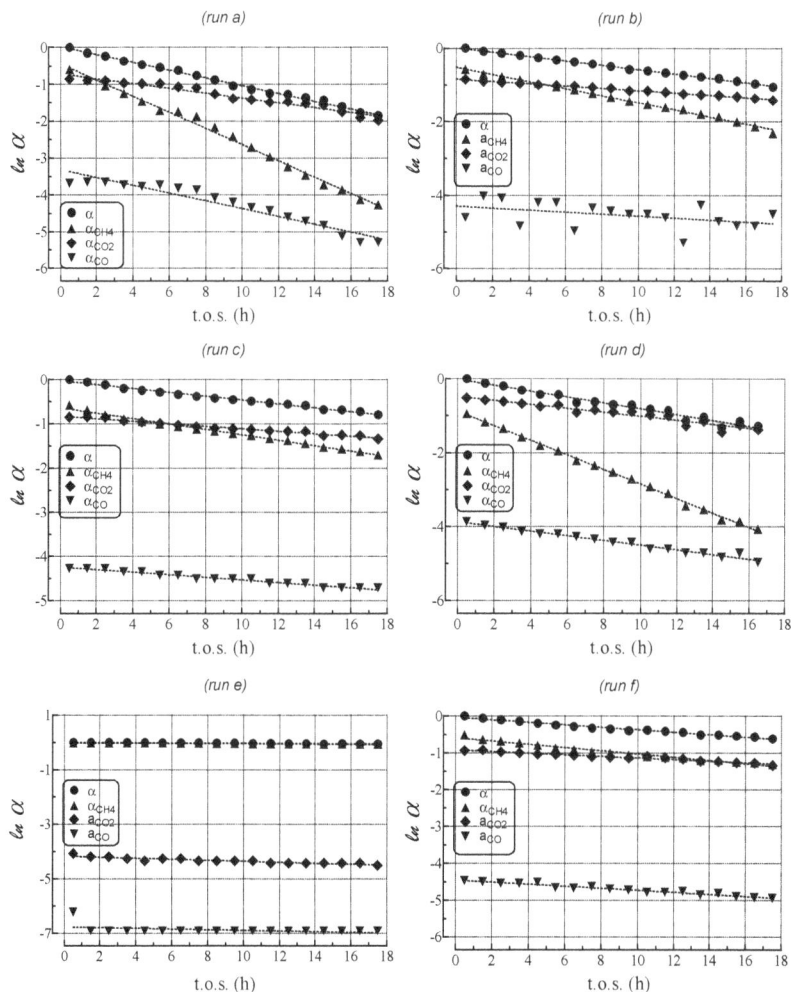

Indeed, direct relationships between k_{MET}, k_{GAS}, and k_{WGS} decay and that of activity loss (k_{deact}) result in straight-line correlations (Figure 9) with slopes equal to 2.7 ± 0.3, 1.1 ± 0.2 and 0.6 ± 0.1, respectively. From this and similar figures it can be shown that MET, GAS, and WGS functionalities depend on different *ensembles* of active sites [11,15,17,19–21,36]; thus, a slope of 1 for k_{GAS}/k_{deact} suggests that hexane conversion involves single sites leading to CO, while MET and WGS require larger and smaller ensemble of active sites for methane and carbon dioxide formation, respectively.

Table 4. Deactivation constants of the various functionalities (T, 450 °C): hexane conversion (k_{deact}), methanation (k_{MET}), gasification (k_{GAS}), water-gas-shift (k_{WGS}).

Run	k_{deact} [a] (h^{-1})	k_{cok} [a] (h^{-1})	k_{MET} (h^{-1})	k_{GAS} (h^{-1})	k_{WGS} (h^{-1})
a	1.1×10^{-1}	1.1×10^{-1}	2.2×10^{-1}	1.1×10^{-1}	6.6×10^{-2}
b	5.9×10^{-2}	5.6×10^{-2}	9.7×10^{-2}	2.7×10^{-2}	3.2×10^{-2}
c	4.4×10^{-2}	4.0×10^{-2}	6.0×10^{-2}	2.8×10^{-2}	2.7×10^{-2}
d	8.1×10^{-2}	7.7×10^{-2}	2.2×10^{-1}	7.1×10^{-2}	6.1×10^{-2}
e	3.5×10^{-3}	0.000	3.5×10^{-3}	1.2×10^{-2}	1.9×10^{-2}
f	3.4×10^{-2}	3.0×10^{-2}	4.4×10^{-2}	2.8×10^{-2}	2.2×10^{-2}

a data taken from Table 2.

In fact, according to Beeckman *et al.* who adopted a stochastic approach to describe deactivation by coking under different mechanisms [34], MET functionality is affected to the maximum extent by the fouling of active sites because of its high number of metal atoms required for each site (e.g., formal site molecularity) in comparison to GAS and WGS ones [11,15,17,19,40,41].

Therefore, in agreement with literature data, kinetic and thermodynamic findings confirm that surface carbon (*i.e.*, C_α) hydrogenation is the crucial step controlling activity, selectivity, and stability against coking in the pre-reforming process. Activity decay is more pronounced at low conversion, mostly under the occurrence of GAS and WGS reactions, as CO formation-decomposition is mainly responsible for carbon build-up *via* an adsorbed C_α intermediate [19,35].

Figure 9. Relationships between k_{MET} (●), k_{GAS} (▼), k_{WGS} (◆) and k_{deact} (*see* Table 4) (adapted from ref. [15] with the kind permission of Wiley, copyright 2006).

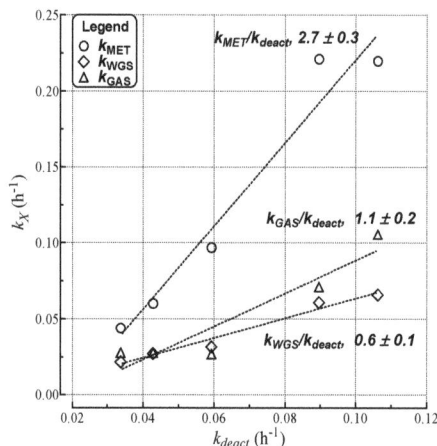

However, with addition of high concentrations of H_2, similar values of k_{deact} and k_{MET} (Table 4), and a much steeper decay of gasification ($k_{GAS}/k_{deact} \approx 3.5$) and WGS ($k_{WGS}/k_{deact} \approx 5$) functionalities [11,17,19] evidence that different reaction and deactivation paths are favored.

3. Experimental Section

3.1. Catalyst Preparation

The 19.1 wt% Ni/MgO catalyst was prepared by incipient wetness impregnation of "smoke" MgO powder (UBE Ltd., Yamaguchi, Japan SA_{BET}, 30 $m^2 \cdot g^{-1}$) with an ethanolic solution of the $Ni(NO_3)_2$ salt [28–31]. The catalyst was dried at 80 °C, calcined at 400 °C (16 h) and pre-reduced for 6 h at 650 °C under H_2 flow.

3.2. Catalyst Testing

Catalytic tests in the pre-reforming of n-hexane were performed at 450 °C and a pressure of 5–15 bar, using an isothermal "fixed bed" stainless steel microreactor (i.d., 6 mm), loaded with 25 mg of catalyst (40–70 mesh), diluted with same-sized SiC in a 1/30 weight ratio [11,15]. Before tests, the catalyst was heated at 450 °C and further reduced *in situ* for 1 h in H_2 flow (100 *STP* mL min^{-1}; *P*, 1 bar). The reaction mixture including n-C_6H_{14} (1.8%), N_2 (8.0%), H_2O (16.2–37.8%) and He (rest) for a variation of the S/C ratio between 1.5 and 3.5 respectively, was fed at the rate of 283 (GHSV, 12,000 h^{-1}) or 566 (GHSV, 24,000 h^{-1}) *STP* mL min^{-1}. The effects of H_2 (11.0%; H_2/C, 1.0) and thiophene (50 ppm in *n*-hexane) were probed at the highest GHSV and a S/C ratio of 2.8. Both water and n-hexane were fed as liquid by HPLC pumps. The reaction temperature was controlled by a thermocouple in contact with the catalytic bed, while the reactor stream was analyzed by a GC equipped with a three-columns analytical system connected to both TCD and FID detectors for permanent gases and hydrocarbons analyses, respectively [11,15].

3.3. Catalyst Characterization

TEM analyses were performed using a PHILIPS CM12 Transmission Electron Microscope (point-to-point resolution, 0.3 nm) on "fresh" and "spent" catalyst samples, dispersed ultrasonically in ethanol and deposited over a thin carbon film supported on a standard copper grid. The Ni particle size distribution (PSD) was obtained from an average of 150–200 particles, and the average volume-area particle size calculated by the conventional statistical formula [29–31]:

$$\overline{d}_{AV} = \frac{\sum_i n_i d_i^3}{\sum_i n_i d_i^2} \qquad (1)$$

The H_2 uptake of "fresh" and "used" catalyst was evaluated by TPD measurements in the range −80–620 °C using Ar as carrier gas (30 *stp* cm$^3 \cdot$min^{-1}) after *in situ* reduction of the samples in H_2 flow at 450 °C for 1 h and subsequent saturation at r.t. (room temperature) and further at −80 °C. The fraction of reduced Ni was evaluated by titration at 450 °C with O_2 pulses (Ni + ½ $O_2 \rightarrow$ NiO) after TPD analysis [31].

TGA-DSC measurements of the "used" catalyst samples (5 mg) in the range of 20–400 °C were performed using a Simultaneous Thermal Analyser (STA 409C, Netzsch), operating in air with a heating rate of 10 °C/min and an accuracy of 0.01 mg.

4. Conclusions

This review summarizes highlights of previous studies (mainly in our laboratory) of reaction and deactivation paths on a "model" Ni/MgO catalyst in the pre-reforming of n-C_6H_{14} with steam at 450 °C. A kinetic and thermodynamic analysis of the data from these studies provides (1) new deactivation rate parameters for carbon deposition and coke formation and (2) insights into the surface reaction network of the pre-reforming process, including hexane cracking, water-gas-shift, and methanation reactions.

Important conclusions from this work can be summarized as follows:

- Deactivation of Ni/MgO by carbon deposition and coking follows a pseudo 1st-order dependence on active site concentration, consistent with an exponential decay with t.o.s. That deactivation by coke and carbon formation is not simple first order in site concentration, is evident from observed decreases in deactivation rate constants with increasing S/C ratio, P, and contact time.
- Comparison of thermodynamic predictions with experimental values of outlet stream composition leads to the conclusion that the kinetics of hexane cracking/reforming, carbon gasification and methanation reactions control the rate of coking; methanation is the critical, rate limiting step responsible for coke deposition in the near absence of H_2. Carbon formation via the Boudouard reaction is apparently thermodynamically and kinetically favored.
- Under simulated industrial pre-reforming conditions at a high H_2 concentration, hydrogen substantially increases reaction throughput by speeding up hydrogenolysis, CO formation via the water-gas-shift, and methanation reactions. H_2 at high concentrations gasifies coke and carbon precursors, thereby largely preventing formation of coke and inactive carbons.
- S-poisoning by thiophene co-feeding causes a sharp drop in catalytic activity due to irreversible poisoning; however, strong pore diffusional resistances in catalyst pellets and bed prevent an accurate analysis of the poisoning mechanism and kinetics.

Author Contributions

This work was conceived by Francesco Arena, who arranged the manuscript, and Giuseppe Trunfio, who carried out the experiments and also conceived and revised the manuscript.

Conflict of Interest

The authors declare no conflict of interest.

References

1. Piel, W.J. Transportation fuels of the future? *Fuel Process. Tech.* **2001**, *71*, 167–179.
2. Nagase, S.; Takami, S.; Hirayama, A.; Hirai, Y. Development of a high efficiency substitute natural gas production process. *Catal. Today* **1998**, *45*, 393–397.

3. Cromarty, B.J.; Hooper, C.W. Increasing the throughput of an existing hydrogen plant. *Int. J. Hydrogen Energy* **1997**, *22*, 17–22.

4. Aitani, A.M. Processes to enhance refinery-hydrogen production. *Int. J. Hydrogen Energy* **1996**, *21*, 267–271.

5. Fierro, J.L.G.; Peña, M.A.; Gómez, J.P. New catalytic routes for syngas and hydrogen production. *Appl. Catal. A* **1996**, *144*, 7–57.

6. Rostrup-Nielsen, J.; Christiansen, L.J. Concept in Syngas Manufacture. *Catalytic Science Series*; Imperial College Press: London, UK, 2011; Volume 10.

7. Bartholomew, C.H.; Farrauto, R.G. *Fundamental of Industrial Catalytic Processes*, 2nd ed.; J. Wiley & sons, Inc.: Hobken, NJ, USA, 2006; pp. 339–371.

8. Wang, X.; Gorte, R.J. A study of steam reforming of hydrocarbon fuels on Pd/ceria. *Appl. Catal. A* **2002**, *224*, 209–218.

9. Zou, X.; Wang, X.; Li, L.; Shen, K.; Lu, X.; Ding, W. Development of highly effective supported nickel catalysts for pre-reforming of liquefied petroleum gas under low steam to carbon molar ratios. *Int. J. Hydrogen Energy* **2010**, *35*, 12191–12200.

10. Shen, K.; Wang, X.; Zou, X.; Wang, X.; Lu, X.; Ding, W. Pre-reforming of liquefied petroleum gas over nickel catalysts supported on magnesium aluminum mixed oxides. *Int. J. Hydrogen Energy* **2011**, *36*, 4908–4916.

11. Arena, F.; Trunfio, G.; Alongi, E.; Branca, D.; Parmaliana, A. Modeling the activity-stability pattern of Ni/MgO catalysts in the pre-reforming of *n*-hexane. *Appl. Catal. A* **2004**, *266*, 155–162.

12. Ayabe, A.; Omoto, H.; Utaka, T.; Kikuchi, R.; Sasaki, K.; Teraoke, Y.; Educhi, K. Catalytic autothermal reforming of methane and propane over supported metal catalysts. *Appl. Catal. A* **2003**, *241*, 261–269.

13. Rostrup-Nielsen, J. Activity of nickel catalysts for steam reforming of hydrocarbons. *J. Catal.* **1973**, *31*, 173–181.

14. Rostrup-Nielsen, J. *Catalysis, Science and Technology*; Springer: Berlin, Germany, 1983.

15. Arena, F. Basic Relationships in the Pre-Reforming of *n*-Hexane on Ni/MgO Catalyst. *AIChE J.* **2006**, *52*, 2823–2831.

16. Christensen, T.S. Adiabatic prereforming of hydrocarbons—An important step in syngas production. *Appl. Catal. A* **1996**, *138*, 285–309.

17. Rostrup-Nielsen, J.; Alstrup, I. Innovation and science in the process industry: Steam reforming and hydrogenolysis. *Catal. Today* **1999**, *53*, 311–316.

18. Sperle, T.; Chen, D.; Lødeng, R.; Holmen, A. Pre-reforming of natural gas on a Ni catalyst. Criteria for carbon free operation. *Appl. Catal. A* **2005**, *282*, 195–204.

19. Bartholomew, C.H. Carbon deposition in methanation and steam reforming. *Catal. Rev. Sci. Eng.* **1982**, *24*, 67–112.

20. Butt, J.B.; Petersen, E.E. *Activation, Deactivation and Poisoning of Catalysts*; Academic Press: Inc.: San Diego, CA, USA, 1998.

21. Forzatti, P.; Lietti, L. Catalyst deactivation. *Catal. Today* **1999**, *52*, 165–181.

22. Bartholomew, C.H. Mechanism of catalyst deactivation. *Appl. Catal. A* **2001**; *212*, 17–60.

23. Kroll, V.C.H.; Swaan, H.M.; Mirodatos, C. Methane Reforming Reaction with Carbon Dioxide Over Ni/SiO$_2$ Catalyst. Deactivation Studies. *J. Catal.* **1996**, *161*, 409–422.

24. Tailleur, R.G.; Davila, Y. Optimal Hydrogen Production through Revamping a Naphtha-Reforming Unit: Catalyst Deactivation. *Energy Fuels* **2008**, *22*, 2892–2901.

25. Sehested, J.; Carlsson, A.; Janssens, T.V.W.; Hansen, P.L.; Datye, A.K. Sintering of nickel steam-reforming catalysts on MgAl$_2$O$_4$ spinel supports. *J. Catal.* **2001**, *197*, 200–209.

26. Ashrafi, M.; Pfeifer, C.; Pröll, T.; Hofbauer, H. Experimental Study of Model Biogas Catalytic Steam Reforming: 2. Impact of Sulfur on the Deactivation and Regeneration of Ni-Based Catalysts. *Energy Fuels* **2008**, *22*, 4190–4195.

27. Gallego, J.; Batiot-Dupeyrat, C.; Barrault, J.; Mondragón, F. Severe Deactivation of a LaNiO$_3$ Perovskite-Type Catalyst Precursor with H$_2$S during Methane Dry Reforming. *Energy Fuels* **2009**, *23*, 4883–4886.

28. Arena, F.; Parmaliana, A. Strategies of design of Ni/MgO catalysts for the reforming of hydrocarbon to hydrogen/syngas. *Curr. Top. Catal.* **2006**, *5*, 69–88.

29. Parmaliana, A.; Arena, F.; Frusteri, F.; Coluccia, S.; Marchese, L.; Martra, G.; Chuvilin, A. Magnesia Supported Nickel Catalysts. Part II: Surface properties and reactivity in CH$_4$ steam reforming. *J. Catal.* **1993**, *141*, 34–47.

30. Frusteri, F.; Spadaro, L.; Arena, F.; Chuvilin, A. TEM evidence for factors affecting the genesis of carbon species on bare and K-doped Ni/MgO catalysts during the dry-reforming of methane. *Carbon* **2002**, *40*, 1063–1070.

31. Arena, F.; Horrell, B.A.; Cocke, D.L.; Parmaliana, A.; Giordano, N. Magnesia-Supported Nickel Catalysts. I: Structure and morphological properties. *J. Catal.* **1991**, *132*, 58–67.

32. Suzuki, T.; Iwanami, H.; Yoshinari, T. Steam reforming of kerosene on Ru/Al$_2$O$_3$ catalyst to yield hydrogen. *Int. J. Hydrogen Energy* **2000**, *25*, 119–126.

33. Ming, Q.; Healey, T.; Allen, L.; Irving, P. Steam reforming of hydrocarbon fuels. *Catal. Today* **2002**, *77*, 51–64.

34. Beeckman, J.W.; Nam, I.-S.; Froment, G.F. Stochastic modeling of catalyst deactivation by site coverage. *Stud. Surf. Sci. Catal.* **1987**, *34*, 365–379.

35. Jackson, S.D.; Thomson, S.J.; Webb, G. Carbonaceous deposition associated with the catalytic steam-reforming of hydrocarbons over nickel alumina catalysts. *J. Catal.* **1981**, *70*, 249–263.

36. Snoeck, J.-W.; Froment, G.F.; Fowles, M.J. Filamentous carbon formation and gasification: thermodynamics, driving force, nucleation, and steady-state growth. *J. Catal.* **1997**, *169*, 240–249.

37. Chen, F.; Zha, S.; Dg, J.; Liu, M. Pre-reforming of propane for low-temperature SOFCs. *Solid State Ionics* **2004**, *166*, 269–273.

38. Sughrue, E.L.; Bartholomew, C.H. Kinetics of carbon monoxide methanation on nickel monolithic catalysts. *Appl. Catal.* **1982**, *2*, 239–256.

39. Rostrup-Nielsen, J. Sulfur Poisoning. In *Progress in Catalyst Deactivation*; Figuerido, J.L., Ed.; Martinus Nijhoff Publisher: The Hague, The Netherlands, 1982; pp. 209–227.

40. Sidjabat, O.; Trimm, D.L. Nickel-magnesia catalysts for the steam reforming of light hydrocarbons. *Top. Catal.* **2000**, *11/12*, 279–282.
41. Fujita, S.; Terenuma, H.; Nakamura, M.; Takezawa, N. Mechanisms of methanation of CO and CO_2 over Ni. *Ind. Eng. Chem. Res.* **1991**, *30*, 1146–1151.

Inhibition of a Gold-Based Catalyst in Benzyl Alcohol Oxidation: Understanding and Remediation

Emmanuel Skupien, Rob J. Berger, Vera P. Santos, Jorge Gascon, Michiel Makkee, Michiel T. Kreutzer, Patricia J. Kooyman, Jacob A. Moulijn and Freek Kapteijn

Abstract: Benzyl alcohol oxidation was carried out in toluene as solvent, in the presence of the potentially inhibiting oxidation products benzaldehyde and benzoic acid. Benzoic acid, or a product of benzoic acid, is identified to be the inhibiting species. The presence of a basic potassium salt (K_2CO_3 or KF) suppresses this inhibition, but promotes the formation of benzyl benzoate from the alcohol and aldehyde. When a small amount of water is added together with the potassium salt, an even greater beneficial effect is observed, due to a synergistic effect with the base. A kinetic model, based on the three main reactions and four major reaction components, is presented to describe the concentration-time profiles and inhibition. The inhibition, as well as the effect of the base, was captured in the kinetic model, by combining strong benzoic acid adsorption and competitive adsorption with benzyl alcohol. The effect of the potassium salt is accounted for in terms of neutralization of benzoic acid.

Reprinted from *Catalysts*. Cite as: Skupien, E.; Berger, R.J.; Santos, V.P.; Gascon, J.; Makkee, M.; Kreutzer, M.T.; Kooyman, P.J.; Moulijn, J.A.; Kapteijn, F. Inhibition of a Gold-Based Catalyst in Benzyl Alcohol Oxidation: Understanding and Remediation. *Catalysts* **2014**, *4*, 89-115.

1. Introduction

As far as the chemical industry is concerned, biomass is forecast to be one of the major successors of oil as a source of carbon for the production of organic molecules [1,2]. However, the chemistry of biomass-derived molecules differs significantly from oil-derived molecules [1,3]. Their higher oxygen content renders them more sensitive to oxidation, requiring milder conditions in selective oxidation processes. Furthermore, the solubility in water is enhanced when organics are oxygenated, which can be either problematic or advantageous. It is now widely recognized that the chemical industry will need to adapt to the new situation [1–3]. Once the catalytic abilities of gold had been discovered [4], its high activity for oxidation reactions at mild conditions, down to room temperature, was quickly noticed. This exceptionally low process temperature allows for a much better control over the selectivity, in particular in selective oxidations. These reactions play an important role in organic synthesis [1,5–11] and, as a consequence, much effort was put into studying and benchmarking gold-based catalysts for selective oxidation of alcohols, ketones, and carboxylic acids.

Benzyl alcohol (BnOH) selective oxidation to benzaldehyde (BnO) is one of these benchmark reactions used extensively to assess the catalytic activity of gold catalysts [1,5–9,11–13]. Alcohol selective oxidation is often carried out in the presence of a base as "promoter" or "co-catalyst" [1–3,7,8,11,12,14,15]. The base enhances the deprotonation of the alcohol, thereby ensuring that the rate-limiting step is the catalytic oxidation step [1,15]. Furthermore, alkaline conditions have also been reported to enhance the selectivity towards benzoic acid (BnOOH) [1,15]. However, the focus is usually on the initial turnover frequency (TOF) [5,7,12], whereas only a few reports mention issues of deactivation and re-usability [1,6,9,11,14].

Deactivation can arise either from catalyst degradation (e.g., sintering) or from catalyst poisoning or fouling [16]. Poisoning and fouling can sometimes be reversed by catalyst regeneration, mainly under oxidative conditions [9]. Sulfur-containing impurities are often responsible for poisoning of noble metal catalysts [16], and desulfurization catalysis has matured for decades to answer this problem, for instance by a combination of hydrodesulphurization (HDS) and guard beds. In a more general perspective, poisoning impurities in the feed can be eliminated by dedicated treatments. In selective oxidation over noble metal catalysts, deactivation can also occur due to the over-oxidation of Pt [14] and Pd [14,17] catalysts when an excess of (molecular) oxygen is present. This over-oxidation means that too much atomic oxygen (a reaction intermediate) is present on the catalytic sites, thereby blocking their accessibility for hydrocarbon adsorption. Because of this, catalysts tested in the oxygen mass-transfer limited regime can exhibit higher activity [1,11,14] than might be predicted based on data in the kinetic regime. It is generally accepted that gold-based catalysts are resistant to over-oxidation [1,3,11], making them promising candidates for oxidation reactions over extended periods of time. However, a reaction intermediate or the product itself can be an inhibiting entity. It should be noted that inhibition is reversible because the inhibition is remediated when the concentration of the inhibitor in the reaction medium is sufficiently reduced, whereas poisoning is irreversible at the reaction conditions [16]. This so-called product inhibition phenomenon is an even greater challenge, as the catalyst creates its own poison while performing the desired reaction. This has been frequently observed both in oxidation and dehydrogenation reactions. For instance, Dimitratos *et al.* [18] attributed deactivation of Au-Pd and Au-Pt catalysts in octanol oxidation to inhibition by the carboxylate formed. They also reported the alleviation of this inhibition when NaOH was present. Zope and Davis [3] reported similar effects for the selective oxidation of glycerol to glyceric acid. They performed the reaction in the presence of 19 different compounds: either products or intermediates in the glycerol oxidation reaction, or species that might be formed from condensation of intermediates and/or products. Ketones, condensation products of ketones or secondary alcohols (forming ketones upon oxidation) were found to be inhibiting compounds. In contrast, simple carboxylic acids such as acetic acid and propionic acid did not show appreciable inhibition, nor did diacids such as malonic or succinic acid, or primary alcohols such as methanol. To the best of our knowledge, and despite reviews mentioning the occurrence of product inhibition during the oxidation of alcohols on gold and platinum group metal catalysts in general [11,14], no detailed study of this phenomenon for BnOH oxidation over gold-based catalysts has been reported.

The current study concerns the Au-catalyzed partial oxidation of BnOH to BnO and benzyl benzoate (BnOOBn) in toluene over the commercial AUROlite™ Au/Al$_2$O$_3$ catalyst. BnOH, which is a primary alcohol, is oxidized to BnO and subsequently to BnOOH, while also BnOOBn can be formed (Figure 1). This system suffers from deactivation, which is particularly observed when the catalyst is re-used in batch-wise operation. This deactivation can be suppressed by the addition of an inorganic base. The aim of the current study is to analyze this deactivation process, identify the possible inhibiting species, elucidate the deactivation mechanism, and evaluate the beneficial effect of the inorganic base. To accomplish this, the reaction was carried out under various reaction conditions, including experiments in the presence of reaction products, bases, and water. Additionally, kinetic modeling was performed in order to confirm the reaction and deactivation mechanisms.

Figure 1. Reaction network: (**a**) oxidation of benzyl alcohol to benzaldehyde; (**b**) oxidation of benzaldehyde to benzoic acid; (**c**) esterification of benzyl alcohol and benzoic acid, and (**d**) esterification of benzyl alcohol and benzaldehyde under oxidative conditions to benzyl benzoate [19].

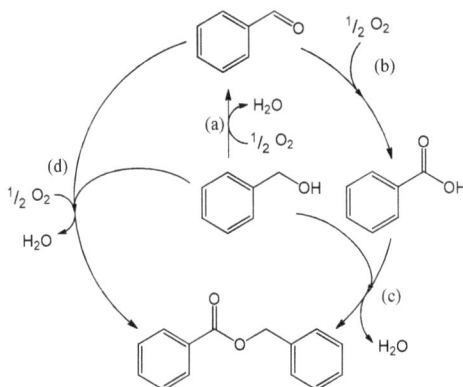

2. Results and Discussion

2.1. Catalyst Deactivation in Base-Free Conditions

Figure 2 displays the concentration profiles for two subsequent experiments performed under the same conditions and in the absence of a base. It should be noted that for all experimental data points, the mass balance of the 4 main components (BnOH, BnO, BnOOH and BnOOBn) closes to 100% within measurement errors, with rare exceptions where discrepancies up to 4% are present. Therefore, disproportionation and dehydration reactions, as reported by Alhumaimess et al. [5], can be neglected. In the first experiment, the initial TOF is 0.7 s^{-1} and conversion levels off at around 55–60% after about 120 min. The selectivity to BnO is 93%, as reported in Table 1. In view of the large discrepancies in TOF reported in literature [1], comparison of our data to previously published ones is not straightforward. It is satisfactory that the values of 0.6 to 0.8 s^{-1} reported in the present study are in the order of magnitude of those of 0.04 to 0.22 s^{-1} reported in the recent review by Davis et al. [1] for nanoparticulate gold and of 2.8 to 4.4 s^{-1} for a gold foil reported in the same review [1]. As indicated by Davis et al., the values span 2 orders of magnitude. The reason for this is most likely the wide range of reaction conditions used in different studies.

Figure 2. Catalyst performance in base-free conditions. Concentration of (♦) benzyl alcohol, (■) benzaldehyde, (●) benzoic acid and (▲) benzyl benzoate *vs.* reaction time for (**a**) first run using fresh AUROlite™ and (**b**) second run using spent AUROlite™. Reaction conditions: T = 80 °C, 0.8 g AUROlite™, $C_{BnOH, t=0}$ = 3.0 × 10^{-4} mol·g^{-1} in 80 mL of toluene, 200 mL·min^{-1} air flow. Concentrations are expressed in moles per unit mass of liquid in the reactor (mol·g^{-1}). The symbols with error bars are the experimental results and the lines represent the kinetic model.

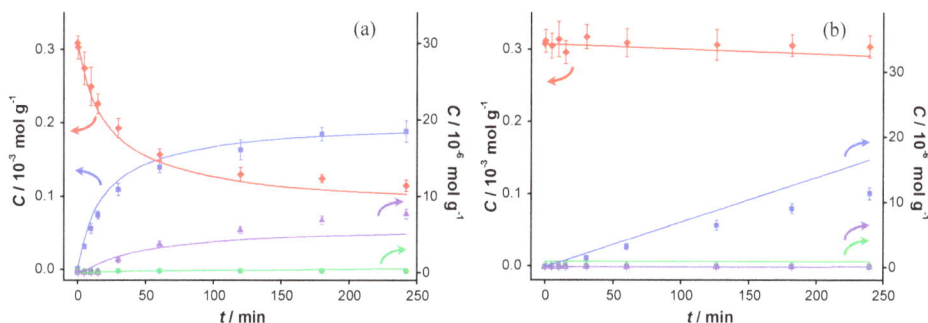

In the second experiment, which is identical to the first one but with re-use of the same catalyst sample after washing with toluene, virtually no conversion is observed, indicating that the catalyst was completely deactivated. Since no other reactants than toluene and benzyl alcohol were present in the reaction mixture, the deactivation must be caused either by catalyst deterioration or by an inhibitor formed during the first experiment.

The potential presence of sulfur-containing impurities was investigated by analyzing the toluene and benzyl alcohol by gas chromatography (GC). No sulfur-containing compounds could be detected, implying that their concentration was below 50 ppb (the pulsed flame photometric detector (PFPD) detection limit). Accounting for the quantities of these chemicals introduced in the reaction mixture, this corresponds to 0.4 nmol of sulfur components at maximum in the reactor. That is five orders of magnitude lower than the total amount of gold present. Since similar concentration profiles as shown in Figure 2 were obtained when using reagents that were pre-treated with activated carbon to remove any strongly adsorbing impurities, we conclude that feed contaminants, including sulfur compounds, are not responsible for the observed deactivation. As a consequence, activated carbon cleaning was deemed not to be necessary and was omitted for the experiments reported here.

Table 1. Benzyl alcohol conversion rate, turn over frequency (*TOF*), conversion at 240 min and selectivity to benzaldehyde at 60% conversion for AUROlite™ catalyst in different test conditions.

Catalytic system	Figure	BnOH conversion rate [a] /mmol·min^{-1}	TOF [b] /s^{-1}	X at 240 min [c] /%	Sel to BnO [d] /%
Base-free	2a	0.37 ± 0.02	0.7 ± 0.1	63 ± 3	93 ± 1
Base-free re-test	2b	0.06 ± 0.04	0.1 ± 0.1	1 ± 5	/
Base-free–BnO	3	0.09 ± 0.04	0.2 ± 0.1	20 ± 10	/
Base-free–BnOOH	4	0	0	2 ± 7	/
K_2CO_3	6a	0.5 ± 0.1	0.9 ± 0.4	78 ± 1	91 ± 1
K_2CO_3 re-test	6b	0.20 ± 0.02	0.4 ± 0.1	72 ± 1	85 ± 1
KF	7a	0.3 ± 0.1	0.6 ± 0.3	81 ± 1	91 ± 1
KF re-test	7b	0.5 ± 0.2	1.0 ± 0.5	95 ± 1	83 ± 1
K_2CO_3–BnOOH	10	0.10 ± 0.03	0.2 ± 0.1	61 ± 7	92 ± 1
K_2CO_3–H_2O	11	0.43 ± 0.03	0.8 ± 0.2	97 ± 1	86 ± 1
Base-free re-test–H_2O	12	0.03 ± 0.05	0.1 ± 0.1	1 ± 4	/

Concentrations are expressed in moles per unit mass of liquid in the reactor (mol·g^{-1}).

[a] $BnOH\ conversion\ rate = \frac{\Delta C_{BnOH}}{\Delta t} \times w_{liq}$, where $\frac{\Delta C_{BnOH}}{\Delta t}$ is the conversion rate of benzyl alcohol (mmol·g^{-1}·min^{-1}), calculated by linear regression of the concentration values between 0 and 15 min and w_{liq} is the total mass of liquid (g); [b] $TOF = \frac{\Delta C_{BnOH}}{\Delta t} \times \frac{0.001}{60} \times w_{liq} \times \frac{M_{Au}}{w_{cat} \times 0.01 \times 0.22}$, where $\frac{w_{cat} \times 0.01}{M_{Au}}$ is the total amount of gold in the reactor (mol) and 0.22 is the amount of edge + corner atoms per amount of gold for 2.5 nm gold nanoparticles [20] (mol·mol^{-1}); [c] $X = \frac{C_{BnOH,t=0} - C_{BnOH,t=240min}}{C_{BnOH,t=0}} \times 100\%$; [d] $Sel_{BnO} = \frac{C_{BnO,X=60\%}}{C_{BnO,X=60\%} + C_{BnOOH,X=60\%} + 2 \times C_{BnOOBn,X=60\%}} \times 100\%$ In parallel and series reaction networks, selectivities have to be compared at the same conversion levels. 60% is chosen here, as it is the level in the base free experiment after 240 min.

These results thus suggest that products or intermediates formed cause the observed deactivation. In order to investigate whether product inhibition is indeed taking place, an experiment was performed with fresh catalyst, where BnO was added to the reactor 20 min prior to the actual start of the catalytic reaction under the same reaction conditions, *i.e.*, before the introduction of BnOH. The 20 min exposure time was selected based on the results of Figure 1a where it can be seen that BnOOBn was already formed at that reaction time, thereby ensuring that all potentially inhibiting products were present. The results in Figure 3 show a very low conversion of 20%, confirming strong product inhibition. It is striking that despite the appreciable amount of BnO introduced at $t < 0$, neither BnOOBn nor BnOOH could be detected until the BnOH was introduced at $t = 0$. Clearly the sites for the sequential reactions of BnO (Figure 1a,b) are fully blocked without the presence of alcohol. It should be noted that for $t < 0$ the BnO concentration shows a slight decrease, indicating that a small amount of BnO is consumed without producing a detectable amount of BnOOH in the liquid phase. We conclude that some product is formed which remains on the catalyst and inhibits further turnovers. Carboxylic acid moieties are well known to interact strongly with gold nanoparticle surfaces, even allowing the stabilization of small gold clusters in colloidal systems [21]. BnOOH is therefore suspected to be the inhibitor, since it is the logical product of BnOH oxidation, although its concentration in the solution stayed below the GC detection limit of $0.20 \ \mu mol \cdot g^{-1}$.

Figure 3. Catalyst performance after pre-addition of benzaldehyde. Concentration of (♦) benzyl alcohol, (■) benzaldehyde, (●) benzoic acid and (▲) benzyl benzoate *vs.* reaction time for catalytic reaction over fresh AUROlite™. Benzaldehyde was introduced 20 min before benzyl alcohol was added. Reaction conditions: $T = 80 \ °C$, 0.8 g AUROlite™, $C_{BnOH, \ t=0} = 3.0 \times 10^{-4} \ mol \cdot g^{-1}$, $C_{BnO, \ t<0} = 1.7 \times 10^{-4} \ mol \cdot g^{-1}$, in 80 mL of toluene, 200 mL·min⁻¹ air flow. Concentrations are expressed in moles per unit mass of liquid in the reactor (mol·g⁻¹). The symbols with error bars are the experimental results and the lines represent the kinetic model.

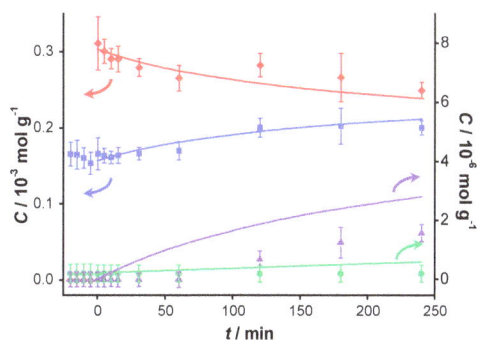

In order to evaluate the effect of BnOOH, a similar experiment was performed in which BnOOH was added 20 min prior to the actual start of the test. The concentration profiles are shown in Figure 4. In this experiment, approximately 40 times less acid was introduced compared to the amount of aldehyde introduced for the experiment in Figure 3. Nevertheless, an even more dramatic inhibiting effect was observed: hardly any conversion of BnOH occurred. The concentration of BnOOH measured by GC at $t < 0$ is 1 µmol·g⁻¹, which is about 25% of what was added, suggesting a strong interaction of BnOOH with the catalyst. The slight increase in concentration of BnOOH upon addition of BnOH at $t = 0$ is attributed to competitive adsorption between the alcohol and the acid. No ester was present at the beginning of the reaction, nor was it detected during the course of the reaction. This suggests that the inhibiting product is either BnOOH or a compound formed from BnOOH. Therefore, product inhibition particularly occurs on catalysts on which BnO can react further to BnOOH. This interpretation is confirmed by a nanostructured gold-based catalyst synthesized in our lab exhibiting 100% selectivity to BnO not showing deactivation for 4 consecutive runs [22].

Figure 4. Catalyst performance after pre-addition of benzoic acid. Concentration of (♦) benzyl alcohol, (■) benzaldehyde, (●) benzoic acid and (▲) benzyl benzoate *vs.* reaction time for catalytic reaction over fresh AUROlite™. Benzoic acid was introduced 20 min prior to the beginning of the reaction. Reaction conditions: T = 80 °C, 0.8 g AUROlite™, $C_{BnOH, t=0}$ = 3.0 × 10^{-4} mol·g^{-1}, $C_{BnOOH, t<0}$ = 4.1 × 10^{-6} mol·g^{-1}, in 80 mL of toluene, 200 mL·min^{-1} air flow. Concentrations are expressed in moles per unit mass of liquid in the reactor (mol·g^{-1}). The symbols with error bars are the experimental results and the lines represent the kinetic model.

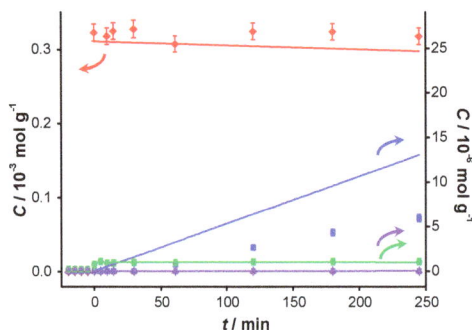

2.2. Influence of a Base on Catalyst Deactivation

Figure 5 shows the reaction mechanism, in basic conditions, of the oxidation of alcohol to aldehyde and the sequential oxidation to carboxylic acid [1,15,23–25]. Au* indicates an active site on the gold surface. A dashed line represents a chemical interaction of a species with an adsorption site on the gold surface. Electron transfers corresponding to bond cleavage or formation are indicated by curved arrows. As indicated in Figure 5, reaction step a, the role of the base is to deprotonate the alcohol, thereby ensuring that the rate limiting step is the reaction on the gold surface, presumably the β-hydride elimination forming the aldehyde (step b yielding R-CH=O) or the carboxylic acid (step d yielding R-COOH) [1,15,23–25]. However, basic conditions have also been reported to enhance the selectivity towards carboxylic acid, by favoring the conversion of aldehyde to the corresponding geminal diol [1,15] as depicted by reaction step c (The diol is shown as R-CH(OH)O⁻, partially deprotonated and adsorbed on Au*). In parallel, the gold site that carries the hydride (Au*-H⁻) is regenerated by adsorbed molecular oxygen (Au*||||O₂) via a peroxyl intermediate (step e yielding Au*-O-OH) [1,15], or via dissociated oxygen from the catalyst support [25]. This also regenerates an OH⁻ (steps f and g) [15] and closes the catalytic cycle as shown on the right-hand side of Figure 5. For stoichiometric reasons, it is clear that two hydrides species must react per O₂ molecule. However, it is unclear whether the second hydride reacts after the O-O bond dissociation (step f), or if the O-O bond dissociation is assisted by the second hydride (in the latter case, step f and g would be simultaneous). The question arises if the presence of a base has an influence on the catalyst stability.

Figure 5. Reaction mechanism of benzyl alcohol oxidation to benzaldehyde and sequential oxidation to benzoic acid over a gold catalyst, co-catalyzed by the base HO⁻ [1,15,23–25].

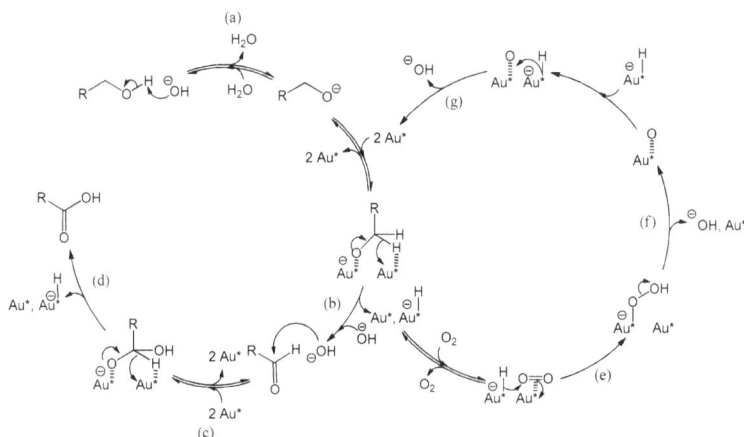

The catalytic reaction was carried out in the presence of two different bases: potassium carbonate (pK_{b1} = 3.68) and potassium fluoride (pK_b = 10.8). Although the pK_b values (defined in aqueous environment) are not directly transferable to the aprotic solvent (toluene), and these bases hardly dissolve in toluene, the pK_b values still indicate the relative basicity of both bases used. The spent catalysts were also re-used. A significant loss of KF was observed during the recovery of the catalyst after the first test, which was compensated for by the addition of another 2.1 g of KF to the spent catalyst, resulting in an estimated total amount of KF present during the second experiment with the spent catalyst of about 3 g. The concentration profiles of these experiments are presented in Figures 6 and 7, respectively.

The results of the first run are similar to those for the base-free experiments, although a slight increase in the initial *TOF*, from 0.7 to 0.9 s⁻¹, is observed (Table 1). Increased activity upon addition of a base has already been reported [1] and is probably due to the enhanced deprotonation of BnOH. The conversion at 240 min is also higher: 78% compared with 63% in base-free conditions. The selectivity to BnO does not change significantly (93% when no base is present *vs.* 91% in the presence of potassium carbonate). However, it should be noted that more ester is formed in the presence of K₂CO₃ (at 240 min: 16 µmol·g⁻¹ compared to 8 µmol·g⁻¹ in base free conditions), and the amount of BnOOH remains below detection limit. The most striking difference, when compared to base-free conditions, is the largely maintained activity when re-using the spent catalyst. The initial *TOF* is lower than that of the fresh catalyst (0.4 s⁻¹ compared with 0.9 s⁻¹), but the conversion at 240 min is comparable (72% compared with 78%). This shows that the addition of a base to the reaction medium largely remediates the strong product inhibition observed under base-free conditions.

Figure 6. Catalyst performance in the presence of K_2CO_3. Concentration of (♦) benzyl alcohol, (■) benzaldehyde, (●) benzoic acid and (▲) benzyl benzoate *vs.* reaction time for catalytic reaction over fresh AUROlite™ in the presence of K_2CO_3. (**a**) First run using fresh AUROlite™ (**b**) second run using spent AUROlite™. Reaction conditions: T = 80 °C, 0.8 g AUROlite™, 3.04 g K_2CO_3, $C_{BnOH, t=0}$ = 3.0 × 10^{-4} mol·g^{-1}, in 80 mL of toluene, 200 mL·min^{-1} air flow. Concentrations are expressed in moles per unit mass of liquid in the reactor (mol·g^{-1}). The symbols with error bars are the experimental results and the lines represent the kinetic model.

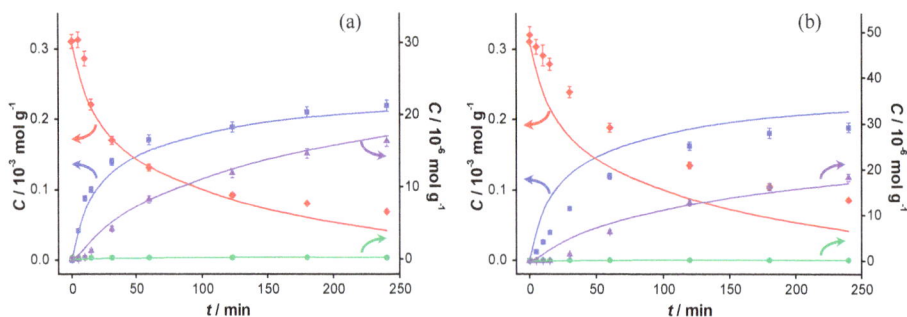

Figure 7. Catalyst performance in the presence of KF. Concentration of (♦) benzyl alcohol, (■) benzaldehyde, (●) benzoic acid and (▲) benzyl benzoate *vs.* reaction time for catalytic reaction over fresh AUROlite™ in the presence of KF. (**a**) First run using fresh AUROlite™ and 2.1 g KF (**b**) second run using spent AUROlite™ and ~3 g KF. Reaction conditions: T = 80 °C, 0.8 g AUROlite™, $C_{BnOH, t=0}$ = 3.0 × 10^{-4} mol·g^{-1}, in 80 mL of toluene, 200 mL·min^{-1} air flow. Concentrations are expressed in moles per unit mass of liquid in the reactor (mol·g^{-1}). The symbols with error bars are the experimental results and the lines represent the kinetic model.

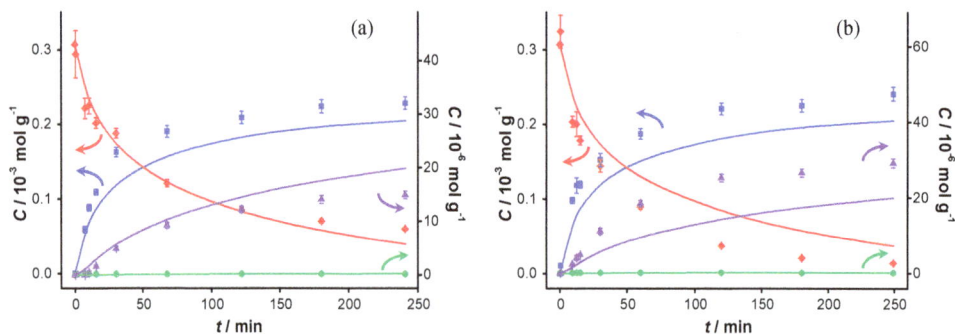

When K_2CO_3 is replaced by an equimolar amount of KF, the results are similar. *TOF* is slightly lower (0.6 s^{-1}) although the difference is within the measurement error. The conversion is 81% after 240 min when KF is used, and 78% when K_2CO_3 is used (Table 1). Apparently, the strength of the base does not affect the activity of the fresh catalyst. Upon re-using the spent catalyst however, a higher initial reaction rate is observed (1.0 s^{-1}) and the final conversion even reaches

95%. This increase is attributed to the additional potassium fluoride added. The preserved catalytic activity proves that no other deactivation mechanism than product inhibition (e.g., sintering or poisoning) is taking place. It is clear that the bases largely neutralize the acid responsible for product inhibition. As this alleviation of the inhibition is always accompanied by an increase of selectivity towards BnOOBn, apparently ester formation is faster in the presence of a base. However, direct esterification of the carboxylic acid with the alcohol catalyzed by a base is mechanistically not likely. After deprotonation of the BnOOH by a base, nucleophilic attack of the alcohol (R-CH$_2$-OH) on the functional carbon of carboxylate anion (R-COO$^-$) is highly unlikely as this would imply that a nucleophile (also seen as a Lewis base) would have to react with an electron rich species. This is illustrated in Figure 8 (steps d and e). The classic acid-catalyzed esterification between carboxylic acid and alcohol is depicted on the left hand-side of Figure 8 for comparison. It shows that under acidic conditions, the functional carbon becomes positively charged (Figure 8 step a, yielding R-C(OH)$_2$$^+$) and thus more prone to nucleophilic attack. Another pathway involving base-catalyzed ester formation from an alcohol and an aldehyde has been suggested by Rodríguez-Reyes *et al.* [19]. This pathway is illustrated in Figure 9, and can explain why in our system a higher ester production is observed under basic conditions.

Figure 8. Reaction mechanism of acid catalyzed esterification of carboxylic acid and alcohol. (**a**) protonation of carbonyl oxygen yielding an electrophilic carbocation; (**b**) nucleophilic attack of the alcohol and (**c**) dehydration yielding the corresponding ester. (**d**) deprotonation of the carboxylic acid by a base yielding a carboxylate anion; (**e**) the nucleophilic attack of the alcohol is then greatly disfavored.

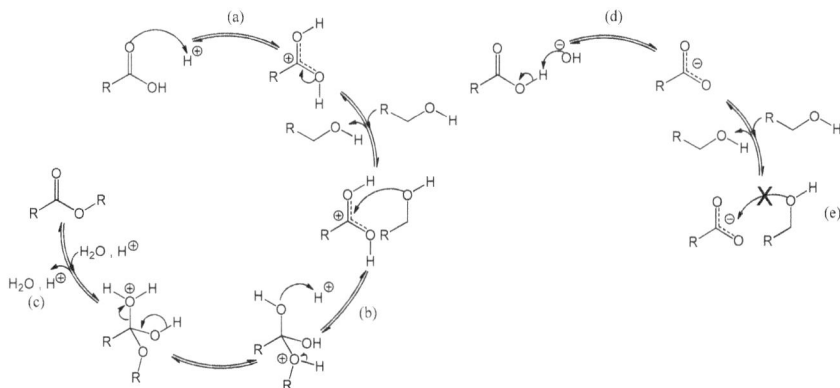

Figure 9. Reaction mechanism of base-catalyzed ester formation from alcohol and aldehyde, adapted from [19]. **(a)** alcohol deprotonation, **(b)** nucleophilic attack of alcoholate on carbonyl, followed by **(c)** β-hydride elimination yields the corresponding ester. The rest of the catalytic cycle consists of the oxidation of the hydride left on the gold surface by molecular oxygen, which also regenerates the base as depicted in Figure 5 (steps f and g).

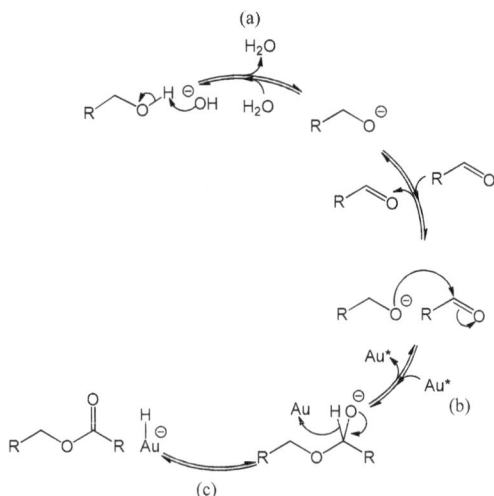

To further elucidate the role of the base on product inhibition, an experiment was performed with potassium carbonate and pre-addition of BnOOH. The concentration profiles are shown in Figure 10 and are directly comparable with the ones shown in Figure 6a where K_2CO_3 was present but no inhibitor was pre-added, and with Figure 4 where no base was present but BnOOH was pre-added.

In comparison with Figure 6a where no BnOOH was added, the initial *TOF* is lower: 0.2 s^{-1} (Table 1). The conversion after 240 min is 61% and the selectivity to BnO is 92%. When compared with the results in Figure 4 (no carbonate added), we can conclude that even though inhibition is still observed, the presence of potassium carbonate greatly reduces it.

Based on our interpretation, the presence of a base results in (partial) neutralization of the BnOOH, thereby alleviating inhibition. In parallel, basic conditions also enhance the subsequent ester formation depicted in Figure 9, leading to more ester production and decreasing aldehyde selectivity. An effect of the amount of potassium on the inhibition is also suspected based on Figure 7a,b, but this has not been further quantified.

Figure 10. Catalyst performance in the presence of K_2CO_3 after pre-addition of benzoic acid. Concentration of (♦) benzyl alcohol, (■) benzaldehyde, (●) benzoic acid and (▲) benzyl benzoate *vs.* reaction time for catalytic reaction over fresh AUROlite™ where benzoic acid was introduced prior to the beginning of the reaction. Reaction conditions: T = 80 °C, 0.8 g AUROlite™, 3.04 g K_2CO_3, $C_{BnOH, t=0}$ = 3.0 × 10^{-4} mol·g^{-1}, $C_{BnOOH, t=0}$ = 4.1 × 10^{-6} mol·g^{-1}, in 80 mL of toluene, 200 mL·min^{-1} air flow. Concentrations are expressed in moles per unit mass of liquid in the reactor (mol·g^{-1}). The symbols with error bars are the experimental results and the lines represent the kinetic model.

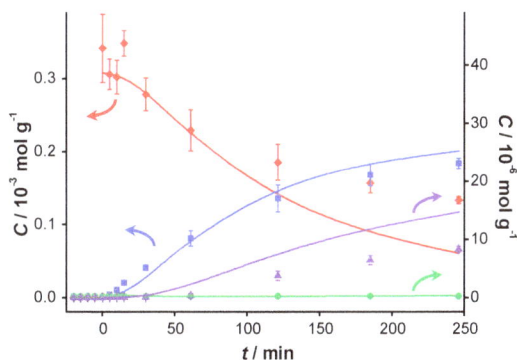

2.3. Influence of Water on Catalyst Deactivation

Water was already recognized to be crucial for Au-catalyzed gas phase CO oxidation by Daté *et al.* [26], who demonstrated its great beneficial role. The mechanism was later proposed by Daniells *et al.* [27]. In many of the reaction mechanisms discussed above, water plays a role. It is produced in an amount equimolar to BnO (see Figures 1a and 5a). The mechanism of the oxidation of aldehyde to carboxylic acid involves a base-catalyzed hydration of aldehyde to geminal diol. Water is also a byproduct of the equilibrium-limited esterification. Yang *et al.* [28] studied the influence of different water contents on the kinetics of oxidation and determined that water has a promoting effect. They observed higher conversions of BnOH and higher selectivities to BnO when an optimal amount of water was used. More recently, Chang and coworkers [29] conducted a computational study to better understand this effect on methanol dehydrogenation/oxidation. They concluded that the promoting effect originates from a facilitated peroxyl formation from O_2 by transfer of hydrogen from the water itself or from the alcohol via the water, where hydrogen bonds are reported to play a key role in this mechanism. Therefore, water could play a role in the deactivation of the catalyst in the present study. With this in mind, the reaction was carried out in the presence of the small amount of water that adheres to the catalyst after immersion in water and filtration. The mass difference before and after this treatment indicates about 0.5 g of water per gram of catalyst, which corresponds to around 550 mol of water per mol of gold. The results of this experiment are presented in Figure 11.

Figure 11. Catalyst performance in the presence of K_2CO_3 and a small amount of water. Concentration profiles of ($\blacklozenge$) benzyl alcohol, ($\blacksquare$) benzaldehyde, ($\bullet$) benzoic acid and ($\blacktriangle$) benzyl benzoate *vs.* reaction time for catalytic reaction over fresh AUROlite™ in the presence of water. Reaction conditions: $T = 80$ °C, 0.8 g AUROlite™, 3.04 g K_2CO_3, ~0.4 g H_2O, $C_{BnOH, t=0} = 3.0 \times 10^{-4}$ mol·g^{-1}, in 80 mL of toluene, 200 mL·min^{-1} air flow. Concentrations are expressed in moles per unit mass of liquid in the reactor (mol·g^{-1}). The symbols with error bars are the experimental results and the lines represent the kinetic model.

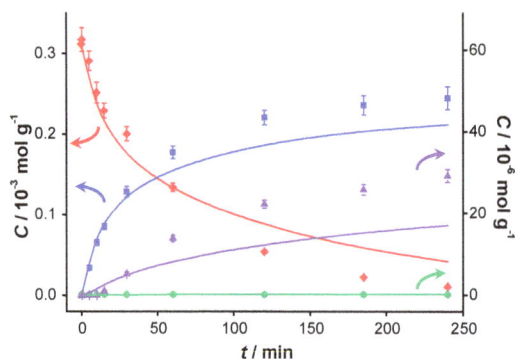

Compared with the dry conditions of Figure 6a the initial *TOF*s are equal within the experimental error, being 0.8 s^{-1} in moist conditions and 0.9 s^{-1} in dry conditions. We do not observe the kinetic effect of water found by Yang *et al.*, but in the presence of water our catalyst does not deactivate and the final conversion is close to 100%. The water thus positively influences the catalyst stability. A possible explanation could be that water interacts with the inhibiting product, thereby diminishing its interaction with the catalyst. However, since potassium carbonate was also present in the reactor, a synergistic effect of base and water cannot be excluded. To address this question, the deactivated catalyst obtained after the experiment of Figure 2b (where no base was present) was tested again in the presence of water and without the addition of any base. Concentration profiles of this experiment are displayed in Figure 12.

Clearly, catalytic activity could not be recovered by this treatment. Thus, water alone does not remove the species responsible for deactivation, and the beneficial influence of water observed in Figure 11 is due to a synergistic effect with the potassium carbonate, e.g., by an enhanced dissolution of the inorganic base. It should also be noted that adding more water than used in these experiments provoked phase separation in which the catalyst agglomerated in the water phase, thereby eliminating the dispersion of the catalyst powder in the organic phase and causing the reaction to proceed in a mass-transport limited regime.

Figure 12. Catalyst performance of the water-washed spent catalyst in base-free conditions from Figure 2 in the presence of water. Concentration of (♦) benzyl alcohol, (■) benzaldehyde, (●) benzoic acid and (▲) benzyl benzoate *vs.* reaction time for catalytic reaction spent AUROlite™ in the presence of water. Reaction conditions: $T = 80\ °C$, 0.8 g AUROlite™, ~1 g H_2O, $C_{BnOH,\ t=0} = 3.0 \times 10^{-4}$ mol·g^{-1}, in 80 mL of toluene, 200 mL·min^{-1} air flow. Concentrations are expressed in moles per unit mass of liquid in the reactor (mol·g^{-1}). The symbols with error bars are the experimental results and the lines represent the kinetic model.

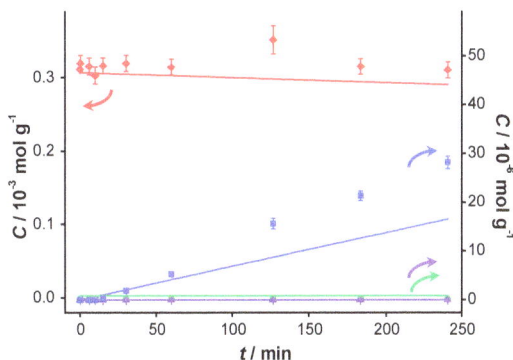

In order to further assess which hypothesis holds, the spent catalyst was analyzed by DRIFTS before and after washing in boiling water for 12 h (Figure 13). The fresh catalyst barely shows any features. In contrast, both the spent catalyst and the spent catalyst after boiling in water show clear absorption features. The catalyst tested in the presence of KF, which did not deactivate after two runs, exhibited different features. Identification of the species on the surface of the catalysts was based on reference spectra recorded of the compounds present in the reactor adsorbed on alumina. The band at 1200 cm^{-1} is attributed to C_{arom}–CHO or C_{arom}–CH_2OH stretching vibrations [30] and originates only from BnOH and BnO. Interestingly, this band is only present for the catalyst tested with KF. A small absorption band at 1390 cm^{-1} is seen in practically all cases and corresponds to the bending vibration of O–H bonds; thus, it cannot be used to differentiate the components of interest. The sharp absorption at 1450 cm^{-1} is due to in-plane bending vibrations of protons at a primary alcohol carbon and is specific to BnOH. Unfortunately, this band is often masked by a broader one due to other O–H vibrations at the same wavenumber. Absorptions between 1500 and 1600 cm^{-1} are attributed to C_{sp2}=C_{sp2} stretching vibrations present in all the compounds of interest. The broad band at 1710 cm^{-1} is specific to benzaldehydes and is due to π-conjugation of bonds throughout the entire molecule. Again, this characteristic feature is only present for the catalyst tested with KF. A smaller feature is also observed for BnOOH for the same reasons, but the carboxyl moiety seems to alter it drastically. The bending mode of water, which should appear at around 1600–1800 cm^{-1}, is hardly visible due to the pre-treatment of the samples at 473 K. Some features are also observed at higher wavenumbers. The small bands centered at around 2740 cm^{-1} and 2820 cm^{-1} arise from wagging and stretching vibrations respectively of the BnO carbonyl proton, but these bands are not detected on any of the spent catalysts. The symmetric and anti-symmetric stretching vibrations of primary alcohol methylene group protons give rise to absorptions at 2870 and 2930 cm^{-1}, respectively. The three bands between 3030 and 3090 cm^{-1} are assigned to C_{sp2}–H vibrations, which stem from any aromatic compound in the reaction medium.

Figure 13. DRIFT spectra of (**a**) fresh AUROlite™ catalyst; (**b**) used with K$_2$CO$_3$; (**c**) used with K$_2$CO$_3$ and washed in boiling water; (**d**) used with KF, (**e**)–(**i**) reference compounds adsorbed on alumina. Catalyst samples were pretreated under He at 473 K before recording.

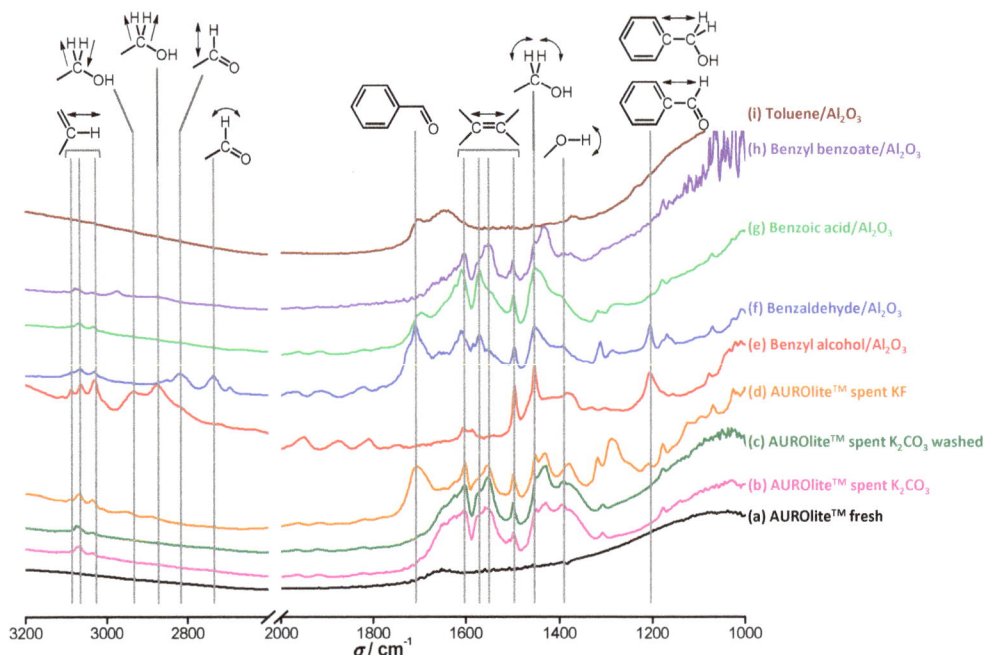

Figure 13 clearly demonstrates that water does not visibly wash off the species involved in the reaction from the catalyst surface. Even after 12 h in boiling water, the intensities of the absorption bands corresponding to aromatic species and oxygenated aromatic species do not show any sign of decrease. Therefore, combined with the results of Figure 12, the hypothesis that water remediates the product inhibition by enhancing desorption of the inhibiting product is refuted. The absorption features on the catalyst tested in the presence of K$_2$CO$_3$ are very similar to those of BnOOH and BnOOBn, although the observed bands are not specific to these compounds. In contrast, the catalyst tested in the presence of KF shows absorption features similar to those of BnO, in particular the characteristic band at 1710 cm^{-1}. Since this catalyst did not deactivate whereas the catalyst tested with K$_2$CO$_3$ showed some deactivation, the attribution of the inhibitor being BnOOH or one of its products is supported. However, in view of the low degree of deactivation shown by the catalyst tested in the presence of K$_2$CO$_3$, a signal corresponding to BnO would be expected. It remains unclear why BnO seems to be absent from this catalyst surface despite the fact that KF and K$_2$CO$_3$ have similar beneficial effects.

2.4. Kinetic Modeling

Since our experiments show that the BnOOH concentration in solution is very low at all times and that the ester formation increases in the presence of a base, it is assumed that the esterification runs entirely through the reaction of BnO with BnOH by H abstraction. The reactions involved in the model are:

$$C_6H_5CH_2OH + \tfrac{1}{2}O_2 \longrightarrow C_6H_5CHO + H_2O \tag{1}$$

$$C_6H_5CHO + \tfrac{1}{2}O_2 \longrightarrow C_6H_5COOH \tag{2}$$

$$C_6H_5CHO + C_6H_5CH_2OH + \tfrac{1}{2}O_2 \longrightarrow C_6H_5COOCH_2C_6H_5 + H_2O \tag{3}$$

The reaction model is accordingly assumed to consist of the set of surface reactions shown in the following set of equations, in which $*$ stands for a catalytic oxidation site:

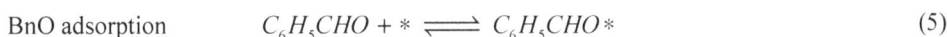

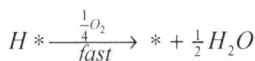

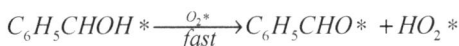

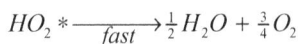

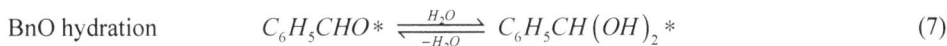

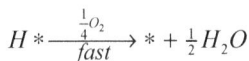

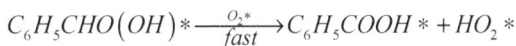

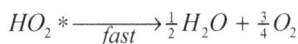

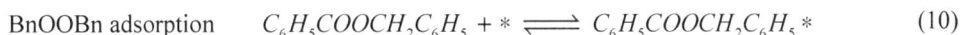

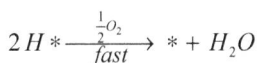

BnOH adsorption $\qquad C_6H_5CH_2OH + * \rightleftharpoons C_6H_5CH_2OH* \tag{4}$

BnO adsorption $\qquad C_6H_5CHO + * \rightleftharpoons C_6H_5CHO* \tag{5}$

BnOH oxidation $\qquad C_6H_5CH_2OH* + * \longrightarrow C_6H_5CHOH* + H* \tag{6}$

$$H* \xrightarrow[fast]{\tfrac{1}{4}O_2} * + \tfrac{1}{2}H_2O$$

$$C_6H_5CHOH* \xrightarrow[fast]{O_2*} C_6H_5CHO* + HO_2*$$

$$HO_2* \xrightarrow[fast]{} \tfrac{1}{2}H_2O + \tfrac{3}{4}O_2$$

BnO hydration $\qquad C_6H_5CHO* \underset{-H_2O}{\overset{H_2O}{\rightleftharpoons}} C_6H_5CH(OH)_2* \tag{7}$

BnO oxidation $\qquad C_6H_5CH(OH)_2* + * \longrightarrow C_6H_5CHO(OH)* + H* \tag{8}$

$$H* \xrightarrow[fast]{\tfrac{1}{4}O_2} * + \tfrac{1}{2}H_2O$$

$$C_6H_5CHO(OH)* \xrightarrow[fast]{O_2*} C_6H_5COOH* + HO_2*$$

$$HO_2* \xrightarrow[fast]{} \tfrac{1}{2}H_2O + \tfrac{3}{4}O_2$$

BnOOH adsorption $\qquad C_6H_5COOH + * \rightleftharpoons C_6H_5COOH* \tag{9}$

BnOOBn adsorption $\qquad C_6H_5COOCH_2C_6H_5 + * \rightleftharpoons C_6H_5COOCH_2C_6H_5* \tag{10}$

BnOOBn formation $\qquad C_6H_5CHO* + C_6H_5CH_2OH* \longrightarrow C_6H_5COOCH_2C_6H_5* + 2H* \tag{11}$

$$2H* \xrightarrow[fast]{\tfrac{1}{2}O_2} * + H_2O$$

It is assumed that all three surface oxidation reactions are irreversible. These reactions require two adjacent sites, one for the adsorbed species originating from the BnOH or BnO and one empty site that was regenerated by the oxygen or, for the esterification, adjacently adsorbed BnOH or BnO. It is also assumed that the hydride species, H*, and the peroxy species. HO₂*, are quickly removed in the excess of oxygen present, leading to the assumption that the occupancy of both species is always very low. Adsorption of water and oxygen is assumed not to inhibit the reaction. The surface reactions are described by the surface species reaction rates shown in the following equations:

BnOH adsorption $\qquad r_{BnOH} = k_{BnOH} C_{BnOH} N_T \theta_* - k_{-BnOH} N_T \theta_{BnOH*}$ (12)

BnO adsorption $\qquad r_{BnO} = k_{BnO} C_{BnO} N_T \theta_* - k_{-BnO} N_T \theta_{BnO*}$ (13)

BnOOH adsorption $\qquad r_{BnOOH} = k_{BnOOH} C_{BnOOH} N_T \theta_* - k_{-BnOOH} N_T \theta_{BnOOH*}$ (14)

BnOOBn adsorption $\qquad r_{BnOOBn} = k_{BnOOBn} C_{BnOOBn} N_T \theta_* - k_{-BnOOBn} N_T \theta_{BnOOBn*}$ (15)

BnOH oxidation $\qquad r_1 = k_1' N_T \theta_{BnOH*} \theta_* s$ (16)

BnO oxidation $\qquad r_2 = k_2' N_T \theta_{BnO*} \theta_* s$ (17)

BnOOBn formation $\qquad r_3 = k_3' N_T \theta_{BnO*} \theta_{BnOH*} s$ (18)

Where: $\quad r_i$ = reaction rate (mol·g$_{cat}^{-1}$·s^{-1})
$\qquad k_i$ = reaction rate constant (unit according to equations)
$\qquad C_i$ = concentration (mol·g^{-1})
$\qquad \theta_i$ = occupancy of the surface sites (-)
$\qquad N_T$ = total number of surface oxidation sites per unit catalyst mass (g$_{cat}^{-1}$)
$\qquad s$ = number of adjacent sites per site (-)

The hydration of BnO followed by the sequential oxidation of the diol species has been lumped here into a single step (17), as the equilibrium constant of the reversible hydration will be highly correlated with the rate constant of the oxidation step.

In general, adsorption reactions are at *quasi*-equilibrium, while the surface reactions are rate-limiting steps. This allows expression of the surface coverages as functions of the concentrations and the adsorption constants:

BnOH adsorption $\qquad \theta_{BnOH*} = K_{BnOH} C_{BnOH} \theta_*$ with $K_{BnOH} = k_{BnOH}/k_{-BnOH}$ (19)

BnO adsorption $\qquad \theta_{BnO*} = K_{BnO} C_{BnO} \theta_*$ with $K_{BnO} = k_{BnO}/k_{-BnO}$ (20)

BnOOH adsorption $\qquad \theta_{BnOOH*} = K_{BnOOH} C_{BnOOH} \theta_*$ with $K_{BnOOH} = k_{BnOOH}/k_{-BnOOH}$ (21)

BnOOBn adsorption $\qquad \theta_{BnOOBn*} = K_{BnOOBn} C_{BnOOBn} \theta_*$ with $K_{BnOOBn} = k_{BnOOBn}/k_{-BnOOBn}$ (22)

Where: K_i = adsorption constant of compound i on a gold active site (g·mol^{-1})
After substitution of all the surface coverage expressions in the site balance:

$$\theta_{BnOH*} + \theta_{BnO*} + \theta_{BnOOH*} + \theta_{BnOOBn*} + \theta_* = 1 \tag{23}$$

$$\theta_* = \frac{1}{\left(1 + K_{BnOH}C_{BnOH} + K_{BnO}C_{BnO} + K_{BnOOH}C_{BnOOH} + K_{BnOOBn}C_{BnOOBn}\right)} \tag{24}$$

The kinetic parameter estimation based on the full model showed that not all adsorption constants could be estimated properly due to a too weak sensitivity. It appeared that only the adsorption term of the most polar species present, BnOOH, was significant allowing the estimation of its adsorption constant and neglecting the adsorption terms of the BnOH, the BnO and the BnOOBn. Since it was experimentally observed that the presence of a base, $i.e.$, K_2CO_3 or KF, results in significant ester formation, it was assumed that this reaction is base-catalyzed. Since hardly any ester is formed in absence of base, it is concluded that the condensation reaction of BnOOH and BnOH does not occur to any significant extent.

By elimination of the coverages of the surface species, the following reaction rate expressions are obtained:

$$r_1 = \frac{k_1}{\left(1 + K_{BnOOH}C_{BnOOH}\right)^2}C_{BnOH} \qquad \text{With: } k_1 = k_1' K_{BnOH} N_T s \tag{25}$$

$$r_2 = \frac{k_2}{\left(1 + K_{BnOOH}C_{BnOOH}\right)^2}C_{BnO} \qquad \text{With: } k_2 = k_2' K_{BnO} N_T s \tag{26}$$

$$r_3 = \frac{S_{base}\,k_3}{\left(1 + K_{BnOOH}C_{BnOOH}\right)^2}C_{BnO}C_{BnOH} \qquad \text{With: } k_3 = k_3' K_{BnO} K_{BnOH} N_T s \tag{27}$$

Where: r_i = reaction rate (mol·kg$_{cat}^{-1}$·s^{-1})

          k_i = lumped rate constant (unit according to equations)

          C_i = concentration (mol·g^{-1})

          S_{base} = base strength (K_2CO_3 or KF) (-)

In the two catalytic experiments in which BnOOH was added prior to the BnOH (Figures 4 and 10), it was observed that the catalyst support Al_2O_3 adsorbs BnOOH more strongly than BnOH. In the first experiment (Figure 4), it was observed that the molar quantity of BnOOH in the liquid was much lower (16 µmol) than that originally added (780 µmol). The amount of BnOOH in solution increased only slightly (to 74 µmol) after subsequent addition of the 22 mmol of BnOH, demonstrating that BnOOH is adsorbed more strongly than BnOH. In the second experiment (Figure 10), it was observed that despite the addition of a similar amount of BnOOH (712 µmol yielding a theoretical concentration of 10.3 µmol·g^{-1}), the concentration in solution remained below detection limit (0.20 µmol·g^{-1}) for the entire reaction. Since the total amount of gold present in the reactor is in all cases 41 µmol, we conclude that adsorption on the catalyst support is responsible for the missing quantities of BnOOH. The adsorption on the support is captured by site coverages assuming to follow Langmuir behavior:

$$\theta_{BnOH, Al_2O_3} = \frac{K^{ads}_{BnOH, Al_2O_3} C_{BnOH}}{1 + K^{ads}_{BnOH, Al_2O_3} C_{BnOH} + K^{ads}_{BnOOH, Al_2O_3} C_{BnOOH}} \qquad (28)$$

$$\theta_{BnOOH, Al_2O_3} = \frac{K^{ads}_{BnOOH, Al_2O_3} C_{BnOH}}{1 + K^{ads}_{BnOH, Al_2O_3} C_{BnOH} + K^{ads}_{BnOOH, Al_2O_3} C_{BnOOH}} \qquad (29)$$

Where: θ_{i, Al_2O_3} = occupancy of the surface sites (-)

K^{ads}_{i, Al_2O_3} = adsorption constant of compound i on the support (g·mol^{-1})

From these experimental data the adsorption constants K^{ads}_{BnOH, Al_2O_3} and K^{ads}_{BnOOH, Al_2O_3} were estimated to be 18.1 g·mol^{-1} and 2.47 × 10^4 g·mol^{-1}, respectively, and the total adsorption site density was found to be 1.12 mmol·g$_{cat}^{-1}$. With a specific surface area of 230 m^2·g^{-1} and using Avogadro's number, the corresponding site density on the support is equivalent to 2.92 sites per nm^2, which is in good agreement with Matulewicz et al. [31] who report a value of 2 sites/nm^2 for their γ-alumina.

This yields a relation between the overall acid concentration in the vessel, $C_{BnOOH, tot}$, and the actual acid concentration in the liquid phase, C_{BnOOH}:

$$C_{BnOOH} = C_{BnOOH, tot} - \sigma_{OH, Al_2O_3} \frac{w_{cat}}{w_{liq}} \frac{K^{ads}_{BnOOH, Al_2O_3} C_{BnOOH}}{1 + K^{ads}_{BnOH, Al_2O_3} C_{BnOH} + K^{ads}_{BnOOH, Al_2O_3} C_{BnOOH}} \qquad (30)$$

Where: $C_{BnOOH, tot}$ = BnOOH concentration if no adsorption would take place (mol·g^{-1})

$\sigma_{OH, Al2O3}$ = adsorption site concentration on the catalyst (mol·g$_{cat}^{-1}$)

w_{cat} = amount of catalyst in the reactor (g$_{cat}$)

w_{liq} = amount of liquid in the reactor (g)

Since the amount of BnOH in the liquid phase is about three orders of magnitude larger than the amount of surface adsorption sites, the influence of BnOH adsorption on the concentration in the liquid phase is neglected. The actual acid concentration in the liquid phase can therefore be calculated directly from this quadratic equation with respect to C_{BnOOH}:

$$C_{BnOOH} = \frac{1}{2}\left(T + \sqrt{T^2 + \frac{4C_{BnOOH, tot}\left(1 + K^{ads}_{BnOH, Al_2O_3} C_{BnOH}\right)}{K^{ads}_{BnOOH, Al_2O_3}}} \right) \qquad (31)$$

Where: $T = C_{BnOOH, tot} - \frac{\left(1 + K^{ads}_{BnOH, Al_2O_3} C_{BnOH}\right)}{K^{ads}_{BnOOH, Al_2O_3}} - \sigma_{OH, Al_2O_3} \frac{w_{cat}}{w_{liq}}$

The strong adsorption of BnOOH could be the main cause of the very low activity of catalysts reused for the same experiment after a test in absence of a base, since it was observed that this acid remains on the catalyst during rinsing with toluene and also during boiling in water (Figure 13). The amount of BnOOH that remains on the catalyst in these 'second runs' (Figures 2b, 6b, 7b and 12) is not known but an estimate can be made. Assuming that the BnOOH concentration in the liquid equals its detection limit of 0.20 µmol·g^{-1}, a coverage of 0.88 is found from our simulations. Thus, it is tentatively concluded that although the amounts of acid formed in the previous runs is very low (often below the detection limit), the coverage is close to 1. Since the inhibition with a reused catalyst was very small or negligible in the presence of K_2CO_3 or KF, it was assumed for simplicity that in these cases all acid was removed, which seems acceptable in view of the time (approximately half an hour, the time needed for heating plus temperature stabilization) that the catalyst particles were in the close vicinity of the K_2CO_3 or KF crystals at reaction conditions (80 °C and well mixed) before the BnOH was added.

The model also accounts for the time allowed between adding the BnO or the BnOOH to the reactor and adding the main reactant BnOH, in all cases about 20 min at reaction conditions (additionally from the heating time). In the case of BnO this causes the formation of significant amounts of BnOOH, inhibiting the reaction (Figure 3).

Besides the three rate constants k_1, k_2, k_3, and the three adsorption equilibrium constants, K_{BnOOH}, $K^{ads}_{BnOH,\,Al_2O_3}$ and $K^{ads}_{BnOOH,\,Al_2O_3}$, there are several other unknown parameters, which are related to the effect of the base present. The experimental results show that the main effects of either K_2CO_3 or KF are (i) a decrease of the inhibition and (ii) an increase in formation of the BnOOBn ester.

In order to account for effect (i), it is assumed that the bases react to potassium benzoate with the BnOOH formed. Although acid-base reactions are typically instantaneous reactions, a finite rate is assumed to account for the transport from the catalyst pores to the insoluble crystals of K_2CO_3 or KF. The reaction rate is assumed to be first-order with respect to the acid and independent of the catalyst concentration:

$$r_{AcBase} = k_{AcBase} C_{BnOOH} \tag{32}$$

Effect (ii) is accounted for by defining parameters to describe the strength of either K_2CO_3 or KF in the catalysis of the esterification of BnO with BnOH:

$$\text{No base: } S_{base} = 1 \tag{33}$$

$$K_2CO_3 : S_{base} = 1 + S_{K_2CO_3} \tag{34}$$

$$KF : S_{base} = 1 + S_{K_2CO_3} S_{KF/K_2CO_3} \tag{35}$$

S_{K2CO3} is defined as the base strength of K_2CO_3 and $S_{KF/K2CO3}$ as the relative base strength of KF compared to K_2CO_3.

Since the experiments were performed in batch operation in a vessel that is assumed to be ideally stirred, the reactor model used to describe the process is the batch reactor model:

$$\frac{dC_{i,L}}{dt} = C_{cat} \sum_j v_{i,j} r_j \tag{36}$$

Where: $C_{i,L}$ = concentration of component i in the liquid phase (mmol·g^{-1})
t = time (s)
C_{cat} = catalyst concentration (mg·g^{-1})
$v_{i,j}$ = stoichiometric coefficient of component i in reaction j (-)
r_j = reaction rate of reaction j (mmol·mg$_{cat}^{-1}$·s^{-1})

The complete experimental dataset used for the parameter estimation contains 140 experimental data points obtained in 14 batch experiments at various conditions. The parameter estimation was carried out using the software package Athena Visual Studio [32], applying Bayesian estimation for multiresponse experiments using the full covariance matrix [33]. The concentrations (expressed in $mol \cdot g^{-1}$) of the four measured liquid components were used as the input for the objective function to be minimized. Since the concentrations of the BnOOH and the ester were typically up to two orders of magnitude smaller than those of the BnOH and BnO, the weight of BnOOBn was set at 10 and that of BnOOH at 100. For experiments with a very low conversion rate of BnOH, the weight of BnO was increased to 10 as well.

With the exception of the experiment of Figure 4, BnOOH concentrations were mostly below the detection limit of 0.20 $\mu mol \cdot g^{-1}$. Since the model predictions are very sensitive to the acid concentration, it was necessary to estimate acid concentrations for experiments where these were not detectable. In those cases, acid concentrations were arbitrarily assumed to be half of the detection limit.

These data lead to fits of the seven kinetic parameters to the experimental dataset and the results are shown in Table 2.

Table 2. Optimal estimates of the kinetic parameters using all experimental data.

Parameter	Unit	Estimate	95% confidence range	
			Value	Relative/%
k_1	$/g \cdot mg_{cat}^{-1} \cdot s^{-1}$	2.69×10^{-3}	$\pm 5.0 \times 10^{-4}$	± 19
k_2	$/g \cdot mg_{cat}^{-1} \cdot s^{-1}$	2.37×10^{-4}	$\pm 5.9 \times 10^{-5}$	± 25
k_3	$/g^2 \cdot mmol^{-1} \cdot mg_{cat}^{-1} \cdot s^{-1}$	6.6×10^{-4}	$\pm 3.5 \times 10^{-4}$	± 54
k_{AcBase}	$/s^{-1}$	0.71	± 0.15	± 21
K_{BnOOH}	$/g \cdot mmol^{-1}$	1.23×10^4	$\pm 1.5 \times 10^3$	± 12
$K_{BnOH, Al_2O_3}^{ads}$	$/g \cdot mmol^{-1}$	18.1	fixed	
$K_{BnOOH, Al_2O_3}^{ads}$	$/g \cdot mmol^{-1}$	2.47×10^4	fixed	
S_{K2CO3}	/-	1.39	± 0.81	± 58
$S_{KF/K2CO3}$	/-	1.37	± 0.39	± 29
SSR*	/-	0.414	-	-

(*) $SSR = \sum_{i=1}^{v} w_i \sum_{k=1}^{n} \left(C_{i,k,exp} - C_{i,k,mod} \right)^2$ (sum of the squared residuals) (w_i = weight factor for response i,

v = number of responses, $C_{i,k,exp}$ = experimental response of component i in experiment k, $C_{i,k,mod}$ = model response of component i in experiment k, n = number of experiments (samples).

While relatively good fits of predicted concentrations of BnOH, benzaldehyde and BnOOBn with time are evident in Figures 2a and 6a for first batch runs with and without K_2CO_3, relatively poor fits of benzaldehyde concentration are observed for second-time runs (Figures 2b and 6b) and the first run with KF (Figure 7a). The poorer fit of the second runs in the presence of K_2CO_3 is understandable since it was assumed in the model that all BnOOH was removed by the base in between the experiments, which is probably not completely justified, indicating that some acid or another inhibiting species remains on the catalyst. In all other experiments, approximate fits of concentrations of one or both products or of all three species (alcohol, aldehyde and ester) are observed (see Figures 3, 4, 7b, and 10–12). Thus, variations in how well the fit follows the data are a logical consequence of attempting to simulate in a single model a wide range of concentrations with and without base and in the absence and presence of strongly inhibiting aldehyde and acid product species. Moreover, the model did not include effects of water and was limited to the three most important reactions and four most important species. The 54% and 58% relative confidence intervals for k_3 and S_{K2CO3} originate from the strong correlation between these two parameters, which is discussed later in this section. The approximate nature of the model can be attributed to:

(1) Assumptions that are only approximately valid, e.g., (a) arbitrary estimates of benzoic acid concentration, and (b) the assumption that benzoic acid is completely removed from the catalyst by interaction with insoluble $K2CO_3$ or KF crystals.

(2) By practical necessity, the limited scope of the mechanistic scheme, e.g., (a) neglecting effects of adsorption or inhibition of some species such as coverages of BnOH of BnOOBn and (b) neglecting the positive effect of water.

Nevertheless, the model provides (1) accurate predictions of initial reaction rate for oxidation of BnOH to BnO on a gold/alumina catalyst and (2) approximate predictions of the effects of BnOOH inhibition and the neutralizing effect of potassium salts to alleviate this inhibition.

The value of the rate constant for the reaction of the BnOOH with the base to potassium benzoate, k_{AcBase}, represents the characteristic time for the transport of BnOOH from a catalytic site to the $K2CO_3$ or KF crystals. The order of magnitude can be compared with an estimate of the characteristic time for diffusion [34] of BnOOH through the catalyst pores to the liquid bulk, obtained from a typical diffusion distance (one third of the catalyst particle size 50 μm, estimated from the sieve mesh size) and an effective diffusivity of 4.8×10^{-10} m$^2 \cdot$s^{-1} (estimated using Wilke and Chang's relationship [35], using porosity-tortuosity ratio of 0.14, based on data of similar aluminas and catalysts):

$$\tau_{diff,\, BnOOH} = \frac{\left(\frac{1}{3} d_p\right)^2}{2 \times D_{eff,\, BnOOH}} \tag{37}$$

Where: $\tau_{diff, BnOOH}$ = diffusion time (s)

d_p = catalyst particle diameter (m)

$D_{eff,\, BnOOH}$ = effective diffusivity (m$^2 \cdot$s^{-1})

This yields a typical diffusion time of 0.3 s, which is about 4.8 times smaller than the typical time $(k_{AcBase})^{-1}$ = 1.41 s. The latter seems a plausible value in view of the additional transport resistance that might be caused by the transfer from the external catalyst surface towards the $K2CO_3$ or KF crystals.

The correlation matrix for the estimated parameters, shown in Table **3**, shows that the strongest correlation occurs between k_3 and S_{K2CO3} with a correlation coefficient of −0.97, which is in line with our conclusion that the ester formation in our system is base-catalyzed. All correlations between the parameters justify maintaining all parameters in the model since these do not exceed the value of 0.99, which is accepted as the limit for a proper parameter estimation [36].

Table 3. Correlation matrix between all the parameters estimated using the optimized kinetic model.

Parameter	k_1	k_2	k_3	k_{AcBase}	K_{BnOOH}	S_{K2CO3}	$S_{KF/K2CO3}$
k_1	1						
k_2	0.52	1					
k_3	0.06	−0.13	1				
k_{AcBase}	0.12	−0.17	−0.02	1			
K_{BnOOH}	0.51	−0.30	0.09	0.64	1		
S_{K2CO3}	0.02	0.12	−0.97	0.06	−0.01	1	
$S_{KF/K2CO3}$	−0.01	−0.02	0.33	−0.05	−0.01	−0.45	1

3. Experimental Section

Toluene (anhydrous, 99.8%), benzyl alcohol (>99%), potassium carbonate (>99.0%), tetradecane (> 99%), potassium fluoride (>99.99%) and phosphorus pentoxide desiccant were supplied by Sigma Aldrich and were used without further purification. AUROlite™ catalyst (Au/Al$_2$O$_3$ 1 wt.%, Au average particle size: 2–3 nm, specific surface area: 200–260 m^2·g^{-1}, from supplier specifications) was supplied by Strem Chemicals in the form of extrudates. The extrudates were crushed and sieved to a particle size <71 μm, thereby excluding diffusion limitations during catalytic experiments as verified using different catalyst particle sizes. The resulting powder was stored in a well-sealed container at 4 °C and in the dark. Catalytic testing under dry conditions was performed with this powder used as such. For the catalytic tests involving water, the desired amount of this powder was suspended in Milli-Q® water (18.2 MΩ·cm) under sonication for 30 min, and then vacuum filtered (using a Büchner funnel). The resulting moist catalyst was collected from the filter with a spatula and used as such for catalytic testing. The mass difference before and after this step indicates that around 0.5 g of water is adsorbed per gram of catalyst.

Catalytic experiments were carried out in a 100 mL round-bottom vessel, the inner diameter of which is 60 mm. The vessel was equipped with a reflux condenser and Teflon baffles, and mechanically stirred at 1300 rpm with a 4-blade Teflon impeller. Upon varying the catalyst quantity in preliminary tests, an initial reaction rate proportional to the catalyst quantity was observed, indicating that mass transport limitations were absent. In a typical catalytic test, 3.04 g of K$_2$CO$_3$ and 0.8 g of AUROlite™ are introduced in the vessel together with 80 mL of toluene. Two complementary tests were performed using 2.07 g of KF instead of K$_2$CO$_3$. The vessel was heated to 80 °C by means of an oil bath, and 200 mL·min^{-1} of air was bubbled through the reaction mixture via a glass frit. When the temperature was stabilized, 2.4 g of BnOH was introduced using a syringe, constituting the beginning of the test ($t = 0$ min). Small samples of 300 μL were taken at recorded times and filtered from catalyst and K$_2$CO$_3$ powders with a 13 mm syringe Teflon filter of 0.2 μm pore size (diameter: 13 mm; pore size: 0.2 μm; PTFE membrane; VWR International) and introduced in a GC sample vial together with 20 μL of tetradecane, the latter being used as internal standard.

GC analyses were performed using a Varian CP-3380 equipped with a FID detector and a CP-Sil 8 CB cat. no. 7453 column (length: 50 m; diameter: 0.25 mm; coating thickness: 0.25 μm). The initial temperature of the GC oven was 150 °C and was maintained for 4 min, then increased with 100 °C·min^{-1} to 220 °C and then maintained at 220 °C for 6.3 min. After testing, the catalyst was recovered by vacuum filtration, washed with 80 mL of toluene at room temperature and stored over P$_2$O$_5$ in an evacuated desiccator. In the case of the water treated catalyst samples, the catalyst was washed with 80 mL of toluene, followed by extensive washing with about 250 mL of Milli-Q$^{®}$ water (18.2 MΩ·cm) at room temperature.

Diffuse Reflectance Infra-Red Fourier Transform Spectroscopy (DRIFTS) spectra were recorded on a Nicolet model 8700 spectrometer, equipped with a high-temperature DRIFTS cell, and a DTGS-TEC detector. The spectra were recorded with 256 scans at 4 cm^{-1} resolution from 4000 to 500 cm^{-1} using potassium bromide (KBr) to perform background subtraction. The samples were pre-treated at 473 K for 1 h in a helium flow of 20 mL·min^{-1}.

4. Conclusions

Our study shows that benzoic acid or compounds formed from benzoic acid cause catalyst inhibition in benzyl alcohol oxidation in toluene and in absence of a base. The introduction of a potassium salt as a base prevents this inhibition by neutralizing the benzoic acid formed. Basic conditions result in a decrease in selectivity to benzaldehyde and in an increase of ester production. The enhanced ester formation probably occurs via condensation of alkoxy species (formed by alcohol deprotonation by the base) with benzaldehyde under oxidative conditions, and is not the result of an increased benzoic acid production followed by esterification, as might be expected under acidic conditions. Water appears to have no influence on inhibition, but may enhance the effect of the base described above by improved dissolution. Although effects of water were not modeled, a kinetic effect for water cannot be excluded.

The concentration *versus* time data of the batch experiments in this study, which covered a wide range and included effects of acid inhibition and base, were fitted to a comprehensive kinetic model for (1) the primary reaction, oxidation of benzyl alcohol to benzaldehyde; (2) secondary oxidation of benzaldehyde to benzoic acid; and (3) secondary esterification of benzyl alcohol and benzaldehyde to benzyl benzoate. Effects of base (potassium salts) were also included in the model. The resulting model predicts concentration-time trends approximately well, including inhibition by benzoic acid and the neutralization of benzoic acid by potassium salts, forming potassium benzoate. A precise fit of the model to experimental data was observed in two first batch runs, with and without K$_2$CO$_3$. Variations in how well the fit follows the data are a logical consequence of attempting to simulate in a single model a wide range of concentrations with and without base and in the absence and presence of product species such as aldehyde and strongly inhibiting acid. Moreover, the model did not include effects of water and was limited to the three most important reactions and four most important species.

Nevertheless, the model provides (1) accurate predictions of initial reaction rate for oxidation of benzyl alcohol to benzaldehyde on a gold/alumina catalyst and (2) approximate predictions of the effects of benzoic acid inhibition and the neutralizing effect of a potassium base to alleviate this inhibition.

Acknowledgments

The Dutch National Research School Combination Catalysis Controlled by Chemical Design (NRSC-Catalysis) is gratefully acknowledged for financial support.

Conflict of Interest

The authors declare no conflict of interest.

References

1. Davis, S.E.; Ide, M.S.; Davis, R.J. Selective oxidation of alcohols and aldehydes over supported metal nanoparticles. *Green Chem.* **2013**, *15*, 17–45.
2. Lilga, M.A.; Hallen, R.T.; Gray, M. Production of oxidized derivatives of 5-Hydroxymethylfurfural (HMF). *Topics Catal.* **2010**, *53*, 1264–1269.
3. Zope, B.N.; Davis, R.J. Inhibition of gold and platinum catalysts by reactive intermediates produced in the selective oxidation of alcohols in liquid water. *Green Chem.* **2011**, *13*, 3484–3491.
4. Haruta, M.; Yamada, N.; Kobayashi, T.; Iijima, S. Gold catalysts prepared by coprecipitation for low-temperature oxidation of hydrogen and of carbon monoxide. *J. Catal.* **1989**, *115*, 301–309.
5. Alhumaimess, M.; Lin, Z.; Weng, W.; Dimitratos, N.; Dummer, N.F.; Taylor, S.H.; Bartley, J.K.; Kiely, C.J.; Hutchings, G.J. Oxidation of benzyl alcohol by using gold nanoparticles supported on ceria foam. *ChemSusChem* **2012**, *5*, 125–131.
6. Deplanche, K.; Mikheenko, I.P.; Bennett, J.A.; Merroun, M.; Mounzer, H.; Wood, J.; MacAskie, L.E. Selective oxidation of benzyl-alcohol over biomass-supported Au/Pd bioinorganic catalysts. *Topics Catal.* **2011**, *54*, 1110–1114.
7. Guo, H.; Kemell, M.; Al-Hunaiti, A.; Rautiainen, S.; Leskelä, M.; Repo, T. Gold-palladium supported on porous steel fiber matrix: Structured catalyst for benzyl alcohol oxidation and benzyl amine oxidation. *Catal. Commun.* **2011**, *12*, 1260–1264.
8. Guo, H.; Al-Hunaiti, A.; Kemell, M.; Rautiainen, S.; Leskelä, M.; Repo, T. Gold catalysis outside nanoscale: Bulk gold catalyzes the aerobic oxidation of π-activated alcohols. *ChemCatChem* **2011**, *3*, 1872–1875.
9. Hao, Y.; Hao, G.P.; Guo, D.C.; Guo, C.Z.; Li, W.C.; Li, M.R.; Lu, A.H. Bimetallic Au-Pd Nanoparticles Confined in Tubular Mesoporous Carbon as Highly Selective and Reusable Benzyl Alcohol Oxidation Catalysts. *ChemCatChem* **2012**, *4*, 1595–1602.
10. Heeskens, D.; Aghaei, P.; Kaluza, S.; Strunk, J.; Muhler, M. Selective oxidation of ethanol in the liquid phase over Au/TiO$_2$. *Phys. Status Solidi. A* **2013**, *250*, 1107–1118.

11. Mallat, T.; Baiker, A. Oxidation of alcohols with molecular oxygen on solid catalysts. *Chem. Rev.* **2004**, *104*, 3037–3058.

12. Quintanilla, A.; Butselaar-Orthlieb, V.C.L.; Kwakernaak, C.; Sloof, W.G.; Kreutzer, M.T.; Kapteijn, F. Weakly bound capping agents on gold nanoparticles in catalysis: Surface poison? *J. Catal.* **2010**, *271*, 104–114.

13. Sá, J.; Taylor, S.F.R.; Daly, H.; Goguet, A.; Tiruvalam, R.; He, Q.; Kiely, C.J.; Hutchings, G.J.; Hardacre, C. Redispersion of gold supported on oxides. *ACS Catal.* **2012**, *2*, 552–560.

14. Mallat, T.; Baiker, A. Oxidation of alcohols with molecular oxygen on platinum metal catalysts in aqueous solutions. *Catal. Today* **1994**, *19*, 247–283.

15. Zope, B.N.; Hibbitts, D.D.; Neurock, M.; Davis, R.J. Reactivity of the gold/water interface during selective oxidation catalysis. *Science* **2010**, *330*, 74–78.

16. Moulijn, J.A.; van Diepen, A.E.; Kapteijn, F. Catalyst deactivation: Is it predictable? What to do? *Appl. Catal.* **2001**, *212*, 3–16.

17. Keresszegi, C.; Bürgi, T.; Mallat, T.; Baiker, A. On the role of oxygen in the liquid-phase aerobic oxidation of alcohols on palladium. *J. Catal.* **2002**, *211*, 244–251.

18. Dimitratos, N.; Villa, A.; Wang, D.; Porta, F.; Su, D.; Prati, L. Pd and Pt catalysts modified by alloying with Au in the selective oxidation of alcohols. *J. Catal.* **2006**, *244*, 113–121.

19. Rodríguez-Reyes, J.C.F.; Friend, C.M.; Madix, R.J. Origin of the selectivity in the gold-mediated oxidation of benzyl alcohol. *Surf. Sci.* **2012**, *606*, 1129–1134.

20. Fang, W.; Chen, J.; Zhang, Q.; Deng, W.; Wang, Y. Hydrotalcite-supported gold catalyst for the oxidant-free dehydrogenation of benzyl alcohol: Studies on support and gold size effects. *Chem. A Eur. J.* **2011**, *17*, 1247–1256.

21. Kimling, J.; Maier, M.; Okenve, B.; Kotaidis, V.; Ballot, H.; Plech, A. Turkevich method for gold nanoparticle synthesis revisited. *J. Phys. Chem. B* **2006**, *110*, 15700–15707.

22. Juan-Alcañiz, J.; Ferrando-Soria, J.; Luz, I.; Serra-Crespo, P.; Skupien, E.; Santos, V.P.; Pardo, E.; Llabrés i Xamena, F.X.; Kapteijn, F.; Gascon, J. The oxamate route, a versatile post-functionalization for metal incorporation in MIL-101(Cr): Catalytic applications of Cu, Pd, and Au. *J. Catal.* **2013**, *307*, 295–304.

23. Fristrup, P.; Johansen, L.B.; Christensen, C.H. Mechanistic investigation of the gold-catalyzed aerobic oxidation of alcohols. *Catal. Lett.* **2008**, *120*, 184–190.

24. Abad, A.; Corma, A.; García, H. Catalyst parameters determining activity and selectivity of supported gold nanoparticles for the aerobic oxidation of alcohols: The molecular reaction mechanism. *Chem. A Eur. J.* **2008**, *14*, 212–222.

25. Pina, C.D.; Falletta, E.; Rossi, M. Update on selective oxidation using gold. *Chem. Soc. Rev.* **2012**, *41*, 350–369.

26. Daté, M.; Okumura, M.; Tsubota, S.; Haruta, M. Vital role of moisture in the catalytic activity of supported gold nanoparticles. *Angew. Chem. Int. Ed.* **2004**, *43*, 2129–2132.

27. Daniells, S.T.; Overweg, A.R.; Makkee, M.; Moulijn, J.A. The mechanism of low-temperature CO oxidation with Au/Fe$_2$O$_3$ catalysts: A combined Mössbauer, FT-IR, and TAP reactor study. *J. Catal.* **2005**, *230*, 52–65.

28. Yang, X.; Wang, X.; Liang, C.; Su, W.; Wang, C.; Feng, Z.; Li, C.; Qiu, J. Aerobic oxidation of alcohols over Au/TiO$_2$: An insight on the promotion effect of water on the catalytic activity of Au/TiO$_2$. *Catal. Commun.* **2008**, *9*, 2278–2281.

29. Chang, C.R.; Yang, X.F.; Long, B.; Li, J. A water-promoted mechanism of alcohol oxidation on a Au(111) surface: Understanding the catalytic behavior of bulk gold. *ACS Catal.* **2013**, *3*, 1693–1699.

30. Socrates, G. *Infrared and Raman Characteristic Group Frequencies. Tables and Charts*, 3rd ed.; John Wiley & Sons Ltd: Heidelberg, Germany, 2004; pp. 50–167.

31. Matulewicz, E.R.A.; Kerkhof, F.P.J.M.; Moulijn, J.A.; Reitsma, H.J. Structure and activity of fluorinated alumina. 1. Determination of the number of protonic sites by an infrared study of adsorbed pyridines. *J. Colloid Interface Sci.* **1980**, *77*, 110–119.

32. *Athena Visual Studio, Software for Modeling, Estimation and Optimization*, Version 14.2. Available online: *www.AthenaVisual.com* 1997–2009 (accessed 25 September 2013).

33. Stewart, W.E.; Caracotsios, M.; Sorensen, J.P. Parameter estimation from multiresponse data. *AIChE J.* **1992**, *38*, 641–650.

34. Atkins, P.W. *Physical Chemistry*, 5th ed.; W.H. Freeman: New York, NY, USA, 1994.

35. Perry, R.H.; Green, D.W. *Perry's Chemical Engineers' Handbook*, 7th ed.; McGraw-Hill: New York, NY, USA, 1997.

36. Stewart, W.E.; Caracotsios M. Computer-Aided Modelling of Reactive Systems; John Wiley & Sons, Inc., Hoboken, NJ, USA, 2008.

Investigation of the Deactivation Phenomena Occurring in the Cyclohexane Photocatalytic Oxidative Dehydrogenation on MoO$_x$/TiO$_2$ through Gas Phase and *in situ* DRIFTS Analyses

Vincenzo Vaiano, Diana Sannino, Ana Rita Almeida, Guido Mul and Paolo Ciambelli

Abstract: In this work, the results of gas phase cyclohexane photocatalytic oxidative dehydrogenation on MoO$_x$/SO$_4$/TiO$_2$ catalysts with DRIFTS analysis are presented. Analysis of products in the gas-phase discharge of a fixed bed photoreactor was coupled with *in situ* monitoring of the photocatalyst surface during irradiation with an IR probe. An interaction between cyclohexane and surface sulfates was found by DRIFTS analysis in the absence of UV irradiation, showing evidence of the formation of an organo-sulfur compound. In particular, in the absence of irradiation, sulfate species initiate a redox reaction through hydrogen abstraction of cyclohexane and formation of sulfate (IV) species. In previous studies, it was concluded that reduction of the sulfate (IV) species via hydrogen abstraction during UV irradiation may produce gas phase SO$_2$ and thereby loss of surface sulfur species. Gas phase analysis showed that the presence of MoO$_x$ species, at same sulfate loading, changes the selectivity of the photoreaction, promoting the formation of benzene. The amount of surface sulfate influenced benzene yield, which decreases when the sulfate coverage is lower. During irradiation, a strong deactivation was observed due to the poisoning of the surface by carbon deposits strongly adsorbed on catalyst surface.

Reprinted from *Catalysts*. Cite as: Vaiano, V.; Sannino, D.; Almeida, A.R.; Mul, G.; Ciambelli, P. Investigation of the Deactivation Phenomena Occurring in the Cyclohexane Photocatalytic Oxidative Dehydrogenation on MoO$_x$/TiO$_2$ through Gas Phase and *in situ* DRIFTS Analyses. *Catalysts* **2013**, *3*, 978-997.

1. Introduction

Photocatalytic oxidation reactions have been widely used in processes such as the decontamination of water and air [1–8]. However, applications of heterogeneous photocatalysis for the synthesis of compounds of commercial importance have been considered only in recent years. The most studied photocatalytic reactions occur in slurry systems. Among them, partial oxidation of cyclohexane is an important commercial reaction, as the resultant products, cyclohexanol and cyclohexanone, are precursors in the syntheses synthesis of adipic acid, in turn intermediates in the production of nylon [9]. In their review, Maldotti *et al.* [10] reported the main aspects concerning the photocatalytic oxidation of cyclohexane. Li *et al.* [11] showed that the quantum size and surface state are key factors governing the selectivity in photoxidation on TiO$_2$ nanoparticles [12,13]. The photo-oxidation of cyclohexane on titanium dioxide was also investigated in neat cyclohexane and in various solvents showing an influence of the solvent media on the cyclohexane oxidation rate and on the selectivity to cyclohexanol and cyclohexanone [14,15].

From theoretical and practical points of view, the ideal solvent for the photo-oxidation of cyclohexane is one that minimizes the strengths of adsorption of the desired products on titanium dioxide and does not compete with cyclohexane and oxygen for adsorption sites. Otherwise, the solvent could be strongly adsorbed but is non-reactive with itself upon forming a radical on the illuminated titanium dioxide surface [16].

Supported transition-metal oxides can absorb light, and the transferred energy can be used to activate C-H bonds in saturated hydrocarbons, a chemical step that is generally unselective [6]. In this context, cyclohexanol, cyclohexanone, and polyoxygenates have been formed from cyclohexane when polyvanadate or polyoxytungstate were supported on several oxides [17,18].

It was shown that attenuated total reflection Fourier transform infrared (ATR-FTIR) is a suitable way to monitor in real time and *in situ* the light-induced heterogeneous oxidations [19], allowing deeper knowledge of the complex phenomena occurring under irradiation to be obtained. ATR-FTIR technique was also performed to study the photodegradation of organic pollutants in water on TiO_2 [20].

The generally low efficiency associated with liquid phase photocatalytic reactions, which typically occur at low conversion, coupled with the difficulty of separating catalyst and products in a liquid has motivated research of gas-solid systems, e.g., catalysts active in the gas-phase partial oxidation of cyclohexane. However, selective photoxidation of cyclohexane yields different reaction products in gas phase relative to liquid phase. In particular, cyclohexene or benzene are selectively obtained through gas phase oxidative dehydrogenation of cyclohexane on MoO_x/TiO_2 photocatalysts, with UV illumination both in fixed [21] and fluidized bed reactors [22–24]. Higher molybdenum loadings improved the benzene selectivity, whereas with only titania, total conversion to carbon dioxide is obtained. The selective formation of benzene is attributed to the poisoning by Mo-species of unselective sites on the titania surface which otherwise totally oxidize cyclohexane to CO_2 and water [24–26].

A mechanism for the catalytic photo-oxidative dehydrogenation of cyclohexane to benzene on MoO_x/TiO_2 was recently proposed by analyzing the gas-phase coming from a photoreactor [21]. This mechanism involves dehydrogenation of cyclohexane to cyclohexene followed by oxy-dehydrogenation to benzene on molybdenum oxide active sites via a detailed sequence of elementary steps [21]. In the same paper, several hypotheses relating to the role of sulfate species in promoting the selectivity to benzene formation were postulated: (1) sulfate present on TiO_2 surface may facilitate hydrogen abstraction from an adsorbed cyclohexane molecule, owing to its strong acid properties, or (2) it may participate in the re-oxidation step of the octahedrally-coordinated polymolybdate surface species [21]. With regard to the influence of sulfate concentration, Ciambelli *et al.* [24,26,27] showed that during photooxidative dehydrogenation of cyclohexane, the presence of sulfate species on the surface of titania favor a high benzene yield. Cyclohexene was produced in low concentration and CO_2 was not detected in gas-phase. Enhanced photooxidative dehydrogenation activity of MoO_x/TiO_2 catalysts was attributed to the increase in surface acidity by sulfate species, which enhances hydrocarbon adsorption coverage on the catalyst surface.

On the other hand, the addition of sulfate to $MoO_x/\gamma-Al_2O_3$ catalysts was found to promote the selective mono-oxidative dehydrogenation of cyclohexane to cyclohexene [28,29]. An optimum in MoO_3 and SO_4 loadings were found to be 8 and 2.4 wt%, respectively (corresponding to MoO_3/SO_4 molar ratio equal to 2.22), while a decrease in the catalytic activity at higher sulfate loading was ascribed to MoO$_x$ decoration by sulfates. By studying the influence of the preparation method and molybdenum loading on sulfated MoO$_x$/γ-Al$_2$O$_3$, it was shown that selectivity to benzene increases with increasing catalyst acidity, as the latter favors cyclohexene adsorption and thus its conversion to benzene. Close proximity of surface sulfates to octahedral polymolybdate appears to be a key parameter promoting photoactivity of these catalysts [28].

While much has been learned from the previous work, an *in situ* study of photocatalyst surfaces under UV irradiation is nevertheless needed to gain a better understanding of the surface phenomena. Typical problems of molybdenum based catalysts that need to be addressed include the role of sulfate and the deactivation occurring during gas-phase photocatalytic oxidative dehydrogenation of cyclohexane. Thus, in this study we present the results of gas phase cyclohexane photocatalytic oxidative dehydrogenation on MoO$_x$/TiO$_2$ through the analysis of products present in gas-phase exit of the reactor, complemented by *in situ* DRIFTS analysis, to monitor the photocatalyst surface in the absence and during UV irradiation.

2. Results and Discussion

2.1. Catalysts Characterization

The list of catalysts investigated and their characterization results are reported in Table 1.

Table 1. List of catalysts and their characterization results.

Catalyst	MoO$_3$[a] loading, wt%	SO$_4$[b], wt%	SSA, m²/g	Mo=O stretching (Raman) [c], cm^{-1}
DT05	0	0.6	67	-
4.7MoDT0.5	4.6	0.6	68	956
DT2	0	2.3	71	-
2MoDT2	1.8	2.2	71	953
8MoDT2	7.6	2.2	63	978

[a] evaluated by ICP-MS; [b] evaluated by TG-MS analysis; [c] these results are obtained from previous studies [21,25,26].

Raman results of all the samples were reported in previous works [21,25,26]. In summary, Raman bands related to Mo-species can be observed in the range 820–1000 cm^{-1}. For all Mo-based catalysts, bands at 956 cm^{-1} on 4.7Mo/DT05, at 953 cm^{-1} on 2MoDT2, at 978 cm^{-1} on 8MoDT2 and at 956 cm^{-1} on 8Mo2S are observed and assigned to terminal Mo=O stretching of octahedral MoO$_x$ species [30]. The increasing of wavenumber was an indication of a higher polymerization degree of Mo species at increasing loadings, corresponding to a Mo-nuclearity likely ranging from 7 to 12 [21].

2.2. Adsorption Measurement in the Absence of UV Irradiation

2.2.1. DRIFTS Analysis

Cyclohexane was admitted in absence of UV irradiation on the catalysts 2MoDT2, 4.7MoDT05, DT2 and DT05, and in Figure 1 the spectra after 1 h in presence of gas mix 2 are reported. The cyclohexane peak at 1450 cm^{-1} appears to be present on the overall tested photocatalysts, indicating the occurring of the adsorption of the hydrocarbon. This first finding was coherently with the occurring of Langmuir–Hinshelwood mechanism [23]. Together with the cyclohexane adsorption, a carbonyl vibration (~1681 cm^{-1}) starts to grow while an organo-sulfur vibration, found at ~1354 cm^{-1}, decreases [31]. The decrease in the organo-sulfur vibration is more evident for the support with higher sulfate amount (DT2). These results suggest that in the absence of irradiation sulfate species initiate a redox reaction through hydrogen abstraction of cyclohexane and formation of sulfate (IV) species. In previous studies, it was concluded that reduction of the sulfate (IV) species via hydrogen abstraction may produce gas phase SO_2 and thereby loss of surface sulfur species [27].

The extent of carbonyl vibration may be related to the content of either molybdenum or sulfate, increasing in the combination of both sulfate and molybdenum. Simultaneously, an isosbestic point on the bridging-OH TiO_2 vibration, probably related to the displacement of water molecules by cyclohexane molecules, can be observed in the range 3600–3750 cm^{-1}. Surface sulfates could be highly hydrated showing a weak absorption at 1635 cm^{-1} and a strong broad absorption ranging from 3000 to 3500 cm^{-1} [32]. However, these signals are not visible because the main phenomenon is the appearance of an organo-sulfur compound (found at ~1354 cm^{-1}) resulting from a surface reaction between sulfate and adsorbed cyclohexane that may have determined the loss of hydroxyls bonded with sulfate.

Figure 2 shows in more detail the behavior of 2MoDT2 catalyst surface during dark adsorption of gas mix 2. While the cyclohexane peak at 1450 cm^{-1} stabilizes in a few minutes, the signals at 1681 and 1354 cm^{-1} grow steadily in adsorption time, indicating a continuous surface modification. Though the peaks evolve at the same time, their behavior is not completely parallel, which indicates different surface phenomena. The 1354 cm^{-1} peak could originate from the oxidation of an organic molecule coupled to the sulfate. Other sulfate related vibrations could not be followed.

Figure 3 shows the spectra after 1 h of gas mix 3 (with H_2O in the feed) during dark adsorption on 2MoDT2, 4.7MoDT05, DT2 and Hombikat. Once again, there is an increased carbonyl vibration and a decreasing in organo-sulfur vibration. In the carbonyl area, only a single peak occurred when H_2O was absent in the gas feed (Figure 1), while when H_2O is present (Figure 3) a different product may be formed leading to 2 different vibrations in this area.

In addition, adsorbed water may be playing a role since it absorbs around 1630 cm^{-1} and an isosbestic point may occur. To verify that the negative peak (~1354 cm^{-1}) was related to sulfate, an unsulfated anatase-TiO_2 (Hombikat) was used as reference. The latter sample showed similar band of adsorbed cyclohexane at 1450 cm^{-1}, weak bands of carbonyl compounds with an intensity similar to that one present on the sample with low sulfate coverage (4.7MoDT05), and different hydroxyl bands. In particular, in the presence of adsorbed cyclohexane, the hydroxyl band at 3680 cm^{-1} is negative, that is hindered by cyclohexane adsorption. The hydroxyls band at 3640 cm^{-1} is instead still present, while organo-sulfur (1350 cm^{-1}) peak is not detectable. When organo-sulfur compounds are present, the band of hydroxyls at 3640 cm^{-1} is negative for the samples at high sulfate content. In the absence of sulfate, on the Hombikat catalyst, the band of hydroxyls at 3680 cm^{-1}, instead, is negative. For this reason, the isosbestic point on the bridging-OH vibration is related to the displacement of water molecules by cyclohexane molecules, preferably interacting with hydroxyls close to the sulfate, or with OH of titania in the absence of sulfate.

Figure 1. DRIFT spectra after 1 h of dark adsorption for gas mix 2 on DT05, DT2, 2MoDT2 and 4.7MoDT05 catalysts. Cyclohexane spectrum is shown as reference.

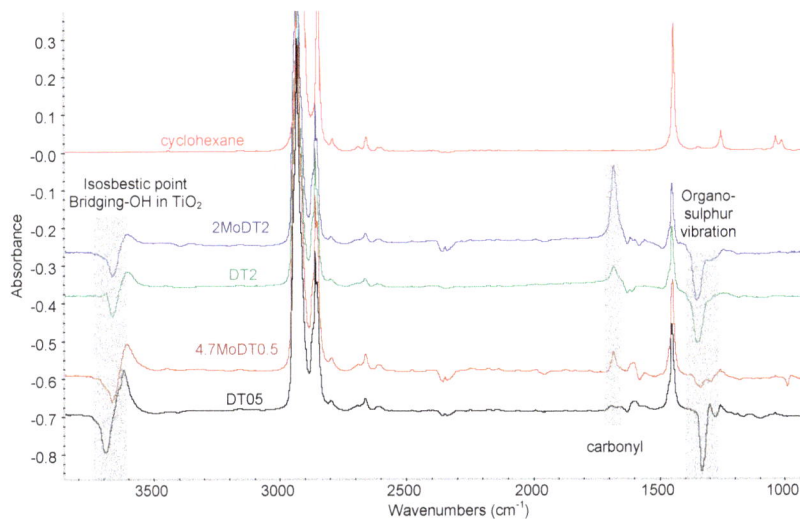

Figure 2. DRIFT spectra during 1 h of dark adsorption for gas mix 2 on 2MoDT2 catalyst.

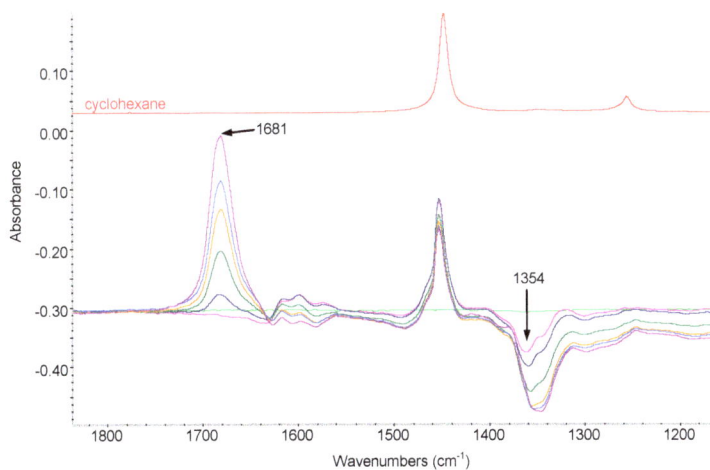

Figure 3. DRIFT spectra after 1 h of dark adsorption for gas mix 3 on DT2, 2MoDT2, 4.7MoDT05 and Hombikat catalysts. Cyclohexane spectrum is shown as reference.

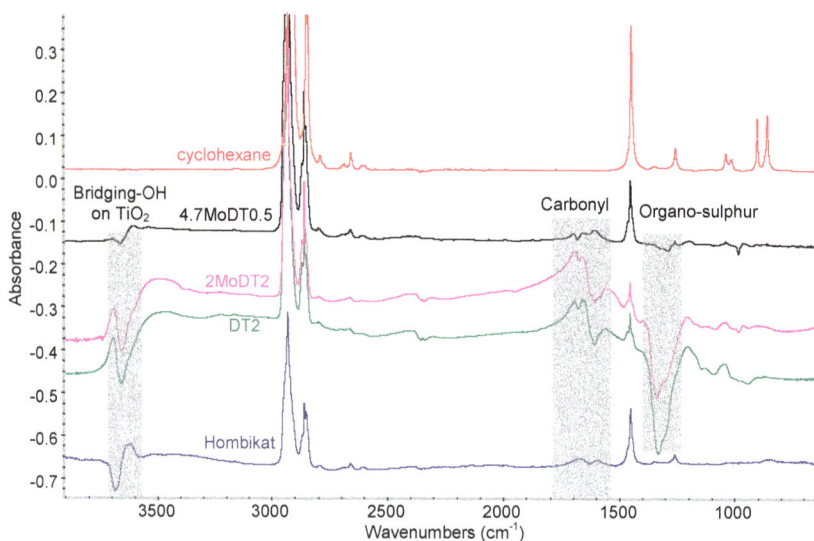

2.2.2. Gas Phase Analysis in the Absence of UV Irradiation

At the run starting time, the gaseous feeding stream was passed through the reactor in the absence of UV irradiation at room temperature. Adsorption of cyclohexane was observed by the decrease in its concentration. Cyclohexane breakthrough time was about 10 min. Thereafter cyclohexane outlet concentration slowly increased to reach the inlet value after about 50 min, indicating that adsorption equilibrium of cyclohexane on the catalyst surface was attained. During this period, no products in gas-phase were detected because they remained adsorbed on catalyst surface (as observed from DRIFT analysis).

A support to the formation of organo-sulfur compounds in dark conditions observed by DRIFTS could be found in an our paper [24] where the amount of cyclohexane adsorbed in the dark on MoO_x/TiO_2 catalysts was linearly correlated with sulfate surface density. The linear increase in the amount of the cyclohexane adsorbed in dark conditions has been determined to increase cyclohexane reaction rate in presence of UV irradiation [24]. In the same paper, it was also indicated that the adsorption of cyclohexane is correlated to the corresponding increase of catalyst acidity. These results suggest that surface sulfate facilitates hydrogen abstraction from adsorbed cyclohexane, increasing its storage on catalyst surface.

2.3. Photocatalytic Activity Tests

2.3.1. Gas Phase Analysis

Photocatalytic tests performed on titania based photocatalysts at low sulfate coverage are reported in Figure 4. Cyclohexane conversion on DT05 and 4.7MoDT05 reached a maximum after about 8 min of irradiation time and then decreased. The strong catalyst deactivation is particularly evident after this time of irradiation.

On DT05, carbon dioxide was the only product detected in gas phase (Figure 5). It started to be formed progressively; after that, the UV light was activated. In this case total carbon mass balance was closed to about 100%. In presence of Mo species, on 4.7MoDT05, the main product was benzene, whose maximum into the production was very delayed with respect both the maximum of cyclohexane conversion, and the achieving of cyclohexane conversion steady state conditions. In fact, the maximum outlet concentration of benzene was 7 ppm after 55 min. CO_2 concentration showed a steady state formation value of 10 ppm. Cyclohexene was formed before benzene in agreement with the mechanism reported in [21], that considered consecutive steps of oxidative dehydrogenation going through cyclohexene, as intermediate, to get finally benzene.

Figure 4. Cyclohexane conversion on DT05 and 4.7MoDT05 during UV irradiation.

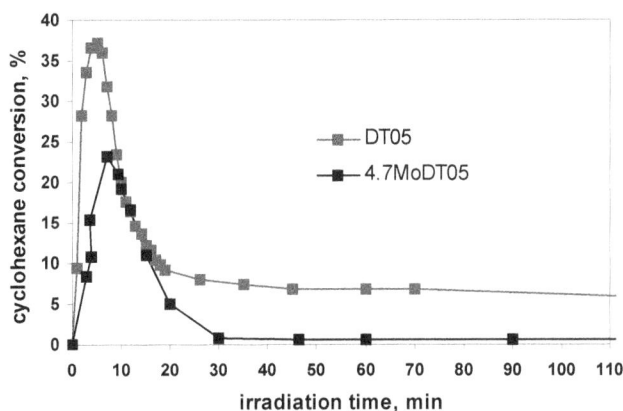

The comparison of cyclohexane conversion over DT2, 2MoDT2 and 8MoDT2 is shown in Figure 6.

A maximum value was reached after about 5 min on all catalysts, then activity decreased approaching a steady state conversion. On 2MoDT2, maximum cyclohexane conversion was about 45%, decreasing to about 10% in 90 min. With the same sulfate content, increasing Mo loading up to 8 wt% MoO_3, the initial maximum conversion was lower, about 15%, while steady state conversion was 2.3% after 30 min. Therefore, the progressive coverage of the titania surface by MoO_x species resulted in a decreased initial and steady state cyclohexane conversions according to our previous results [21]. On DT2 the initial maximum conversion was higher with respect to 8MoDT2 (about 25%), while steady state conversion was similar. Thus, with lower sulfate coverage, steady state photocatalyst activity was smaller, so underlining the relevance of sulfate presence on the photocatalysts performances.

Figure 5. (**a**) Benzene outlet concentration; (**b**) carbon dioxide outlet concentration and (**c**) cyclohexene outlet concentration on DT05 and 4.7MoDT05 catalysts during UV irradiation.

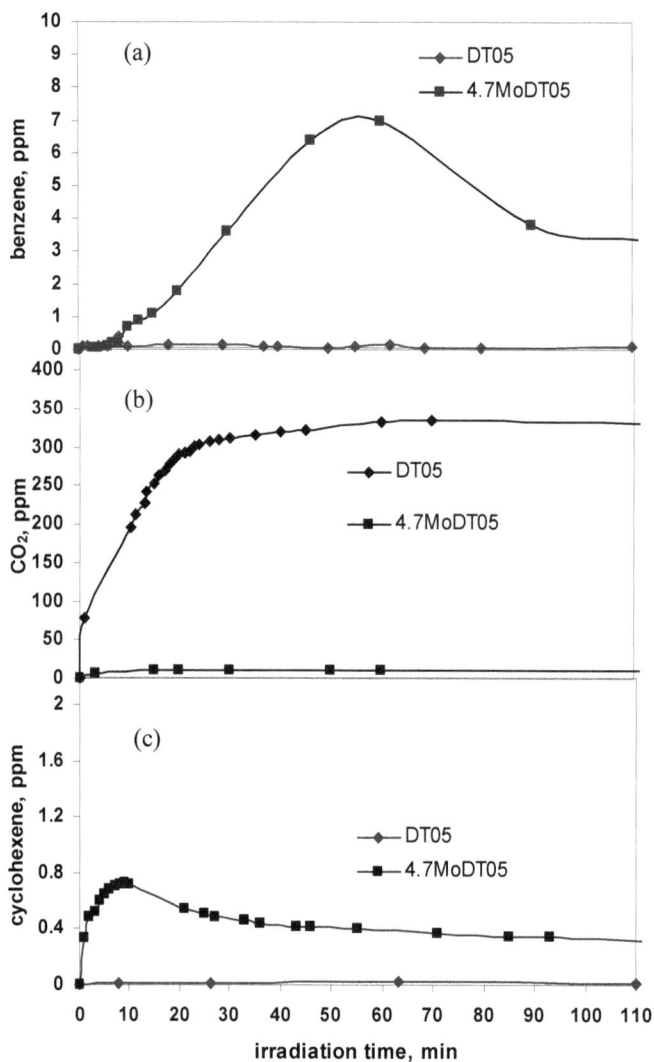

Figure 6. Cyclohexane conversion on DT2, 2MoDT2 and 8MoDT2 catalysts during UV irradiation.

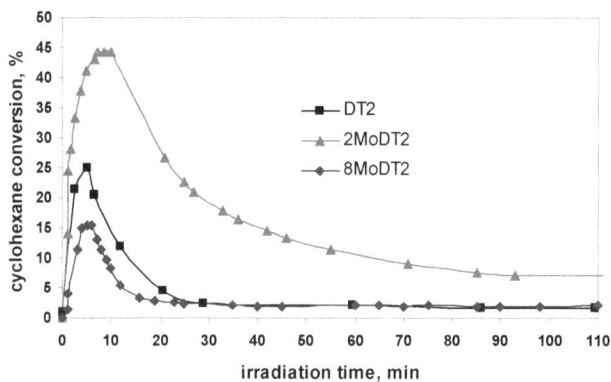

Figure 7. (**a**) Benzene outlet concentration, (**b**) carbon dioxide outlet concentration and (**c**) cyclohexene outlet concentration on DT2, 2MoDT2 and 8MoDT2 catalysts during UV irradiation.

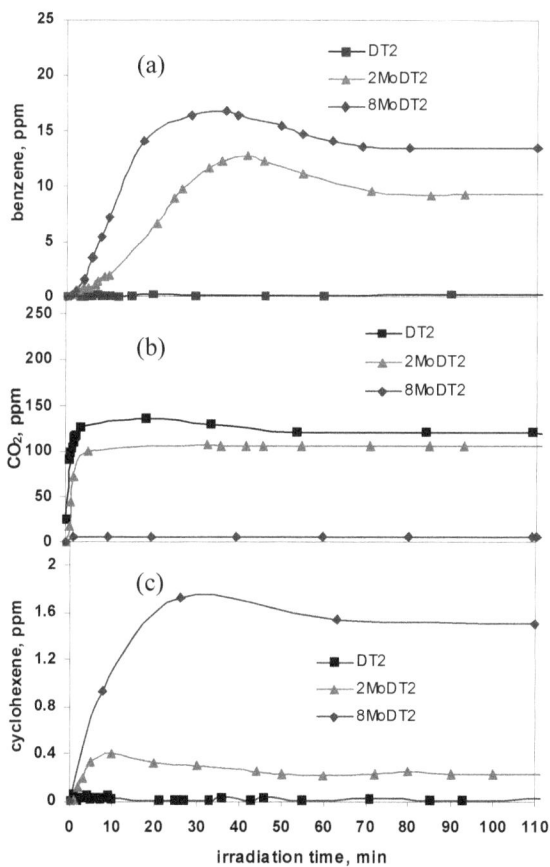

On all MoO$_x$/TiO$_2$ catalysts, reaction products were benzene, CO$_2$ and few amount of cyclohexene (Figure 7). Benzene concentration showed a maximum (17 ppm after 36 min on 8MoDT2 and 13 ppm after 45 min on 2MoDT2), while CO$_2$ concentration was 100 ppm on 2MoDT2 and 6 ppm on 8MoDT2. As observed for DT05 catalysts, the only product observed on DT2 was carbon dioxide and its concentration was the highest.

The higher cyclohexene concentration was recorded for 8MoDT2 whereas its formation was negligible on 2MoDT2, reaching a value less than 1 ppm.

In summary, gas phase analysis of cyclohexane and reaction products evidenced that the presence of MoO$_x$ species on the surface of titania, at same sulfate content, changes the selectivity of the catalyst with increasing molybdenum loading. These results indicate that the interaction between titania and supported molybdenum oxide plays an essential role in the catalyst selectivity. In addition also the amount of surface sulfate influenced benzene yield, decreasing when the sulfate content is lower.

2.3.2. DRIFTS Analysis

The results of photocatalytic reaction are shown in Figure 8a,b for gas mix 2. The CH stretching vibration of cyclohexane increases steadily during illumination probably because cyclohexane adsorption was still occurring, possibly as the result of light induced dehydration of the surface, making free new adsorption sites. So besides dehydration and the continued adsorption of cyclohexane on TiO$_2$ sites, also depletion of hydroxyl groups occurs during photoreaction, as showed by the increasing negative absorption in the range 3200–3800 cm^{-1}. This has been also observed performing the reaction in liquid phase.

During UV irradiation, the peaks, likely to be ascribed to adsorbed cyclohexanone (1691 and 1671 cm^{-1}) were found (Figure 4a), indicating the occurrence of an oxidation step in the reaction. The formation of two different peaks could be related to the presence of different active sites at the surface, but also to a different ketone. The peak at 1671 cm^{-1}, in liquid phase cyclohexane photoxidation, was attributed to a stronger adsorption site, while the 1691 cm^{-1} to a weaker cyclohexanone adsorption site [19]. The position of these peaks is, however, different from the one observed during cyclohexanone dark adsorption (1681cm^{-1}), so during this step, the formation of different surface products could be supposed. An examination of the literature was then performed with the aim to support the latter hypothesis.

In a paper concerning the cyclohexene photo-oxidation over V/TiO$_2$ catalysts [33], cyclohexenone was formed during irradiation. One of the most intense band of this unsaturated carbonyl compound is located at 1692 cm^{-1}. So the signal at 1691 cm^{-1} could be ascribed also to the presence of cyclohexenone, in turn formed from cyclohexene [33]. In addition, it should be considered that C=C stretching frequency of alkenes lies in the range between 1680–1620 cm^{-1} [34], therefore it overlaps to the signals due to adsorbed ketones. The formation of cyclohexene was also found in cyclohexane oxidation at low temperatures using copper chloride in pyridine as catalyst [35] and by liquid phase photocatalytic oxidation of cyclohexane on TiO$_2$ in various solvents [15]. In particular, by using dichloromethane as solvent, the presence of cyclohexene and cyclohexenone was found. Thus it is not possible to exclude the formation of an unsaturated cyclic hydrocarbon in the presence of molybdenum, taking into account also that the cyclohexene adsorbs on the surface through the C=C double bond, and the contribution of the -HC=CH- stretching would therefore be absent [36].

With regard to benzene detected in gas phase, it is not observed in DRIFTS analysis, in agreement with the low affinity evinced in [21] for the titania surface. Both the two kind of hydroxyls disappear during the photoreaction.

Figure 8. DRIFT spectra during 90 min of UV irradiation for gas mix 2 on 2MoDT2 catalyst in the range 1100–1925 cm^{-1} (**a**) and in the range 2750–3900 cm^{-1} (**b**).

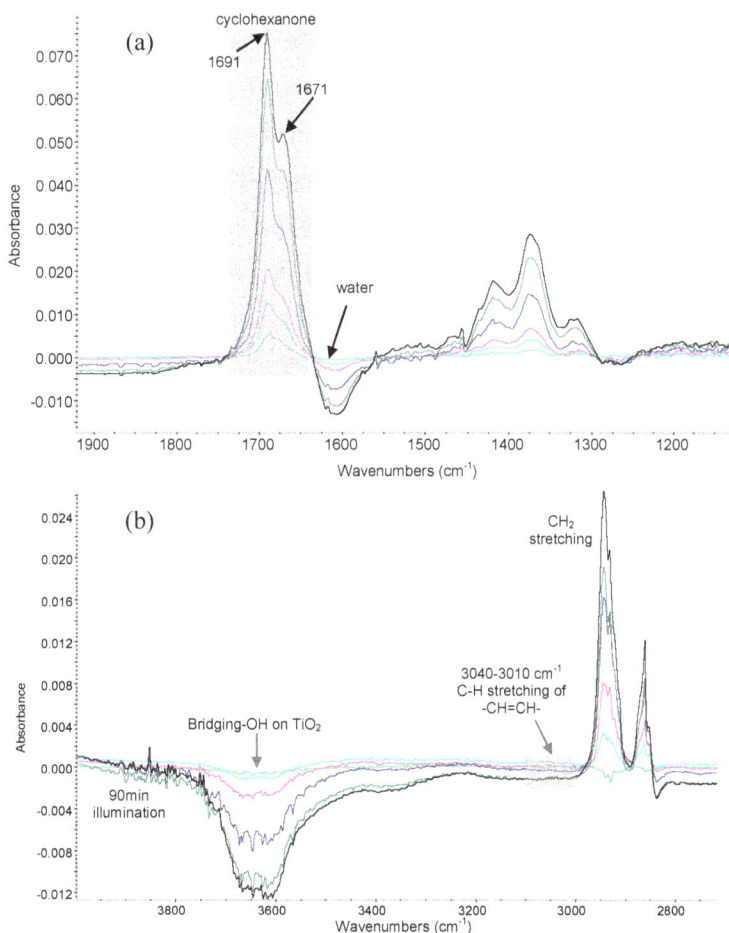

Figure 9 compares the photocatalytic activity of 8MoDT2, 4.7MoDT05 and 2MoDT2 with DT2 and DT05 supports. The DT05 and DT2 supports show similar carbonyl absorptions, with stronger 1671 cm^{-1} vibration. When molybdenum is added to the catalyst, a decrease in this peak is observed and the 1691cm^{-1} peak becomes more prominent. Similar to 2MoDT2, 8MoDT2 photocatalyst shows two bands at 1671 and 1691 cm^{-1}. The latter signal could be ascribed also in this case to cyclohexenone. The contribution at about 1690 cm^{-1} is not visible for DT2 and DT05, indicating that the formation of this compound is due probably correlated to the presence of molybdenum. The band at 1690 cm^{-1} is also not present on 4.7MoDT05. This result could be explained considering that gas phase results on 4.7MoDT05 (Figures 4 and 5) showed the lowest activity and benzene production with together a strong deactivation during UV irradiation. Therefore, in this case, the formation of carbonylic compounds is predominant. On the other hand, the formation of oxygenated products (stronger 1671 cm^{-1} vibration) is observed for all catalysts, and a decrease in the formation of these products occurs with increasing molybdenum loading confirmed by the decrease of the band at 1671 cm^{-1} when molybdenum is added on TiO$_2$ surface.

Figure 9. DRIFT spectra after 90 min of UV irradiation for gas mix 2 on 8MoDT2, 4.7MoDT05, 2MoDT2, DT2 and DT05 catalysts.

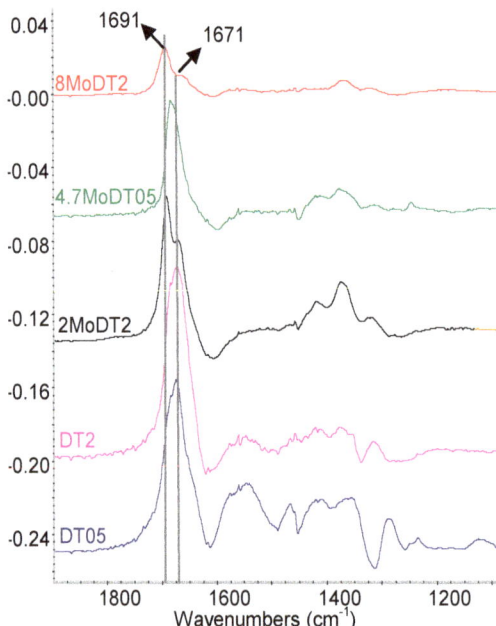

The other data to consider are the bands in the region from 1300 cm^{-1} to about 1450 cm^{-1}, ascribed to the carbon deposits on the catalyst surface during the photoreaction, which determine the deactivation of catalyst observed in the literature [37]. These observations suggest that these species are formed on bare titania, where total oxidation occurs preferentially, while Mo-species hinder titania surface sites active for total oxidation. The photocatalytic tests performed on all catalysts revealed that a rapid decays of activity occurred in the first minutes on stream (Figures 4 and 6). This initial decrease of activity is probably due to a poisoning of the surface by carbon deposits, blocking of a part of the catalytic surface sites.

2.4. Further Test to Assess Intermediates

With the aim to confirm if carbonylic compounds are the responsible of catalyst deactivation, a photocatalytic test on 8MoDT2 was carried out by feeding cyclohexanone with the same operating conditions used for cyclohexane.

The obtained results are reported in Figure 10.

Figure 10. Cyclohexanone conversion and CO_2 concentration during UV irradiation.

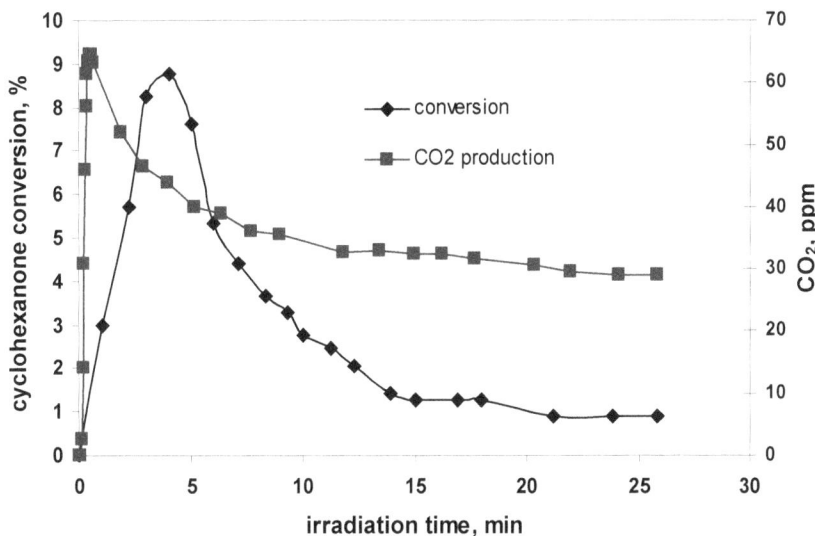

Maximum cyclohexanone conversion was about 9%, and then activity decreased to 1% in 15 min. A steady state condition was obtained after about 21 min of irradiation with a conversion of approximately 0.9%. Carbon dioxide was formed immediately after the UV sources were switched on and reached a concentration of about 60 ppm after an irradiation time of 0.5 min. CO_2 was the only observed product and no other reaction products were detected in gas phase. These results evidenced that cyclohexanone (carbonyl compounds shown by DRIFTS spectra) is the precursor for CO_2 production. According to DRIFTS spectra, the accumulation on surface of intermediates formed by cyclohexanone oxidation during irradiation is the responsible of catalyst deactivation.

With regard to the influence of sulfate, a considerable decrease in the formation of carboxylates is observed with higher molybdenum and sulfate content confirming that the simultaneous presence of Mo-species and sulfate sharply increases photocatalytic activity and selectivity. The acidity induced by the sulfate on TiO_2 surface [38] furnishes hydrocarbon activation towards partial oxidation and supplying the initial step in the absence of irradiation [24]. However, if the sulfate is lost during dark adsorption, it can induce a deactivation of catalyst under illumination.

For gas mix 3, when H_2O is also present, almost no reaction has been observed for the molybdenum catalysts. The decrease in reactivity due to the hydration of metal oxides has been discussed in literature [39].

An example is shown in Figure 11, for the 4.7MoDT05 catalysts, in which very small peaks were observed under UV irradiation. The increase of CH stretching vibration indicates that cyclohexane adsorption on the catalyst surface is still occurring evidencing a deactivation phenomenon. The contribution of the reaction products to the spectra is very low, only visible by the formation of small peaks between 1800–1000 cm^{-1} and the decrease of the TiO$_2$ bridging-OH.

Figure 11. DRIFT spectra during 90 min of UV irradiation for gas mix 3 on 4.7MoDT05 catalyst.

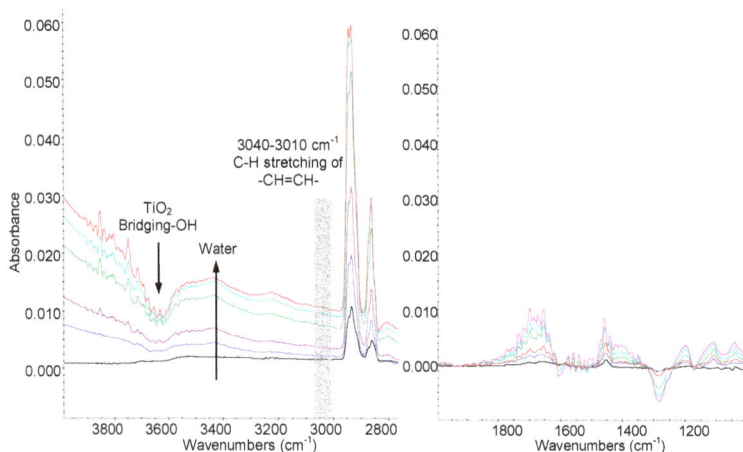

Figure 12. DRIFT spectra after 90 min of UV irradiation for gas mix 3 on DT2 and Hombikat catalysts.

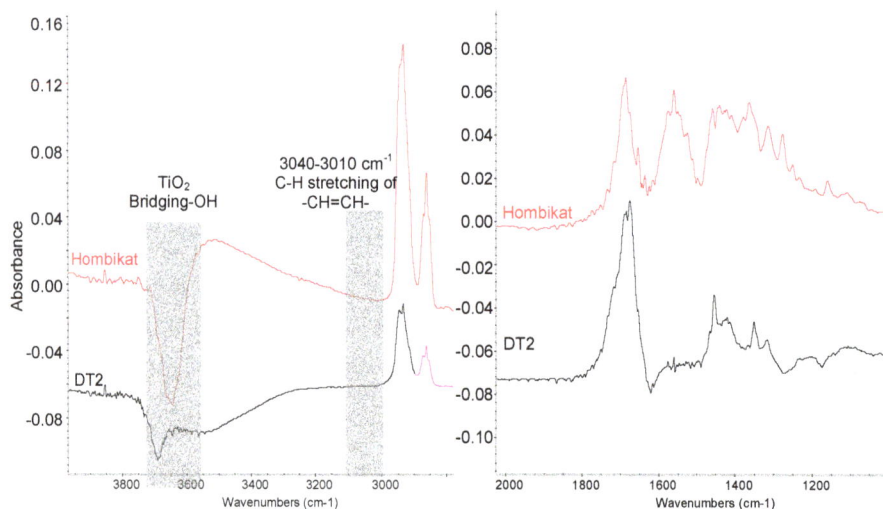

Under gas mix 3 flow better results were observed when there was no molybdenum on the TiO$_2$ surface, as can be seen in Figure 12. While on MoO$_x$/TiO$_2$ catalysts, the water presence in the gas feed greatly decreased the product formation, in DT2 this effect was not observed. The reactivity of DT2 under gas mix 2 (without H$_2$O) and 3 (with H$_2$O) showed similar reaction products and spectral intensity. As expected from previous work, no peaks related to unsaturated hydrocarbons formation were observed on DT2 [21]. Hombikat is also shown as reference, which appears to catalyze the formation of more surface products and higher bridging-OH deactivation than the ones observed for sulfated TiO_2. This last result is a further confirmation of the role of sulfate in the deactivation of photocatalysts.

3. Experimental Section

3.1. Catalysts Preparation and Characterization

Two titanias were used as supports: two commercial titania samples (DT and DT51 by Millenium Inorganic Chemicals) with different sulfate content (respectively 0.5 wt% and 2 wt% expressed as SO_3) contributed from the experimental procedure used for the synthesis of samples [40]. The samples are named, respectively, DT2 and DT05 with reference to the sulfate content. MoO_x-based catalysts were prepared by wet impregnation of titania with aqueous solution of ammonium heptamolybdate $(NH_4)_6Mo_7O_{24}$ $4H_2O$, followed by drying at 120 °C and calcination at 400 °C for 3 h. Unsulfated titania (Hombikat) was used as reference.

Thermogravimetric analysis (TG-DSC-MS) was carried out in air flow on powder samples with a thermoanalyzer (Q600, TA) in the range 20–1000 °C with an heating rate of 10 °C/min. Sulfate content has been evaluated from the weight loss in the range in which the release of SO_2 occurred. Specific surface area was obtained by N_2 adsorption-desorption isotherm at −196 °C with a Costech Sorptometer 1040. Powder samples were treated at 150 °C for 2 h in He flow (99.9990%) before testing. Laser Raman spectra of powder samples were obtained with a Dispersive MicroRaman (Invia, Renishaw), equipped with 785 nm diode-laser, in the range 100–2500 cm^{-1} Raman shift. Chemical analysis of molybdenum loading was performed by inductive coupled plasma-mass spectrometry (7500c ICP-MS, Agilent) after microwave digestion (Ethos Plus from Milestone) of samples in HNO_3/HCl and HF/HCl mixtures.

3.2. Gas phase Analysis

Photocatalytic tests were carried out feeding 1000 ppm cyclohexane, 1500 ppm oxygen and 1600 ppm water in N_2 (total flow rate: 830 (stp) cm^3/min) to a continuous gas-solid annular photocatalytic fixed-bed reactor. The tests were carried out in presence of water in order to minimize catalyst deactivation as reported in literature [37].

Oxygen and nitrogen were fed from cylinders, nitrogen being the carrier gas for cyclohexane and water vaporized from two temperature controlled saturators containing pure cyclohexane and water. The gas flow rates were measured and controlled by mass flow controllers (Brooks Instrument). The annular section of the reactor [27] (reactor volume: 7 l) was realized with two axially mounted 500 mm long quartz tubes of 140 and 40 mm diameter, respectively. The reactor was equipped with seven 40 W UV fluorescent lamps providing photons wavelengths in the range from 300 to 425 nm, with primary peak centered at 365 nm. One lamp (UVA Cleo Performance 40 W, Philips) was centered inside the inner tube while the others (R-UVA TLK 40 W/10R flood lamp, Philips) were located symmetrically around the reactor.

Although light sources vary in their intensity, they have the same emission spectrum and in the case of photocatalytic oxidative dehydrogenation of cyclohexane, the selectivity of the reaction was not influenced by light intensity [23].

The overall system composed of UV lamps and photoreactor are covered by aluminum foils to minimize the dispersion of the photons in the space surrounding the photoreactor. In order to avoid temperature gradients in the reactor caused by irradiation, the temperature was controlled to 35 ± 2 °C by cooling fans. The catalytic reactor bed was prepared *in situ*, by coating quartz flakes previously loaded in the annular section of a quartz continuous flow reactor with aqueous slurry of catalysts powder. The coated flakes were dried at 120 °C for 24 h in order to remove the excess of physisorbed water. This treatment resulted in uniform coating well adhering to the quartz flakes surface. The gas composition was continuously determined by on line analyzers connected to a PC for data acquisition. CO and CO_2 concentration is measured by an on line non dispersive IR analyzer (Uras 10, Hartmann & Braun), working on the basis of specific adsorption of IR radiation (wavelength from 2 to 8 μm). Oxygen, cyclohexane and reaction products composition is determined by an on line quadrupole mass detector (MD800, ThermoFinnigan) that can analyze the outlet reactor gas, introduced into a heated silica capillary, up to $m/z = 800$.

3.3. DRIFTS Analysis

A layer of KBr powder was introduced in the DRIFTS (3 window cell) holder [41], over which each of the different catalysts and supports were deposited and pressed. The catalysts were heated up to 120 °C for 30 min in He flow (app. 30mL/min) to remove weakly adsorbed water from the surface.

Different gas mixtures (100mL/min flow) were prepared for the catalytic tests:

Gas mix 1: 1000ppm Cyclohexane, 1500ppm O_2

Gas mix 2: 1000ppm Cyclohexane, 5000ppm O_2

Gas mix 3: 1000 ppm Cyclohexane, 250 ppm H_2O, 5000ppm O_2

Further, for DRIFTS analysis, nitrogen is the carrier gas for cyclohexane and water vaporized from two temperature controlled saturators containing pure cyclohexane and water.

A background (128 scan averages) of the dried catalyst at room temperature was taken and used for the adsorption step, which was followed with IR during 40–60 min, until surface stabilization was achieved. A spectrum (128 scan averages) was taken after adsorption stabilization, which was used as background for the reaction step. Reaction was continued for 90 min; the first minute of collection was performed in the dark, and the rest was done under irradiation by a 150 W Xe light with a light diffuser. All spectra, except for background spectra, consisted of 64 averaged scans and water vapor correction has been applied to most of them.

Only the background spectrum was collected at room temperature, while during the irradiation the temperature increased up to a value very similar to that one of the fixed bed reactor used for gas phase analysis. In this way, the selectivity of the reaction was not different.

4. Conclusions

From coupling of gas phase and DRIFTS analysis, a deeper knowledge of phenomena occurring during photocatalytic selective oxidation of cyclohexane on MoO_x/TiO_2 was obtained.

During adsorption measurements in the absence of UV irradiation, there is a clear indication of organo-sulfur compound formation accompanied by H-abstraction by the acidic sulfate species. The accompanying stepwise reduction of the sulfate by protons causes formation of SO_2 and thereby loss of surface sulfur. The decrease in surface sulfur during the dark adsorption experiment leads to a subsequent initial decrease in photocatalytic activity during irradiation since the acidity induced by the sulfate facilitates hydrocarbon activation in partial oxidation of cyclohexane after supplying the initial step in the absence of irradiation. The further decrease of photoactivity during irradiation is due to a poisoning of the surface by carbonaceous species derived from carbonylic compounds formed on bare titania, blocking of a portion of the catalytic surface sites. A considerable decrease in the rate of poisoning by carbonaceous species is observed at higher molybdenum and sulfate contents, confirming that the simultaneous presence of Mo-species and sulfate enhances photocatalytic activity and selectivity. Gas phase analysis of cyclohexane and reaction products evidenced that the presence of MoO_x species at same sulfate coverage increases the selectivity of the catalyst with increasing molybdenum content indicating that the interaction between titania and supported molybdenum oxide plays an essential role in changing the catalyst selectivity. In addition increasing coverage of surface sulfate influenced benzene yield.

Conflicts of Interest

No author of the present manuscript has a direct financial relation with the commercial identities mentioned in the paper that might lead to a conflict of interest.

References

1. Sannino, D.; Vaiano, V.; Sacco, O.; Ciambelli, P. Mathematical modelling of photocatalytic degradation of methylene blue under visible light irradiation. *J. Environ. Chem. Eng.* **2013**, *1*, 56–60.
2. Sacco, O.; Stoller, M.; Vaiano, V.; Ciambelli, P.; Chianese, A.; Sannino, D. Photocatalytic Degradation of Organic Dyes under Visible Light on N-Doped TiO_2 Photocatalysts. *Int. J. Photoenergy* **2012**, Article ID 626759:1–Article ID 626759:8.
3. Murcia, J.J.; Hidalgo, M.C.; Navío, J.A.; Vaiano, V.; Sannino, D.; Ciambelli, P. Cyclohexane photocatalytic oxidation on Pt/TiO_2 catalysts. *Catal. Today* **2013**, *209*, 164–169.
4. Stoller, M.; Movassaghi, K.; Chianese, A. Photocatalytic degradation of orange II in aqueous solutions by immobilized nanostructured titanium dioxide. *Chem. Eng. Trans.* **2011**, *24*, 229–234.
5. Ibrahim, H.; de Lasa, H. Photo-catalytic conversion of air borne pollutants: Effect of catalyst type and catalyst loading in a novel photo-CREC-air unit. *Appl. Catal. B* **2002**, *38*, 201–213.

6. Augugliaro, V.; Bellardita, M.; Loddo, V.; Palmisano, G.; Palmisano, L.; Yurdakal, S. Overview on oxidation mechanisms of organic compounds by TiO₂ in heterogeneous photocatalysis. *J. Photochem. Photobiol. C* **2012**, *13*, 224–245.

7. Sannino, D.; Vaiano, V.; Isupova, L.A.; Ciambelli, P. Heterogeneous Photo-Fenton Oxidation of Organic Pollutants on Structured Catalysts. *J. Adv. Oxid. Technol.* **2012**, *15*, 294–300.

8. Sannino, D.; Vaiano, V.; Ciambelli, P.; Isupova, L.A. Structured catalysts for photo-Fenton oxidation of acetic acid. *Catal. Today* **2011**, *161*, 255–259.

9. Molinari, A.; Amadelli, R.; Mazzacani, A.; Sartori, G.; Maldotti, A. Tetralkylammonium and sodium decatungstate heterogenized on silica: Effects of the nature of cations on the photocatalytic oxidation of organic substrates. *Langmuir* **2002**, *18*, 5400–5405.

10. Maldotti, A.; Molinari, A.; Amadelli, R. Photocatalysis with organized systems for the oxofunctionalization of hydrocarbons by O₂. *Chem. Rev.* **2002**, *102*, 3811–3836.

11. Li, X.; Chen, G.; Po-Lock, Y.; Kutal, C. Photocatalytic oxidation of cyclohexane over TiO₂ nanoparticles by molecular oxygen under mild conditions. *J. Chem. Technol. Biotechnol.* **2003**, *78*, 1246–1251.

12. Stoller, M.; Miranda, L.; Chianese, A. Optimal Feed location in a Spinning Disc Reactor for the Production of TiO₂ Nanoparticles. *Chem. Eng. Trans.* **2009**; *17*, 993–998.

13. de Caprariis, B.; Di Rita, M.; Stoller, M.; Verdone, N.; Chianese, A. Reaction-precipitation by a spinning disc reactor: Influence of hydrodynamics on nanoparticles production. *Chem. Eng. Sci.* **2012**, *76*, 73–78.

14. Boarini, P.; Carassiti, V.; Maldotti, A.; Amadelli, R. Photocatalytic oxygenation of cyclohexane on titanium dioxide suspensions: Effect of the solvent and of oxygen. *Langmuir* **1998**, *14*, 2080–2085.

15. Almquist, C.B.; Biswas, P. The photo-oxidation of cyclohexane on titanium dioxide: an investigation of competitive adsorption and its effects on product formation and selectivity. *Appl. Catal. A* **2001**, *214*, 259–271.

16. Berg, O.; Hamdy, M.S.; Maschmeyer, T.; Moulijn, J.A.; Bonn, M.; Mul, G. On the wavelength-dependent performance of Cr-doped silica in selective photo-oxidation. *J. Phys. Chem. C* **2008**, *112*, 5471–5475.

17. Maldotti, A.; Amadelli, R.; Varani, G.; Tollari, S.; Porta, F. Photocatalytic processes with polyoxotungstates: Oxidation of cyclohexylamine. *Inorg. Chem.* **1994**, *33*, 2968–2973.

18. Teramura, K.; Tanaka, T.; Yamamoto, T.; Funabiki, T. Photo-oxidation of cyclohexane over alumina-supported vanadium oxide catalyst. *J. Mol. Catal. A* **2001**, *165*, 299–301.

19. Almeida, A.R.; Moulijn, J.A.; Mul, G. *In situ* ATR-FTIR study on the selective photo-oxidation of cyclohexane over anatase TiO₂. *J. Phys. Chem. C* **2008**, *112*, 1552–1561.

20. Xu, W.; Raftery, D.; Francisco, J.S. Effect of irradiation sources and oxygen concentration on the photocatalytic oxidation of 2-propanol and acetone studied by *in situ* FTIR. *J. Phys. Chem. B* **2003**, *107*, 4537–4544.

21. Ciambelli, P.; Sannino, D.; Palma, V.; Vaiano, V.; Bickley, R.I. Reaction mechanism of cyclohexane selective photo-oxidation to benzene on molybdena/titania catalysts. *Appl. Catal. A* **2008**, *349*, 140–147.

22. Ciambelli, P.; Sannino, D.; Palma, V.; Vaiano, V.; Mazzei, R.S. Improved Performances of a Fluidized Bed Photoreactor by a Microscale Illumination System. *Int. J. Photoenergy* **2009**, Article ID 709365:1– Article ID 709365:7.

23. Palma, V.; Sannino, D.; Vaiano, V.; Ciambelli, P. Fluidized-Bed Reactor for the Intensification of Gas-Phase Photocatalytic Oxidative Dehydrogenation of Cyclohexane. *Ind. Eng. Chem. Res.* **2010**, *49*, 10279–10286.

24. Ciambelli, P.; Sannino, D.; Palma, V.; Vaiano, V. The effect of sulfate doping on nanosized TiO_2 and MoO_x/TiO_2 catalysts in cyclohexane photooxidative dehydrogenation. *Int. J. Photoenergy* **2008**, Article ID 258631:1–Article ID 258631:8.

25. Ciambelli, P.; Sannino, D.; Palma, V.; Vaiano, V. Photocatalysed selective oxidation of cyclohexane to benzene on $MoO_x/TiO2$. *Catal. Today* **2005**, *99*, 143–149.

26. Ciambelli, P.; Sannino, D.; Palma, V.; Vaiano, V. Cyclohexane photocatalytic oxidative dehydrogenation to benzene on sulfated titania supported MoO_x. *Stud. Surf. Sci. Catal.* **2005**, *155*, 179–187.

27. Sannino, D.; Vaiano, V.; Ciambelli, P.; Eloy, P.; Gaigneaux, E.M. Avoiding the deactivation of sulfated MoO_x/TiO_2 catalysts in the photocatalytic cyclohexane oxidative dehydrogenation by a fluidized bed photoreactor. *Appl. Catal. A* **2011**, *394*, 71–78.

28. Ciambelli, P.; Sannino, D.; Palma, V.; Vaiano, V.; Eloy, P.; Dury, F.; Gaigneaux, E.M. Tuning the selectivity of MoO_x supported catalysts for cyclohexane photo oxidehydrogenation. *Catal. Today* **2007**, *128*, 251–257.

29. Ciambelli, P.; Sannino, D.; Palma, V.; Vaiano, V.; Mazzei, R.S.; Eloy, P.; Gaigneaux, E.M. Photocatalytic cyclohexane oxidehydrogenation on sulfated $MoO_x/gamma-Al_2O_3$ catalysts. *Catal. Today* **2009**, *141*, 367–373.

30. Cheng, C.P.; Schrader, G.L. Characterization of supported molybdate catalysts during preparation using laser Raman spectroscopy. *J. Catal.* **1979**, *60*, 276–294.

31. Bellamy, L.J.; Mayo, D.W. Infrared frequency effects of lone pair interactions with antibonding orbitals on adjacent atoms. *J. Phys. Chem.* **1976**, *80*, 1217–1220.

32. Bezrodna, T.; Puchkovska, G.; Shimanovska, V.; Chashechnikova, I.; Khalyavka, T.; Baran, J. Pyridine-TiO_2 surface interaction as a probe for surface active centers analysis. *Appl. Surface Sci.* **2003**, *214*, 222–231.

33. Mul, G.; Wasylenko, W.; Hamdy, M.S.; Frei, H. Cyclohexene photo-oxidation over vanadia catalyst analyzed by time resolved ATR-FT-IR spectroscopy. *Phys. Chem. Chem. Phys.* **2008**, *10*, 3131–3137.

34. Coates, J. Vibrational spectroscopy: Instrumentation for infrared and Raman spectroscopy. *Appl. Spectrosc. Rev.* **1998**, *33*, 267–425.

35. Schuchardt, U.; Pereira, R.; Rufo, M. Iron(III) and copper(II) catalysed cyclohexane oxidation by molecular oxygen in the presence of tert-butyl hydroperoxide. *J. Mol. Catal. A* **1998**, *135*, 257–262.

36. Manner, W.L.; Girolami, G.S.; Nuzzo, R.G. Sequential dehydrogenation of unsaturated cyclic C5 and C5 hydrocarbons on Pt(111). *J. Phys. Chem. B* **1998**, *102*, 10295–10306.

37. Einaga, H.; Futamura, S.; Ibusuki, T. Heterogeneous photocatalytic oxidation of benzene, toluene, cyclohexene and cyclohexane in humidified air: Comparison of decomposition behavior on photoirradiated TiO2 catalyst. *Appl. Catal. B* **2002**, *38*, 215–225.

38. Ciambelli, P.; Fortuna, M.E.; Sannino, D.; Baldacci, A. The influence of sulfate on the catalytic properties of V_2O_5-TiO_2 and WO_3-TiO_2 in the reduction of nitric oxide with ammonia. *Catal. Today* **1996**, *29*, 161–164.

39. Amano, F.; Yamaguchi, T.; Tanaka, T. Photocatalytic oxidation of propylene with molecular oxygen over highly dispersed titanium, vanadium, and chromium oxides on silica. *J. Phys. Chem. B* **2006**, *110*, 281–288.

40. Taranto, J.; Frochot, D.; Pichat, P. Photocatalytic treatment of air: Comparison of various tiO2, coating methods, and supports using methanol or *n*-Octane as tst pollutant. *Ind. Eng. Chem. Res.* **2009**, *48*, 6229–6236.

41. Carneiro, J.T.; Moulijn, J.A.; Mul, G. Photocatalytic oxidation of cyclohexane by titanium dioxide: Catalyst deactivation and regeneration. *J. Catal.* **2010**, *273*, 199–210.

MDPI AG
Klybeckstrasse 64
4057 Basel, Switzerland
Tel. +41 61 683 77 34
Fax +41 61 302 89 18
http://www.mdpi.com/

Catalysts Editorial Office
E-mail: catalysts@mdpi.com
http://www.mdpi.com/journal/catalysts

www.ingramcontent.com/pod-product-compliance
Lightning Source LLC
Chambersburg PA
CBHW051923190326
41458CB00026B/6392